Facilities Management and Corporate Real Estate Management as Value Drivers

Facilities Management (FM) and Corporate Real Estate Management (CREM) are two closely related and relatively new management disciplines with developing international professions and increasing academic attention. From the outset both disciplines have had a strong focus on controlling and reducing costs for real estate, facilities and related services. In recent years there has been a shift towards putting more focus on how FM and CREM can add value to the organisation. This book is driven by the need to develop a widely accepted and easily applicable conceptual framework for adding value by FM and CREM. It presents state of the art theoretical knowledge and empirical evidence about the impact of buildings and facilities on twelve value parameters, as well as how to manage and measure these values. The findings are connected to a new Value Adding Management model.

The book is research based with a focus on guidance to practice. It offers a transdisciplinary approach, integrating academic knowledge from a variety of different fields with practical experience. It also includes twelve interviews with practitioners, shedding light as to how they manage adding value in practice. This is a much needed resource for practitioners, researchers and teachers in the field of FM and CREM, as well as students at both undergraduate and postgraduate levels.

Per Anker Jensen is Professor in Facilities Management and head of the externally funded Centre for Facilities Management – Realdania Research, Technical University of Denmark (DTU), Denmark. He holds the degrees of MSc in Civil Engineering, PhD and MBA. Besides research and teaching he has twenty years of experience in practice as a consultant, a project manager and a facilities manager. He was a member of the board of EuroFM and chairman of EuroFM's Research Network Group in 2007 and 2008. He is currently project manager of the joint EuroFM research project on the Added Value of FM, which started in January 2009.

Theo van der Voordt is Associate Professor in Corporate and Public Real Estate Management at the Faculty of Architecture at the Delft University of Technology, the Netherlands. His current research focuses on workplace performance, experience and the use of new work environments, the design and management of health care real estate, and value adding management of FM and CREM. His research aims to develop and test workplace strategies, conceptual models and practical tools to support data collection and decision making processes. This work is conducted in close cooperation with the Centre for People and Buildings (CfPB) in Delft, a knowledge centre that specialises in the relationship between people, working processes and the working environment.

Facilities Management and Corporate Real Estate Management as Value Drivers

How to manage and measure adding value

**Edited by
Per Anker Jensen and
Theo van der Voordt**

Routledge
Taylor & Francis Group

LONDON AND NEW YORK

First published 2017 by Routledge

2 Park Square, Milton Park, Abingdon, Oxon, OX14 4RN

605 Third Avenue, New York, NY 10017

Routledge is an imprint of the Taylor & Francis Group, an informa business

First issued in paperback 2020

British Library Cataloguing-in-Publication Data
A catalogue record for this book is available from the British Library

Library of Congress Cataloging in Publication Data
Names: Jensen, Per Anker, editor. | Voordt, D. J. M. van der, editor.
Title: Facilities management and corporate real estate management as value drivers : how to manage and measure adding value / edited by Per Anker Jensen and Theo van der Voordt.
Description: Abingdon, Oxon ; New York, NY : Routledge, 2017. | Includes bibliographical references and index.
Identifiers: LCCN 2016009862| ISBN 9781138907188 (hardback : alk. paper) | ISBN 9781315695150 (ebook : alk. paper)
Subjects: LCSH: Real estate management. | Facility management. | Value.
Classification: LCC HD1394 .F225 2017 | DDC 658.2--dc23
LC record available at https://lccn.loc.gov/2016009862

ISBN: 978-1-138-90718-8 (hbk)
ISBN: 978-0-367-73688-0 (pbk)

Typeset in Galliard
by Saxon Graphics Ltd, Derby

Contents

Preface

This book is a result of many years of collaboration in a research group that is part of the European Facilities Management (FM) network, EuroFM. The research group launched a first book on "The Added Value of FM" at the European FM Conference 2012 in Copenhagen, where eighteen researchers representing seven nationalities explored the concept of added value and various conceptual frameworks in connection to research data from different sectors. The reviewers from practice as well as academia were very positive. However, they also told us that the book should not be seen as the end, but more like the beginning of investigating this important topic. This response gave us a strong incentive to continue our collaborative research across European borders.

One of the learning points from our book from 2012 was that developing conceptual frameworks for mapping, analysing and visualising added value was useful, but it also had limitations in being rather static and not very action oriented. Therefore, in our follow-up research we put more focus on investigating how to manage and measure added value in practice and developing a practical management tool, which we call Value Adding Management. For instance, we have investigated how practitioners in FM and Corporate Real Estate Management (CREM) deal with value adding by interviewing professionals from companies in Denmark and the Netherlands. We also conducted a cross-paper analysis of eighteen value added related papers presented at the European Facility Management Conferences 2013–2015 in Prague, Berlin and Glasgow, and three papers from the CIB Conference 2014 in Copenhagen.

The current book is a follow-up to our book from 2012. This second book is driven by the need to develop a widely accepted and easily applicable conceptual framework for adding value by FM and CREM, and to support FM and CREM managers and other decision makers in how to manage and measure added value. Its intent is to compile and create an overview of evidence from the many research activities that are going on in this area. As part of this process we observed that companies increasingly work on documenting added value of FM and CREM. In the European research group we have created a strong network, which has made it possible to engage contributors from different countries and disciplines in developing the book.

The purpose of the book is to raise awareness of the opportunities and risks of adding value by FM and CREM, and the need for a more holistic approach with balanced interests between benefits and costs of interventions. We defined twelve specific value parameters, and the invited authors discuss how to manage and measure them based on state of the art theoretical knowledge, practical experience and empirical data for each value parameter and across the different parameters. Furthermore, the authors discuss ways to select and prioritise (conflicting) values and connected impact factors such as the vision, mission and strategy of the organisation, its maturity, and the external context.

The book is research based with a focus on guidance to practice. The target groups are practitioners, researchers and teachers from the field of FM and CREM, as well as students at both undergraduate and postgraduate level. It is not intended as a core textbook, but it will be suitable for supplementary reading for courses in FM and CREM.

The ambition of the book is to be transdisciplinary by integrating the two closely related disciplines and professions, FM and CREM, with a view to examining how they can be value drivers for the organisations they support and their surroundings. Therefore, the editors have asked Ron van der Weerd, chairman of EuroFM, former Dean at the School of Facility Management and current project director at the Hanze University of Applied Sciences, Olav Egil Sæbøe, FM/CREM practitioner and external lecturer in higher education in Norway, and Hans de Jonge, CREM practitioner and professor in Real Estate Management at the Delft University of Technology in the Netherlands, to write forewords.

Twenty-three authors from universities in five European countries have been involved in writing the eighteen chapters in the book. We have established teams of two or more authors to write the chapters, thereby ensuring that the topics are presented with diverse inputs and perspectives within a common framework. The book also includes twelve interviews with practitioners from FM and CREM in six European countries about their perception and practice in relation to adding value. The background, purpose and structure of the book are explained further in Chapter 1. The work on this book by Per Anker Jensen, Susanne Balslev Nielsen and Giulia Nardelli, Centre for Facilities Management – Realdania Research (www.cfm.dtu.dk), Technical University of Denmark, has been financially supported by the Danish private foundation Realdania (www.realdania.dk). As editors we want to thank all the contributors and the people from Routledge for the collaboration on the book. We also thank Ron van der Weerd (EuroFM), Olav Egil Sæbøe (Pro-FM) and Professor Hans de Jonge (Brink Group) for their kind words and for recommending this book to an international audience.

Per Anker Jensen and Theo van der Voordt

Forewords

EuroFM's mission is the advancement of knowledge in Facility Management in Europe and its application in practice, education and research. An understanding of how FM can add value to organisations is crucial for professional development. The topic has for some years obtained increasing attention in research, education and practice. This book provides a great contribution to the current body of knowledge on how to manage and measure added value in organisations.

The former chairman of EuroFM's Research Network Group, Professor Per Anker Jensen, Technical University of Denmark, proposed the formation of a research group on "The Added Value in FM". He has chaired the group from its establishment in 2009. The group has organised a number of workshops and has been an active part of the annual European FM Conferences (EFMC) with many dedicated sessions, papers and presentations as part of the research symposia as well as workshops and panel debates with practitioners and researchers. Since 2013 the group has been chaired jointly by Per Anker Jensen and Theo van der Voordt, Delft University of Technology.

EuroFM is pleased to see and to promote the publication of this book. We are certain that it will be an important element in the advancement of knowledge in FM, and will be of inspiration for practitioners, students, teachers and other researchers for the activities in their organisations. We would like to thank the editors for developing the book and the authors for their contributions.

R.M.D. (Ron) van der Weerd
LL.M MSc, Project Director Reconstruction ZP7
Hanze University of Applied Sciences, Groningen, the Netherlands
Chair EuroFM, European Facility Management Network, www.EuroFM.org

When I was appointed Corporate Real Estate Manager in a Norwegian corporate business in the 1990s, the key management issue was definitely not "added value". "Lowest cost of operations and maintenance" was the main benchmarking objective and the planning basis for improvements and Real Estate development. The various aspects of user satisfaction were not – or only vaguely – associated with the core business. Value creation and service qualities were not given proper attention in this respect. "Added value" was not a professional term in our FM vocabulary. Similar situations were at that time frequent in other Nordic and European business environments.

Thanks to universities and networks like EuroFM and NordicFM as well as peers in the European FM industry, the value issue in Real Estate and Facilities Management has increasingly been given more professional attention, particularly over the last decade. In the Nordic countries the Danish Technical University (DTU) and its Centre for Facilities Management has stood out as a front runner with a series of well documented RE&FM research projects. Already in 2008 a FM Value Map was presented in a book/collection of thirty-six cases of FM best practice in the Nordic countries. In my own country, since 2005 the Norwegian University of Science and Technology (NTNU) has offered a Master's course in Real Estate and Facilities Management, where RE&FM issues with adding value potential are generally addressed throughout the programme.

In this second volume on the Added Value of Facilities Management I am particularly pleased to note the multiple identifications of close relations between Corporate Real Estate Management and Facilities Management from a value adding perspective. These relations appear the more evident – and obvious – when focusing on expectations for added value as achievements of management excellence on a broad scale. I feel that this book has at least a couple of important missions in top corporate level environments. Primarily, by analysis, research and practical examples it will convince organisations' strategic levels about opportunities for added value that may be achieved for their particular requirements in terms of successful and well managed holistic CRE/FM concepts. Second – but no less important – by the same impulse it will inspire the further development of common understanding, professional excellence and maturity in the field of Corporate Real Estate Management and Facilities Management on all levels and across borders.

Olav Egil Sæbøe
Owner and Senior Adviser, Pro-FM Consulting, Oslo, Norway
Management Adviser, Lecturer and External Examiner in FM Master's
programmes at the Norwegian University of Science and Technology (NTNU),
Lecturer in FM Bachelor programmes at Oslo and
Akershus University College of Applied Sciences (HiOA)

In the late 1990s I presented a keynote address on added values in Corporate Real Estate at a EuroFM meeting in Rotterdam. The focus was on seven value parameters: increasing productivity, reducing costs, facilitating flexibility, increasing the value of assets, controlling risk, supporting organisational culture, and enhancing marketing/PR. This led to a fierce debate, and we would later find out why this topic is so complex. The performance of organisations is dependent on so many variables that it is very difficult to establish causal relationships between the nature of facilities and the performance of organisations. That is exactly why I admire the tenacity of Per Anker Jensen and Theo van der Voordt. Together they have been doing research in this field for several years, and they keep on going. Together with Theo van der Voordt I have supervised various PhD candidates working on this subject at Delft University of Technology, including Jackie de Vries (2007) on *Performance by Real Estate*, Chaiwat Riratanaphong (2014) on *Performance Measurement of Workplace Change in Two Different Cultural Contexts*, and Johan van der Zwart (2014) on *Building for a Better Hospital: Value Adding Management and Design of Health Care Real Estate*. There I saw Theo's determination to find more evidence for the added value of FM and Real Estate. In between Theo also co-supervised Alexandra den Heijer (2010) who wrote a PhD thesis on *Managing the University Campus*, with Hugo Priemus being her promotor. I am happy that outside Delft several people are also involved in research in this field. It is good to notice that EuroFM has put the added value of FM and CREM high on the research agenda.

My combination of functions – CEO of a group of companies dealing with management, consultancy and software development for the real estate and construction sector and professor of Real Estate Management at Delft University of Technology – allows me to see which theoretical concepts are useful in daily practice. I see a need for more evidence on how FM/CREM can add value to organisations and their primary processes. Knowing the costs and benefits of different interventions can support decision makers to find solutions that provide value for money and fit with the different needs, preferences and conditions of all stakeholders. I use the added value concept very often in discussions with my own staff and with our clients, for instance to discuss the added value of a new office concept or an investment in new university buildings. It convinces me that research in this field is valuable for both science and practice. A lot still needs to be done…

I am really pleased to see that the earlier book by Per Anker Jensen and Theo van der Voordt on the Added Value of Facilities Management has now got a successor that integrates insights from both FM and CREM disciplines and elaborates twelve value parameters by presenting a conceptual analysis, and state of the art theory, research and practice. It shows us the ways to measure and manage added value.

Hans de Jonge
CEO of Brink Groep, The Netherlands
Professor of Real Estate Management at the Faculty of Architecture
and the Built Environment, Delft University of Technology,
the Netherlands

Abbreviations

AV	Added Value
BCM	Business Continuity Management
BIFM	British Institute of Facilities Management
BREEAM	Building Research Establishment's Environmental Assessment Method
BSC	Balanced Scorecard
CAFM	Computer Aided Facility Management
CEN	Comité Européen de Normalisation (European Committee for Standardization)
CEO	Chief Executive Officer
CFM	Centre for Facilities Management, Technical University of Denmark
CFO	Chief Financial Officer
CfPB	Center for People and Buildings
COO	Chief Operation Officer
COSO	Committee of Sponsoring Organizations of the Treadway Commission
CR	Corporate Responsibility
CRE	Corporate Real Estate
CREM	Corporate Real Estate Management
CRERM	Corporate Real Estate Risk Management
CSR	Corporate Social Responsibility
DALY	Disability Adjusted Life Years
DGNB	German Building Sustainability Certification (Deutsche Gesellschaft für Nachhaltiges Bauen)
DTU	Technical University of Denmark (Danmarks Tekniske Universitet)
Ebit	Earnings before interest and taxation
EFMC	European Facility Management Conference
EPC	Engineering, Procurement and Construction
EVA	Economic Value Added
FM	Facilities/Facility Management
FS	Facilities/Facility Service
fte	full time equivalent
GFA	Gross Floor Area
HR	Human Resources

H&S	Health and Safety
HRM	Human Resource Management
HSSEQ	Health, Safety, Security, Environment, Quality
HSSE	Health, Safety, Security, Environment
HVAC	Heating, Ventilating and Air Conditioning
I-FM	Integrated FM
IAQ	Indoor Air Quality
ICT	Information and Communication Technology
IFMA	International Facility Management Association
ISO	International Organization for Standardization
IT	Information Technology
KPI	Key Performance Indicator
LCC	Life Cycle Costing
LEED	Leadership in Energy and Environmental Design (American Environmental Certification Scheme)
MTM	Motion and Time Methods
NFA	Net Floor Area
OCAI	Organisational Culture Assessment Instrument
PDCA	Plan, Do, Check, Act
P&O	People and Organisation
POE	Post-Occupancy Evaluation
PPP	Public Private Partnership
RA	Risk Assessment
RE	Real Estate
RI&E	Risk Inventory and Evaluation
RM	Risk Management
RMAA	Repair, Maintenance, minor Alteration and Addition
ROI	Return On Investment
SBS	Sick Building Syndrome
S&I	Space and Infrastructure
SIA	Sustainability Impact Assessment
SLA	Service Level Agreement
SMART	Specific, Measurable, Assignable, Realistic and Time-bound
SWOT	Strengths, Weaknesses, Opportunities, Threats
TVA	Total Value Added
UN	United Nations
UNCED	United Nations Conference on Environment and Development
VAM	Value Adding Management
VSM	Value Survey Module
WCED	World Commission on Environment and Development
WODI	Work Environment Diagnostic Instrument
WPI	Workplace Innovation
ZHAW	Zürich University of Applied Science

Contributors

Rianne Appel-Meulenbroek is Assistant Professor in Corporate Real Estate Management and co-editor of the *Journal of Corporate Real Estate*. She holds the degrees of MSc and PhD in Real Estate Management. Besides research and teaching she has four years of experience from practice as a consultant. She is a member of the board of the European Real Estate Society (ERES) and track chair at ERES' yearly conferences. Her research focuses on Corporate Real Estate (CRE) management and the ways that building design can support an organisation, through satisfaction of employees, innovation, knowledge sharing etc.
Affiliation: Eindhoven University of Technology, Faculty of Architecture, Building and Planning, Urban Science and Real Estate Unit, www.tue.nl

Feike Bergsma is a Senior Lecturer and Researcher in Facility Management at the Hospitality Business School of the Saxion University of Applied Sciences. He holds an MSc in Business Administration from the University of Twente. As a facility manager, he was responsible for the introduction of innovative ways of working in educational buildings, the integration of facility organisations and the development of new service concepts. Currently Feike is involved in developing study and research programmes. Within the FM Added Value Programme he and a growing team of researchers, master's and bachelor students are exploring new ways of making FM more visible and tangible to various stakeholders.
Affiliation: Saxion University of Applied Sciences, Hospitality Business School, Facility Management, www.saxion.edu/site/about-saxion/research-centres/hospitality/

Sandra Brunia is Researcher and Project Leader at the Center for People and Buildings (CfPB), a not-for-profit scientific knowledge centre that develops knowledge and decision support tools regarding the relations between people, work processes and work environments. She holds an MSc degree in Culture, Organisation and Management from the Faculty of Social Sciences, VU University Amsterdam. Her research focuses on the work environment from an end-user perspective, including use and satisfaction, cultural aspects and how to match the work environment with the working processes in the organisation.
Affiliation: Center for People and Buildings, Delft, the Netherlands, www.cfpb.nl

Iris de Been works at the Center for People and Buildings (CfPB) in Delft as a Researcher. She graduated from the University of Amsterdam with a master's degree in Psychology, specialising in Cognitive and Environmental Psychology. Through applied research she focuses on the evaluation of (office) buildings from a user's perspective. In her work she also aims to bring case specific results together in order to develop more generic knowledge about the experience of office buildings.
Affiliation: Center for People and Buildings, Delft, the Netherlands, www.cfpb.nl

Rob Geraedts is Associate Professor in Design and Construction Management (DCM) at the Faculty of Architecture, Department of Management in the Built Environment (MBE) at the Delft University of Technology. His present research focuses on Design and Construction Management, the flexibility or adaptability of the product (buildings), the flexibility of the process (design and construction), the transformation of vacant buildings into new functions, and Open Building to meet continuously changing individual user demands. Since 1996 he has been an active member and scientific reviewer of CIB W104, Open Building Implementation.
Affiliation: Delft University of Technology, Faculty of Architecture, Department of Management in the Built Environment, www.mbe.bk.tudelft.nl

Brenda Groen is Associate Professor (Lector) in Experience and Service Design at the Hospitality Business School, Saxion University of Applied Sciences, the Netherlands. She has twenty years of experience as a Lecturer in Facility Management. For several years she was the course director of the Master's in Facility and Real Estate Management, and of an MBA programme. Her current research interests are hospitality in healthcare, the visual representation of hospitality, host-guest interactions in the hospitality industry, new ways of working, and service design (thinking).
Affiliation: Saxion University of Applied Sciences, Research Group Experience and Service Design, www.saxion.nl/hospitality/

Geir Karsten Hansen is Professor in Architectural and Facilities Management at the Department of Architectural Design and Management and Head of the Centre for Real Estate and Facilities Management at NTNU. His main field of research is programming, the evaluation of processes and buildings, and the usability of buildings related to a user and organisational perspective where theory, methods and tools are developed through several projects and case studies. Hansen has previously worked with topics like adaptability and flexibility and the new use and transformation of existing buildings, and has run several architectural design courses on this.
Affiliation: Department of Architectural Design and Management/Centre for Real Estate and Facilities Management, Faculty of Architecture and Fine Art, Norwegian University of Science and Technology, www.ntnu.no/ansatte/geir. hansen/

Barry Haynes is a Principal Lecturer at Sheffield Hallam University. His main teaching includes Facilities Management (FM) and Corporate Real Estate Management (CREM). He is currently a member of the editorial advisory board for the *Journal of Corporate Real Estate*. In addition, he has acted as guest editor for the *Journal of Corporate Real Estate* and *Facilities*. He is co-author of the text book *Corporate Real Estate Asset Management: Strategy and Implementation*. His research focuses on the impact the workplace has on occupier performance. His research aims to evaluate how FM and CREM can add value through strategic alignment.
Affiliation: Sheffield Hallam University, Faculty of Development and Society, Department of the Natural and Built Environment, www.shu.ac.uk/faculties/ ds/nbe/staff/barry-haynes.html

Jan Gerard Hoendervanger works as a Senior Researcher and Lecturer at Hanze University of Applied Sciences in Groningen, the Netherlands, with a focus on various aspects of Corporate Real Estate Management. He is co-author of a (Dutch) textbook that integrates Facility Management and Real Estate Management in the whole life cycle of accommodating organisations. In his PhD research he analysed the psychological aspects of activity-based work environments. Jan Gerard collaborates with practitioners, students and fellow researchers from various disciplines within the Hanze 'NoorderRuimte' knowledge centre, focusing on healthy work environments. He has a background as a management consultant in the field of Corporate Real Estate and New Ways of Working, and was primarily educated in Management Science at the University of Twente.
Affiliation: Hanze University of Applied Sciences in Groningen, http:// nl.linkedin.com/in/jangerardhoendervanger/

Aart Hordijk is a retired part-time Professor of Real Estate Valuation at Tilburg University. As of 2005 he holds a PhD from Maastricht University on "Valuation and Construction Issues in Real Estate Indices". As guest professor he was also involved in Real Estate Valuation courses at Delft University of Technology, and was guest Lecturer at the Corporate Real Estate Management programme. He has been working in real estate commercial business for thirty years – first as a private investor in Real Estate, and then later he started DTZ Research and worked at ABP-Real Estate. For fifteen years he was director of the ROZ IPD property index. He has also been the representative for the Netherlands on the IVSC (International Valuation Standards Committee) for more than ten years.
aartchordijk@gmail.com

Per Anker Jensen is Professor in Facilities Management and Head of the externally funded Centre for Facilities Management – Realdania Research, Technical University of Denmark (DTU). He holds the degrees of MSc in Civil Engineering, PhD and MBA. Besides research and teaching he has twenty years of experience from practice as a consultant, project manager and facilities manager. He was a member of the board of EuroFM and chairman of EuroFM's Research Network Group in 2007 and 2008. He is currently project manager of the joint EuroFM research project on The Added Value of FM, which started in January 2009.

Affiliation: Centre for Facilities Management – Realdania Research, DTU Management Engineering, Technical University of Denmark, www.cfm.dtu.dk, www.man.dtu.dk

Keith Jones is Professor of Facilities Management and Head of the Department of Engineering and the Built Environment at Anglia Ruskin University. His expertise covers a wide range of built environment and management issues, including sustainability analyses of existing social, economic and technical systems within the built environment, the design and development of new sustainable systems/practices for facilities management, the impacts of climate change on the vulnerability, resilience and adaptive capacity of local communities, and occupant behaviour and carbon reduction in buildings. He is the leader of the CIB Working Commission W070 on Facilities Management and Maintenance.

Affiliation: Department of Engineering and the Built Environment, Anglia Ruskin University, www.anglia.ac.uk

Antje Junghans is Professor in Facilities Management at the Department of Architectural Design and Management, Faculty of Architecture and Fine Art, NTNU. She has initiated and is currently leading research on Sustainable Facilities Management. Ongoing research is aimed at the improvement of energy efficiency in non-residential buildings, highlighting different approaches towards sustainability in FM from building, user and management perspectives. Additional research activities intend to develop a general perspective on the scope of FM and efforts towards understanding the FM discipline and its development. She is an active member of international research groups and scientific committees, and working as scientific reviewer.

Affiliation: Department of Architectural Design and Management/Centre for Real Estate and Facilities Management, Faculty of Architecture and Fine Art, Norwegian University of Science and Technology, antje.junghans@ntnu.no

Frans Melissen is Professor (Lector) in Sustainable Business Models at the Academy of Hotel and Facility Management, NHTV Breda University of Applied Sciences, the Netherlands. In his research he focuses on the link between sustainability and human behaviour, with special emphasis on mitigating the social dilemma by means of sustainable business models. Frans has contributed chapters to various (text)books, co-authored the management book *Workin' Wonderland* and (co)authored a number of papers in refereed journals, including the Highly Commended Paper Literati Award-winning paper "Facilities management: lost, or regained?" in *Facilities*.

Affiliation: NHTV Breda University of Applied Sciences, Academy of Hotel and Facility Management, www.hospitalityatwork.nl

Giulia Nardelli is a post-doctoral Researcher at the Centre for Facilities Management – Realdania Research, Technical University of Denmark (DTU). In her current research she focuses on the management of stakeholder relations throughout innovation and organisational change processes in facility management services. Giulia holds a PhD in Innovation and Facility Management from Roskilde University (Denmark) and received the European FM Researcher of the Year Award 2014.
Affiliation: Centre for Facilities Management – Realdania Research, DTU Management Engineering, Technical University of Denmark, www.cfm.dtu.dk, www.man.dtu.dk

Susanne Balslev Nielsen is Associate Professor at DTU Management Engineering, Division of System Analysis, and a Deputy at the Centre for Facilities Management Realdania Research, Technical University of Denmark (DTU). Her research addresses the needs of professionals and advances knowledge about the challenges and benefits of integrating the sustainability perspective in context specific engineering. Susanne is educated as a Civil Engineer (1993) and received her PhD (1998) for her study of the sustainable transformation of urban infrastructure. She has been Professor II in Facilities Management at Oslo and Akershus University College of Applied Sciences in Norway since August 2014.
Affiliation: Centre for Facilities Management – Realdania Research, DTU Management Engineering, Technical University of Denmark, www.cfm.dtu.dk, www.man.dtu.dk

Nils O.E. Olsson is a Professor in Project Management at the Norwegian University of Science and Technology (NTNU). He has a PhD from NTNU and a Master of Science degree from Chalmers in Sweden. Recently his main research focus has been on project users and clients, as well as on project flexibility. His current research includes project owner perspectives on large projects and big data applications in project management. He has extensive experience as a consultant, research scientist and manager. He has served as research director of the Concept research programme and director of the programme for continuing education in project management at NTNU.
Affiliation: Department of Production and Quality Engineering, Norwegian University of Science and Technology, Trondheim, www.ntnu.edu/employees/nils. olsson

Alexander Redlein is a Professor in Real Estate and Facility Management, President of REUG and a past president of IFMA Austria. After his inter-disciplinary studies at the Vienna University of Technology and at the Vienna University of Economics and Business Administration, he has been engaged in research, education and consultancy in the area of FM for more than twenty years. In numerous projects he has acted as a strategic advisor, setting up FM concepts for international companies. As a researcher he conducted international studies about the status quo of FM in Central and Eastern

Europe, and about the value added by FM/RE. In addition, he heads a MBA course in FM at the TU Wien, as well as several FM certification courses in CEE and in India.

Affiliation: Institute for Real Estate and Facility Management (IFM), Vienna University of Technology, www.ifm.tuwien.ac.at

Hilde Remøy is Assistant Professor in Real Estate Management. Her focus of research is the adaptive reuse of real estate. In 2010 she completed her PhD research on the adaptive reuse and conversion of offices into housing at Delft University of Technology. Hilde teaches, does scientific and contract research on adaptive reuse, and frequently publishes articles on the topic. She is a supervisory board member of SHS Delft, a student organisation developing student housing by temporary adaptive reuse, and a board member of the European Real Estate Society (ERES). Hilde got her MSc degree in Architecture and Urbanism at the Norwegian University of Science and Technology and at the Politecnico di Milano. From 1998 to 2005 she worked as an architect at different Dutch architecture offices.

Affiliation: Delft University of Technology, Faculty of Architecture and the Built Environment, Department of Management in the Built Environment, www. bk.mbe.tudelft.nl

Arrien Termaat (MSc) is a Senior Lecturer and Researcher in Facilities Management at the Hospitality Business School, Saxion University of Applied Sciences, the Netherlands. He has eight years of experience as a Lecturer in Facility Management and chairs the education committee for the programme. In his role as Lecturer and Researcher he aims to connect research, practice and education to find new solutions to optimise the use of buildings and workspace. Prior to his academic work he worked at a national Energy Company within the Facility Management and Real Estate department. Arrien has presented at various conferences and is one of the initiators of the Dutch 'Knowledge Network Optimization of Buildings and Workspace' group in the Netherlands.

Affiliation: Saxion University of Applied Sciences, Hospitality Business School, Facility Management programme, www.saxion.nl/fm

Theo van der Voordt is Associate Professor in Corporate and Public Real Estate Management at the Faculty of Architecture of the Delft University of Technology. His current research focuses on workplace performance, the experience and use of new work environments, the design and management of health care real estate, and value adding management of FM and CREM. His research aims to develop and test workplace strategies, conceptual models and practical tools to support data collection and decision making processes. This work is conducted in close co-operation with the Centre for People and Buildings (CfPB) in Delft, a knowledge centre that specialises in the relationship between people, working processes and the working environment.

Affiliation: Delft University of Technology, Faculty of Architecture, Department of Management in the Built Environment, www.mbe.bk.tudelft.nl/, www.tudelft.nl/ djmvandervoordt, www.cfpb.nl

Juriaan van Meel is a Senior Researcher at the Centre for Facilities Management – Realdania Research (CFM), the Technical University of Denmark (DTU). He is also a partner at the Dutch consultancy firm ICOP. As a practitioner, Juriaan has been involved in office projects in the Netherlands, Scandinavia and the Middle East. He has written several books about workplace design, including *Workplaces Today* (2015) and *Planning Office Spaces* (2010).
Affiliation: Centre for Facilities Management – Realdania Research, DTU Management Engineering, Technical University of Denmark, www.cfm.dtu.dk

Martine Vonk is Professor (Lector) in Ethics and Technology at Saxion University of Applied Sciences, Deventer, the Netherlands. She received her PhD from the Institute for Environmental Studies at the Vrije Universiteit Amsterdam. Her research focuses on ethical questions that come along with technological developments, with a focus on clean tech and the circular economy. She has worked as researcher and consultant for organisations and corporations aiming to develop a support base for sustainability, shared visions and practices in the organisation.
Affiliation: Saxion University of Applied Sciences, Research Group Ethics and Technology, www.saxion.nl/hospitality/site/onderzoek/Lectoraten/Lectoraat_ Ethiek_en_Technologie

Abstracts

PART I: INTRODUCTORY CHAPTERS

1. Introduction and overall framework

Per Anker Jensen and Theo van der Voordt

Chapter 1 explains the background and purpose of the book and gives an overview of the structure in its three different parts. The current book is a follow-up to a book on the same topic published in 2012 by the same research group. It aims to integrate the two related disciplines of Facilities Management (FM) and Corporate Real Estate Management (CREM) and to discuss how to manage and measure added value in FM and CREM practice. The most important conceptual frameworks from earlier research are briefly summarised and analysed. Chapter 1 finishes with the presentation of a generalised "Value Adding Management Model", which condenses a common underlying cause-effect model in the earlier frameworks and follows the triplet of input-throughput-output. This new model provides the overall framework for the book, which is explained further in the following chapters in Part I.

Keywords: Background, Purpose, Conceptual frameworks, Process model, Value Adding Management

2. FM and CREM interventions

Per Anker Jensen and Theo van der Voordt

This chapter explains the first part of the generalised Value Adding Management model, i.e. "Intervention" or "Decision on type of change", in a FM and CREM context. The chapter presents a typology with six types of FM and CREM interventions concerning the physical environment, facilities services, interface with core business, supply chain, internal processes, and strategic advice and planning. All types of interventions are explained based on the literature and with examples from practice. The examples include both text boxes detailing small cases and interventions from interviews with practitioners from Denmark and the

Netherlands, where the interventions are related to the prioritised values and KPIs mentioned by the interviewees.

Keywords: Interventions, Typology, Decision on change, Cases, Interviews

3. Value Adding Management

Per Anker Jensen and Theo van der Voordt

This chapter explains the throughput part of the generalised Value Adding Management model, i.e. "Management" or "Implementation", in a FM and CREM context. It introduces the concept of Value Adding Management and presents various conceptual models and research findings on strategic alignment between FM/CREM and core business. It also explores the topics of stakeholder management and relationship management. Stakeholder management and relationship management are closely related. A distinction is made between stakeholder management as related to management of multiple stakeholders, and relationship management concerning a dyadic relationship between two parties. The chapter finishes with a presentation of some results from recent research into the current practice of adding management.

Keywords: Value Adding Management, Implementation, Stakeholder management, Relationship management

4. Value parameters

Theo van der Voordt and Per Anker Jensen

This chapter explains the third part of the generalised Value Adding Management model, i.e. "Added Value" or "Outcome", in a FM and CREM context. It gives an overview of the different added value parameters that have been included in various studies on the added value of FM and CREM and presents the results from our recent research on how FM and CREM practitioners in Denmark and the Netherlands prioritise added value parameters. Based on the literature and our recent research we have made a selection of twelve added value parameters, which form the basis for the chapters in Part II. The chapter finishes by showing how added value can be measured according to the literature and how it is measured in practice.

Keywords: Performance measurement, Prioritised values, Added value parameters, Measuring added value, Performance indicators

PART II: VALUE PARAMETERS

5. Satisfaction

Theo van der Voordt, Sandra Brunia and Rianne Appel-Meulenbroek

This chapter presents some findings from surveys on employee satisfaction in different work environments in the Netherlands and various other European countries. It first discusses why employee satisfaction is relevant for organisations and which factors may influence employee satisfaction. Then the chapter discusses empirical data about employee satisfaction with various building characteristics, facilities and services, and which items are perceived as most important. Based on these analyses typical interventions are presented with related benefits and costs. Furthermore, the chapter discusses how to measure employee satisfaction and suggests a list of eleven topics that should be included in employee surveys to measure employee satisfaction: opportunities to communicate and to concentrate, meeting rooms, seclusion rooms, personal storage facilities, indoor climate, noise levels, chairs, desks and other office equipment, office leisure (e.g. tea/coffee, washroom/shower, restaurant/canteen), general cleanliness and IT services. The chapter ends with a number of questions for future research.

Keywords: Employee satisfaction, Important aspects, Interventions, KPIs, Future research

6. Image

Theo van der Voordt

This chapter explores the concept of image in connection to FM and CREM. It presents state of the art knowledge about the influence of buildings and building-related facilities and services on the image of an organisation and how FM and CREM can be used to support a corporation's identity and express brand values. It also discusses a number of typical interventions to support a positive corporate image. Finally this chapter presents various ways to measure corporate image and possible Key Performance Indicators. The chapter ends with perspectives on further research. The chapter is based on a review of the literature with a focus on FM and CREM related journals and various graduation theses.

Keywords: Image, Identity, Brand values, Corporate Real Estate, Facilities, Management

7. Culture

Theo van der Voordt and Juriaan van Meel

Facilities are not just a functional means of production, but also 'cultural artefacts'. They tell a story about a company culture and corporate identity. Vice versa, facilities can help to shape cultural values and support cultural change, provided that physical change goes hand-in-hand with organisational change and management commitment. This chapter tries to operationalise the relationship between the physical environment and organisational culture. It also discusses typical interventions such as the implementation of new workplace concepts. Benefits can be found in the very visible and tangible nature of the physical environment, making it a powerful means of communication. Costs may come from a 'culture clash' when the chosen interventions are too 'alien' to the organisation. The intangible nature of culture makes it hard to formulate and measure KPIs. The chapter suggests conducting surveys and asking people about their organisation's culture and the extent to which there is a 'cultural fit' with the work environment. Concerning the future, it will be interesting to assess how organisational cultures evolve, if new generations of workers will adopt other cultural values, and how these cultural changes may affect or be affected by buildings, facilities and services.

Keywords: Organisational culture, Workplace design, Facilities Management, CREM

8. Health and safety

Per Anker Jensen and Theo van der Voordt

Buildings, facilities and services can have a substantial influence on health and safety (H&S). Ergonomic furniture may help to prevent or reduce complaints of the arm, neck and/or shoulder. Insufficient lighting, too much noise and unhealthy indoor air may result in work fatigue, headaches, irritation of the eyes, nose or throat, and increased blood pressure. Hazardous materials, harmful substances and radiation may lead to severe diseases. Slippery floors and stairs may cause fall accidents. In industrial plants, ill-considered production processes and poorly designed machines may even kill people. Therefore H&S are relevant values in themselves, but they also have an impact on other values such as productivity, employee satisfaction, Corporate Social Responsibility, sustainability, profitability and risk. This chapter presents state of the art FM/CREM related research on H&S, with a focus on the indoor climate and workplace layout. It also presents the benefits and costs of various interventions to improve H&S. The chapter ends with examples of input KPIs to measure H&S characteristics of the supply side, output KPIs to measure the actual and perceived impact of facilities on health and safety of the end users, and suggestions for further research.

Keywords: Health, Safety, Indoor climate, Lighting, Noise, Spatial layout

9. Productivity

Iris de Been, Theo van der Voordt and Barry Haynes

The economy of developed countries is strongly based on the productivity of knowledge workers, both quantitatively and qualitatively. Measuring the productivity of knowledge workers can be quite a challenge, let alone measuring the specific impact of the building, facilities and services on labour productivity. Research has shown that occupant surveys are effective and cost efficient methods to gain insight into the impact of these factors on (perceived) labour productivity. More objective measures, such as the amount of absenteeism, can complement the more subjective outcomes. Various studies show that physical conditions such as indoor climate and greenery, spatial layout, ergonomics and aesthetics can have a substantial effect on the productivity of knowledge workers. In particular, support for conducting focused work, concentration and communication is essential. It is therefore recommended to at least measure the extent to which people perceive the work environment as supportive to these activities.

Keywords: Labour productivity, Knowledge work, Communication, Concentration, Spatial layout

10. Adaptability

Rob Geraedts, Nils O.E. Olsson and Geir Karsten Hansen

Due to its long technical lifetime a building must be flexible in order to be able to cope with qualitative and quantitative changes in demands. The added value of flexibility is the ability to adapt the building to changing market or user demands, the reduction of the risk of future vacancy, lower adaptation costs of buildings-in-use, higher rental income, happier users, a longer lifespan and as such a more sustainable building. A potential risk is that costly provisions made for future adaptability will not actually be used in a given period. This chapter explores the concept of adaptive capacity, which can be split into three different factors: organisational flexibility, process flexibility, and product flexibility. A first version of a method to define the demand for and to assess the supply of adaptive capacity included 143 indicators. To make it more practically applicable a light version (called Flex 2.0) has been developed with the seventeen most important key performance indicators.

Keywords: Adaptable, Flexible, Sustainable, Life Cycle, Added Value, Circular Economy

11. Innovation and creativity

Rianne Appel-Meulenbroek and Giulia Nardelli

This chapter outlines how dedicated FM and CREM practices in workplace management may contribute to the added value of FM and CREM by sustaining innovation across all the layers of the served organisation. The chapter presents the core benefits and costs of interventions for the innovation and creativity of the served organisation, and proposes a list of the related KPIs and how to measure them. In addition, the chapter highlights how innovation in FM and CREM processes/services may be specifically managed to contribute even more to added value for clients and end users, by increasing the effectiveness and efficiency of FM and CREM practices. The chapter finishes with a presentation of perspectives and reflections for future research.

Keywords: Innovation, Performance measurement, Knowledge sharing, Workplace management, Creativity

12. Risk

Per Anker Jensen and Alexander Redlein

Risk control in FM/CREM should definitely be seen as a benefit, while risks constitute potential costs. Risk Management in practice is a strategic management activity and is usually based on tactical or operational activities concerning Risk Assessment. The chapter identifies seven types of interventions to control risks in FM/CREM related to business continuity, analyses how these interventions can be managed in order to avoid, reduce or transfer risks, and discusses what costs the interventions might incur. Typical general KPIs for RM are cost related, but in relation to business continuity KPIs often concern time, for instance uptime and recovery time. In recent years awareness of climate change has increased the focus on risk in terms of the resilience of the built environment.

Keywords: Risk, Risk Management, Risk Assessment, Business Continuity, Resilience

13. Cost

Alexander Redlein and Per Anker Jensen

Since 2005 the Vienna University of Technology has analysed the demand side of FM on a yearly basis in different European countries such as Austria, Germany, Bulgaria, Italy, Romania and Spain. The goal is to determine the value added of FM and FM departments and the parameters influencing its magnitude. Areas of savings and increase of productivity, and reasons for these effects, are derived directly from a statistically sound sample. The populations for the surveys were the Top 500 companies in the different countries. The research is based on a mixed method approach. The studies provide information about which specific areas are responsible for costs, why they cause costs, and why different cost drivers require differentiated cost planning and cost control. The chapter mainly focuses on cost savings, particularly concerning whether organisations with a FM department have more facility services with savings than organisations without a dedicated FM department, and whether outsourcing can be seen as a cost-saving approach.

Keywords: Cost, Value added, Demand side, Mixed method approach, Statistical analysis

14. Value of assets

Hilde Remøy, Aart Hordijk and Rianne Appel-Meulenbroek

This chapter focuses on the financial side of Corporate Real Estate (CRE). First, the effect of ownership or leasing on the balance sheet is discussed. In addition, the lifecycle effects of ownership are looked at in connection with renovation, restructuring or alternative use. Particular attention is paid to the importance of regularly valuing CRE at market value and the financial risks of not valuing and not strategically managing CRE. The chapter shows that CRE is not always easy to value. Buildings may have specific characteristics without any particular market value, or which are only valuable for similar enterprises and specific use. The value might also be influenced by industry trends or labour costs, followed by shifts of the company's activities to other locations or even other countries. Consequently, active CRE financial management should have a high priority. Involvement in business plans and decisions is essential to fulfil that role.

Key words: CRE ownership and lease, Valuation, Alternative use, Value monitoring, CRE, Financial risk

15. Sustainability

Susanne Balslev Nielsen, Antje Junghans and Keith Jones

This chapter introduces the societal goal of sustainable development and offers a range of indicators to set targets and measure the sustainable performance of FM and CREM. Through an understanding of sustainability theory and practice, FM and CREM professionals will be able to formulate strategic goals and set performance targets that measure the sustainability of their service provision and support the development of alternative service solutions. The indicators can be set against generic FM standards or building certification schemes, or they may be bespoke, reflecting an individual organisation's specific priorities, challenges and aspirations. Whichever approach is adopted a whole lifecycle perspective is essential for assessing the sustainable performance of products and services, and a balanced scorecard approach is recommended to evaluate and compare alternative service delivery options. The indicators and approach presented in this chapter can be applied at the single service delivery level, to the design of new building developments and refurbishment projects, to single buildings, or to a portfolio of buildings to provide FM and CREM support for organisational transition to a more sustainable future.

Keywords: Sustainable Facilities Management, Building assessment, Qualitative and quantitative data collection, Environmental performance

16. Corporate Social Responsibility

Brenda Groen, Martine Vonk, Frans Melissen and Arrien Termaat

This chapter discusses the contribution of Facilities Management (FM) and Corporate Real Estate Management (CREM) to Corporate Social Responsibility (CSR). After a brief introduction to the definitions of and guidelines for CSR, the chapter discusses the choices companies may make regarding the required maturity level of CSR, touching on the level of ambition of the company as a whole and the responsibilities of FM. Next, an input-throughput-output model for CSR is discussed, showing both the prerequisites and outcomes of CSR. In order to determine the contribution of FM and CREM to CSR, a long list of KPIs is described and illustrated by means of examples in annual reports. The chapter ends with a discussion of the underestimation of the contribution of FM and CREM to CSR and calls for further development of relevant KPIs.

Keywords: Corporate Social Responsibility, Sustainability, Triple-P, Annual reports

PART III: EPILOGUE

17. Tools to measure and manage adding value by FM and CREM

Jan Gerard Hoendervanger, Feike Bergsma, Theo van der Voordt and Per Anker Jensen

This chapter connects the simple Value Adding Management (VAM) model from Chapter 1 to the Plan-Do-Check-Act cycle, also known as the Deming cycle. The Plan phase is related to the input part of the VAM model. It includes a strategic analysis to identify the drivers to change, to define the organisational and related FM/CREM objectives, and to explore internal and external conditions. Furthermore, strategic choices have to be made about the interventions that are expected to add most value to the organisation. The Do phase relates to the throughput part and is called 'strategy-in-action'. Its focus is on the implementation of change and change management. The Check phase elaborates the output/ outcome part of the VAM model. It includes checks to see whether the interventions have resulted in improved performance of FM and CREM (output), whether the output improves organisational performance (outcome), whether the outcome fits with the organisational objectives and as such adds value to the organisation, and what side effects (positive or negative) have come to the fore. This step includes performance measurement and the evaluation of appropriate KPIs. The findings form the basis for the Act phase, where decisions on further actions are considered, e.g. extending the current interventions, reconsidering other interventions, or redefining the organisational objectives.

Keywords: Value Adding Management, PDCA-cycle, Tools, Strategy, Interventions, KPIs

18. Reflections, conclusions and recommendations

Theo van der Voordt and Per Anker Jensen

This final chapter reflects on the comprehensive analyses of how to manage and measure the twelve different value parameters that were presented in Chapters 4 to 16. Furthermore it links the findings to the background and purpose of this book (Chapter 1) and the input-throughput-output components of the Value Adding Management model (Chapters 2 to 4). A cross-chapter analysis of the State of the Art sections for each value parameter shows that much theoretical and empirical work has been conducted to operationalise the twelve value parameters, to develop ways to measure and manage performance and added value, and to collect evidence about input-output/outcome relationships. The overview of appropriate interventions, ways to measure their impact, and a shortlist of Key Performance Indicators per value parameter (Chapter 17) can be used to support decision makers in selecting appropriate interventions to solve current problems and to add value to organisations. The cross-chapter analysis also shows that it is difficult to quantify cause-effect relationships, due to the many factors that affect organisational performance and adding value by FM and CREM. In addition to the impact of interventions in buildings, facilities and services, the way interventions are implemented plays an important role as well. The internal context (e.g. leadership, staff, culture, resources) and the external context (e.g. legislation, labour market, benchmark with competitors) may also affect the relationship between facilities and organisational performance. Further research is still needed to disentangle the complex relationships between input factors and organisational outcomes, and to develop new ways to measure the outcomes, for instance by using sensors and apps, narratives, longitudinal observations, focus group discussions and "big data".

Keywords: Value parameters, Evidence, Measuring, Data collection, Cause-effect

Part I
Introductory chapters

1 Introduction and overall framework

Per Anker Jensen and Theo van der Voordt

Introduction

Facilities Management (FM) and Corporate Real Estate Management (CREM) are two closely related and relatively new management disciplines with developing international professions and increasing academic attention. Both disciplines have from the outset had a strong focus on controlling and reducing costs for property, work space and related services. In recent years there has been a change towards putting more focus on how FM/CREM can add value to the organisation. A EuroFM research group established in 2009 and chaired by Per Anker Jensen, Professor in Facilities Management and head of the Centre for Facilities Management (CFM) at the Technical University of Denmark (DTU), has studied this development.

A main result until now was the publication of the anthology *The Added Value of Facilities Management – Concepts, Findings and Perspectives* (Jensen et al., 2012a), which was launched during the European Facilities Management Conference (EFMC) in Copenhagen, May 2012. That publication gave an overview of the many different concepts that have been developed, and contained many important insights on the topic. The present book is a follow-up to the book from 2012 and will focus on instrumental issues, i.e. how to measure added value and how to manage value adding facilities and corporate real estate. Therefore, in the following we will for convenience call the book from 2012 "the first book" without further reference, and the new book "the second book".

The research group has since the first book focused on developing a more practical applicable framework and methods that can provide guidance on how to manage and measure added value. The first activity was a joint workshop for researchers and practitioners during the EFMC in Prague in May 2013. The workshop confirmed that the concept of Added Value is interpreted in many ways and linked to a huge variety of different topics. Prioritisation of different types of added value proved to be highly subjective and dependent on the participant's position, experience and personal beliefs. Most prioritised values included the contribution of FM and CREM to quality of life, the productivity of the core business, user satisfaction and sustainability. The participants found it difficult to mention concrete measures of *how* to add value, partly due to different interpretations of the term "measures" as "interventions" vs. "ways to measure". The answers ranged from concrete measures such as evaluate happiness, satisfaction, work support, creating energy savings in building retrofitting, and take care of shuttle buses and parking facilities for bikes, to abstract measures such as steering on economics, efficiency and effectiveness, or "good price and value for the client".

The editors of the first book were, besides Per Anker Jensen from the DTU, Theo van der Voordt, Associate Professor in CREM at Delft University of Technology, and Christian Coenen, Professor of Marketing and Service Management at Zürich University of Applied Science. This group also organised the workshop in 2013 mentioned above. Since then the research group has been chaired jointly by Per Anker Jensen and Theo van der Voordt. Their next activity was to explore in more depth how people cope in practice with added value, and if and how they incorporate this concept into their daily practice. For this reason they conducted a survey consisting of ten interviews with experienced senior facility managers, corporate real estate managers, consultants and service providers – five in Denmark and five in the Netherlands. The findings confirmed the huge variety in interpretations of added value and the application of concrete measures, with user satisfaction (of employees and clients), cost reduction and improved productivity as the most frequently prioritised values. This survey resulted in a conference paper, which was presented at a session on "The Added Value of FM" at the research symposium during the EFMC in Berlin in June 2014 (Van der Voordt and Jensen, 2014a). The authors were afterwards contacted by the editor of the Australian professional FM Magazine, where a shorter version of the paper was later published as an article (Van der Voordt and Jensen, 2014b) with subsequent articles in a Danish and a European FM magazine (Jensen and Van der Voordt, 2015a, 2015b). Several of the interviews included in Part II of the second book are based on this survey from autumn 2013.

Another important activity by the editors of the second book was to write a critical review of eighteen papers on the Added value of FM and CREM that were presented at the Research Symposia of the EFMC conferences from 2013 to 2015 in Prague, Berlin and Glasgow respectively, and three papers from the CIB 2014 Conference in Copenhagen (Jensen and Van der Voordt, 2015c, 2015d). This review showed that during the last five years much good research on the added value of FM and CREM has been conducted. However, the cumulative creation of a shared body of knowledge is still limited. Furthermore, it appears that most research focuses on one or two value parameters such as usability, end-user satisfaction, productivity or safety. The twenty-one papers primarily discuss the benefits of FM/CREM interventions, and to a much lesser extent or not at all the sacrifices and costs. Most research on added value hardly pays any attention to the implementation process and seldom includes data from both an ex-ante and ex-post evaluation of FM/CREM performance before and after change.

This book

The planning of the second book started in January 2014, when the editors met at Delft University in connection with a PhD defence. The plans were presented to the EuroFM Research Network Group at the member meetings in Helsinki in February 2014. The preparation continued when Theo van der Voordt was guest researcher at DTU in spring 2014, including meetings with Per Anker Jensen and a workshop on the Added Value of FM with researchers from CFM. An open meeting for everybody interested was arranged during EFMC 2014 to involve twelve participants, where an outline of the book was presented.

Over the summer of 2014 the editors sent out invitations to interested contributors. The editors preferred that each chapter should be written in collaboration between two or more authors representing different backgrounds and affiliations, to ensure that the topic was covered comprehensively. Those who expressed an interest in contributing were asked to suggest possible co-authors and afterwards they were asked to write a chapter outline, including an annotated list of content and main sources as well as a work plan with the distribution of contributions and responsibilities of each author of the chapter. The chapter outlines were reviewed by the editors and discussed at the first author workshop in February 2015 in connection with EuroFM member meetings in the Hague. The authors then wrote full draft chapters, which were reviewed by the editors and discussed at the second author workshop in a pre-conference meeting at the EFMC in Glasgow 2015. The authors revised their chapters and the book was finalised and submitted for publication in autumn 2015.

The book addresses a topic of importance for both academics and practitioners. It presents an exemplary application of research-based knowledge to guidance for practitioners. It is based on research results and practical experiences from a number of academics and professionals in different countries. It is the first book that explicitly addresses both FM and CREM and their related communities. It combines a systematic presentation of knowledge and methodology and interviews with highly experienced practitioners building on many years of cumulative international knowledge development.

The book is divided into three parts. Part I – consisting of Chapters 1 to 4 – explains the background and presents a framework for understanding and analysing added value, as well as a list of twelve value parameters. Part II includes a chapter for each of the twelve value parameters, giving a state of the art description of the benefits and costs of typical interventions, an overview of possible KPIs, and recommendations for which and how to use such indicators in practice. Each chapter finishes with perspectives on the need for new knowledge and development. Between these chapters, Part II also includes twelve interviews with practitioners about their experience with managing and measuring added value. Part III is an epilogue, which presents tools for value adding management of a spectrum of value parameters, and overall reflections, conclusions and recommendations.

Definition of added value

In the first book we came up with a generic definition for Added Value of FM based on an extensive list of value terminology (Table 17.3, pp. 274–278 in the first book) and combined with the definition of FM in the European standard. According to the EN15221-1 definition, FM can be defined as "*the integration of processes (...) to maintain and develop (...) services which support and improve the effectiveness of its primary activities*" (CEN, 2006).

Thus, creating *Value* for the core business translates for FM into – as a minimum – delivering and maintaining services that support the effectiveness of the primary activities at a competitive level. Regarding CREM the same holds true, as shown in the widely cited definition of CREM as "*the alignment of the real estate portfolio of a corporation or public authority to the needs of the core business, in order to obtain maximum added value for the business and to contribute optimally to the overall performance of the organisation*" (Dewulf et al., 2000). To create *Added Value* it is necessary for CREM and FM also to develop accommodation, facilities and services so that they improve the effectiveness of the primary activities in an efficient way. As such, added value is a perceived trade-off between the benefits and sacrifices of FM and CREM, i.e. between on one side their contributions to improving the performance of the organisation regarding people, primary and supporting processes, the economy and the surroundings, and on the other side the costs, time and risks connected with achieving these benefits.

Existing conceptual frameworks and process models

An important starting point for the EuroFM research group on the Added Value of FM was the FM Value Map developed by Per Anker Jensen and presented in Figure 1.1. Chapter 3 in the first book describes the making of the FM Value Map. Besides the FM Value Map, from the start of the research group's work there were two main conceptual frameworks for mapping added value from CREM. One framework was developed by Anna-Liisa Sarasoja – then Lindholm (Lindholm and Leväinen, 2006) – at Helsinki University of Technology (today part of Aalto University) during her PhD studies. A simplified version is shown in Figure 1.2. The other framework shown in Figure 1.3 was developed by Jackie de Vries at the Delft University of Technology during her PhD research. Anna-Liisa Sarasoja was a member of the original research group and Theo van der Voordt had been involved in the PhD research of Jackie de Vries, acting as her daily supervisor.

The FM Value Map and the framework of Jackie de Vries both include a basic process model based on input → throughput → output. However, it is used in a different way. In the FM Value Map the process model specifically refers to processes in FM and not to the core business, with input being FM resources, throughput being FM processes and output being FM provisions. The logic of the FM Value Map is that the FM provisions as outputs can lead to different

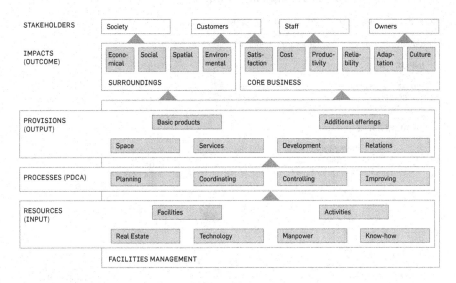

Figure 1.1 FM Value Map, generic version, levels 1 and 2 (Jensen et al., 2008; Jensen, 2010)

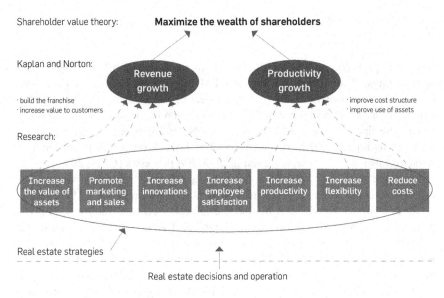

Figure 1.2 Conceptual framework of Anna-Liisa Sarasoja (Jensen et al., 2012b)

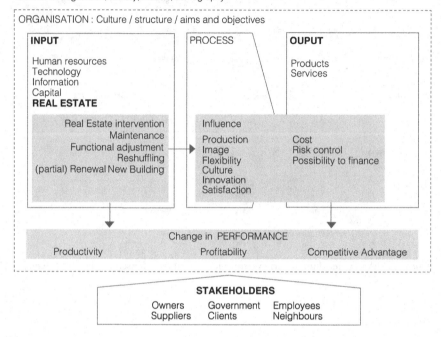

CONTEXT: Legislation, society, market, demography

Figure 1.3 Conceptual framework of Jackie de Vries (De Vries et al., 2008)

types of impacts (outcome/added value parameters) on core business and on the surroundings. The distinction between FM as a support function to a core business is a fundamental part of a good deal of the existing theory on FM – although this distinction is not undisputed. It is even included in the definition of FM in the first European FM standard (CEN, 2006) using the term primary activities to represent the core business. The set of now seven European FM standards also includes a separate standard on FM processes (CEN, 2011).

In Jackie de Vries' framework the process model is related to the overall business organisation and there is no distinction of a separate CREM process as such. The inputs are divided in the five general business resources: Human Resources, Technology, Information, Capital and Real Estate, with real estate viewed as the fifth resource. This is in line with the seminal CREM work by Joroff et al. (1993). Embedded in the process model is a model of real estate intervention that leads to different types of influences (added values) on the business process and business outputs. This embedded model can be seen as a cause-effect model similar to the notion that outputs lead to impacts in the FM Value Map. The framework of Anna-Liisa Sarasoja does not include a process model in a similar way; rather, it is basically structured as a cause-effect model with real estate decisions and operations leading to different types of real estate strategies (added values).

The first book also presented a new framework, developed by Alexandra den Heijer, in Chapter 11 on "Linking Decisions and Performance: Adding Value Theories Applied to the University Campus" (Den Heijer, 2012); see Figure 1.4. This framework was partly based on the framework of Jackie de Vries, but it was redesigned in a different form and extended with various other value parameters.

In addition, the first book presented a further development of the framework of Anna-Liisa Sarasoja, which includes aspects of environmental sustainability (Green FM). At EFMC 2014 another new framework was presented by Gerritse et al. (2104), based partly on the FM Value Map but with inspiration from other frameworks as well. We will not present all these frameworks in this second book, because we want to focus on the essential aspects that unite the existing frameworks in the attempt to develop a very simple model, which is easier to apply in practice and hopefully will become commonly accepted – particularly by practitioners.

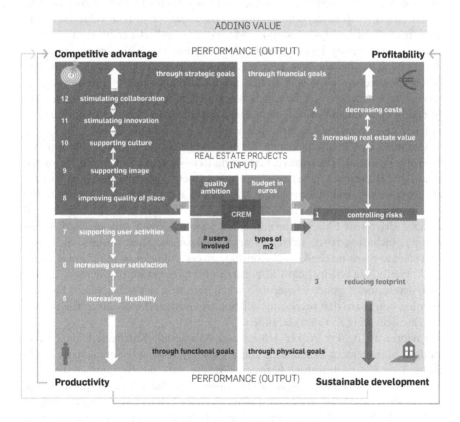

Figure 1.4 Conceptual model of Alexandra den Heijer (2012)

From added value towards value adding management

One of the essential learning points from the first book was that adding value needs to be managed. Interventions in FM and CREM will always have consequences, but to ensure a positive impact on the organisation the interventions need to be managed in a proper and professional way. All the three mentioned conceptual models include stakeholders that benefit from the added value. The model of Anna-Liisa Sarasoja is focused on Shareholders, while the FM Value Map includes Owners, Employees, Customers and Society as stakeholders. The model of Jackie de Vries distinguishes six different stakeholders: Owners, Government, Employees, Suppliers, Clients and Neighbours. The model of Den Heijer clusters all stakeholders in four types: Policy Makers (strategic decisions), Controllers (focus on finance), Users (focusing on usability and experience value) and Technical Managers (facilitate all values in new or existing buildings).

Several chapters in the first book discussed stakeholder and relationship management and the subjective nature of value, making clear that the added value of FM cannot be created without cooperation and understanding the different value perspectives. The different stakeholders have their own roles in the co-creation of value, and they might also perceive values differently. Chapter 5 by Christian Coenen, Keith Alexander and Herman Kok about "FM as a Value Network: Exploring Relationships amongst FM Stakeholders" (Coenen et al., 2012) proposed considering FM as a network of relationships which create perceived value among key stakeholders, i.e. clients, customers and end users. Furthermore, perceived value can only exist and be produced within this specific network of relationships. Chapter 5 in the first book extended the idea of the FM Value Map by taking up a demand-driven, co-creating and subjective perspective of value and differentiating between various dimensions or perceived value in FM.

Chapter 7 by Christian Coenen, Daniel Von Felten and Doris Waldburger on "Beyond Financial Performance: Capturing Relationship Value in FM" (Coenen et al., 2012) presented a relationship management approach designed to gain insights into the field of relationship value in FM and to provide a contrast to the dominant financial perspective of value in FM. Various value dimensions and relevant drivers of FM relationship value were described and analysed in the chapter, including trust, reliability and adaptability. These insights help FM providers to focus on the key relationship drivers and on how to optimise them. Additionally, the results help FM clients to be more aware of the main drivers that determine a good relationship when selecting a FM supplier. The key learning points are that success in collaborative relationships leads to the success of value delivering to the stakeholders.

Chapter 10 by Per Anker Jensen and Akarapong Katchamart on "Value Adding Management: A Concept and a Case" (Jensen and Katchamart, 2012) focused on the relationships between FM and the core business at strategic, tactical and operational levels, and argued that the relationships with the stakeholders should be managed differently at each level. At the strategic level FM should have a business orientation, where considerations for the whole corporation are in focus. This calls for joint decision making involving all main stakeholders at management level, which can take the form of a coalition. At the tactical level FM should have a customer orientation, where the specific needs of each business unit are in focus. This calls for bilateral negotiation and decision making. At the operational level FM should have a service orientation, where the individual users' needs are in focus and the services are either provided based on price per order or based on a service charge. The chapter also presents an interesting case from the Danish construction toy producer LEGO, where such principles of Value Adding Management (VAM) have been implemented. The concept of VAM will be dealt with in more detail in Chapter 3.

The Value Adding Management model

In the presentation of the conceptual frameworks above a general process model can be recognised:

Input → Throughput → Output → Outcome = Impact = Added Value

We also identified an underlying cause-effect model that is included in all the four conceptual frameworks with different wordings, as shown in Table 1.1.

By combining the general process model with the cause-effect model and including value adding management as the intermediary between cause and effect we can define a generalised Value Adding Management model:

Intervention → Management → Added Value

Intervention is used as the general term for cause, and Added Value is used as the general term for effect. This model is very simple and combines essential aspects of the different conceptual frameworks supplemented with management of the implementation of the intervention, to ensure that the FM/CREM interventions lead to added value for the organisation. In relation to the general process model the focus in the generalised Value Adding Management model is on how output by appropriate management can lead to outcome.

This is equivalent to:

Decision on type of change → Implementation → Outcome

Table 1.1 Cause–effect model in the four conceptual frameworks

Framework	Cause	Effect
FM Value Map	Provisions (Output)	Impact (Outcome)
Anna-Liisa Sarasoja	Real estate decisions and operation	Added Value
Jackie de Vries	Real estate intervention	Influence
Alexandra den Heijer	Real estate projects (Input)	Performance (Output)

And also to:

What → How → Why

What is the kind of change and the improvement FM/CREM intends to make to add value; *how* is the way FM/CREM manages the change and implements the improvement; and *why* is the benefit the core business organisation is expected to achieve, i.e. the positive outcome of benefits minus sacrifices in terms of costs, time and risks.

In this book we will use this simple model as a generic framework to further develop our understanding of how to manage and measure the added value of FM/CREM. The three elements in the Value Adding Management model as presented above can be seen as "black boxes". In Chapters 2, 3 and 4 we will open each of these black boxes and reveal what they contain in a FM and CREM context.

References

CEN (2006) *Facility Management – Part 1: Terms and definitions.* European Standard EN 15221-1. European Committee for Standardization.

CEN (2011) *Facility Management – Part 5: Guidance on the Development and Improvement of Processes.* European Standard EN 15221-5. European Committee for Standardization.

Coenen, C, Alexander, K. and Kok, H. (2012) 'FM as a Value Network: Exploring Relationships amongst FM Stakeholders'. Chapter 5 in Jensen, P.A., Van der Voordt, T. and Coenen, C. (Eds.) *The Added Value of Facilities Management – Concepts, Findings and Perspectives.* Centre for Facilities Management – Realdania Research, DTU Management Engineering, and Polyteknisk Forlag, pp. 75–91.

Coenen, C., Von Felten, D. and Waldburger, D. (2012) 'Beyond Financial Performance: Capturing Relationship Value in FM'. Chapter 7 in Jensen, P.A., Van der Voordt, T. and Coenen, C. (Eds.) *The Added Value of Facilities Management – Concepts, Findings and Perspectives.* Centre for Facilities Management – Realdania Research, DTU Management Engineering, and Polyteknisk Forlag, pp. 105–122.

Den Heijer, A. (2012) 'Linking Decisions and Performance: Adding Value Theories Applied to the University Campus'. Chapter 11 in Jensen, P.A., Van der Voordt, T. and Coenen, C. (eds.) *The Added Value of Facilities Management – Concepts, Findings and Perspectives.* Centre for Facilities Management – Realdania Research, DTU Management Engineering and Polyteknisk Forlag, pp. 177–194.

De Vries, J., De Jonge, H. and Van der Voordt, D.J.M. (2008) 'Impact of real estate interventions on organisational performance', *Journal of Corporate Real Estate*, 10 (3), pp. 208–223.

Dewulf, G., Krumm, P. and De Jonge, H. (2000) *Successful corporate real estate strategies.* Nieuwegein: Arko Publishers.

Gerritse, D., Bergsma, F.H.J. and Groen, B.H. (2014) 'Exploration of added value concepts in facilities management practice: Learning from financial institutes'. Paper in Alexander, K. (Ed.) *Promoting Innovation in FM.* Research Papers: Advancing knowledge in FM, *International Journal of Facilities Management, EuroFM Journal,* March.

Jensen, P.A. (2010) 'The Facilities Management Value Map: A conceptual framework', *Facilities*, 28 (3/4), pp. 175–188.

Jensen, P.A. and Katchamart, A. (2012) 'Value Adding Management: A Concept and a Case'. Chapter 10 in: Jensen, P.A., Van der Voordt, T. and Coenen, C. (Eds.) *The Added Value of Facilities Management – Concepts, Findings and Perspectives*. Centre for Facilities Management – Realdania Research, DTU Management Engineering, and Polyteknisk Forlag, pp. 104–176.

Jensen, P.A., Nielsen, K. and Nielsen, S.B. (2008) *Facilities Management Best Practice in the Nordic Countries – 36 cases*. Centre for Facilities Management – Realdania Research, DTU Management Engineering. Technical University of Denmark.

Jensen, P.A., Van der Voordt, T. and Coenen, C. (Eds.) (2012a) *The Added Value of Facilities Management – Concepts, Findings and Perspectives*. Centre for Facilities Management – Realdania Research, DTU Management Engineering, and Polyteknisk Forlag.

Jensen, P.A., Van der Voordt, T., Coenen, C, Von Felten, D., Lindholm, A.-L., Nielsen, S.B., Riratanaphong, C. and Pfenninger, M. (2012b) 'In Search for the Added Value of FM: What we know and what we need to learn' *Facilities*, 30 (5/6), pp. 199–217.

Jensen, P.A. and Van der Voordt, T. (2015a) 'Merværdi for rigtige mennesker (Added value for real people).' *FM Update*, 1, March.

Jensen, P.A. and Van der Voordt, T. (2015b) 'How can facility managers add value?' *EuroFM Insight*, 33, June.

Jensen, P.A., and Van der Voordt, T. (2015c) *How can FM create value to organisations?* Naarden: EuroFM report.

Jensen, P.A. and Van der Voordt, T. (2015d) 'Added Value of FM – A critical review'. Conference paper. *European Facility Management Conference EFMC 2015*, Glasgow, 1–3 June.

Joroff, M.L., Louargand, M., Lambert, S. and Becker. F. (1993) *Strategic Management of the Fifth Resource*. Report no. 49, Atlanta: Industrial Development Research Foundation.

Lindholm, A.-L. and Leväinen, K.I. (2006) 'A framework for identifying and measuring value added by corporate real estate', *Journal of Corporate Real Estate*, 8 (1), pp. 38-46.

Van der Voordt, D.J.M. and Jensen, P.A. (2014a) 'Adding Value by FM: Exploration of management practice in the Netherlands and Denmark'. Paper in Alexander, K. (Ed.) *Promoting Innovation in FM: Research Papers Advancing knowledge in FM. International Journal of Facilities Management, EuroFM Journal*, March.

Van der Voordt, D.J.M. and Jensen, P.A. (2014b) 'Adding value by facility management: A European perspective', *FM Magazine*, October–November 2014, pp. 15–18.

2 FM and CREM interventions

Per Anker Jensen and Theo van der Voordt

Introduction

The purpose of this chapter is to explain what the first part of the generalised Value Adding Management model, called "Intervention" or "Decision on type of change", is in a FM and CREM context. The chapter presents a typology of FM and CREM interventions, which includes both product- and process-related interventions.

The typology consists of the following six types of FM and CREM interventions:

1 Changing the physical environment (on different scale levels: portfolio, building, space)
2 Changing facilities services
3 Changing the interface with core business
4 Changing the supply chain
5 Changing the internal processes
6 Strategic advice and planning.

In our interviews with ten practitioners in Denmark and the Netherlands we asked an open question about the respondents' top five prioritised values (to be presented in Chapter 4), and asked them to specify concrete interventions to attain each value (Van der Voordt and Jensen, 2014). We have classified the total of twenty-nine mentioned interventions according to the typology. The results are shown in Table 2.1. The most mentioned interventions concerned strategic advice and planning, changing the physical environment, and changing the facilities services. There were surprisingly few examples of interventions concerning changing the supply chain and changing the internal processes, and there were no examples of interventions concerning changing the interface with core business.

Table 2.1 Interventions mentioned in ten interviews from Denmark and the Netherlands

Type of intervention	Denmark	The Netherlands	Total
Changing the physical environment	2	6	8
Changing facilities services	5	3	8
Changing the interface with core business	0	0	0
Changing the supply chain	2	0	2
Changing the internal processes	1	1	2
Strategic advice and planning	6	3	9
Sum	**16**	**13**	**29**

In the following six sections each type of intervention is presented with a number of examples.

Changing the physical environment

The physical environment is essential to both FM and CREM. It includes buildings, internal and external spaces, technical services (installations), indoor climate, fitting out, furniture, workplaces, technology, artwork and ambience. Brand (1997) referred to the so-called "six Ss" model to discuss the different life spans of various building layers:

- *Site*: the location of the building
- *Structure*: the construction consisting of columns, loadbearing walls and floors
- *Skin*: the separation between outside and inside, i.e. the façade and the roof
- *Services*: technical installations that are needed to create nice and attractive indoor air quality, and electrical equipment
- *Space plan*: i.e. location of spaces, connections and separations, for instance resulting in a cellular office, a group office, an open office or a combi-office
- *Stuff*: interior design of furniture, floor and wall finishing, kitchen attributes and so on.

For CREM the main focus is on acquisition, the development of real estate and the management of buildings-in-use, while for FM the main focus is on supporting workplaces, the development of working environments, and the delivery of services such as reception, catering, cleaning and ICT. Typical examples of changing the physical environment include:

- Moving to another location (new or existing building); see example in Box 2.1
- New building
- Rebuilding, refurbishment or adaptive re-use, i.e. conversion to new functions
- Changing workplace layout, e.g. conversion of a cellular office with personal desks to an activity-based work setting with shared use of a variety of task-related workspaces
- Changing appearance; see example in Box 2.2.

Box 2.1   Philips' move from Eindhoven to Amsterdam

Before the Second World War, Philips was accommodated in a number of different buildings scattered in the south of the Netherlands. Employees were picked up from home with small shuttle buses and brought back at the end of the day. In order to stimulate employment in the whole region, and also because of various corporate takeovers, the number of locations was increasing. Due to organisational changes, globalisation, technological developments and the move of work to low income countries, the demand for space has drastically changed in recent decades. In the early 1990s Philips started the so-called Centurion Operation. Almost all locations in Eindhoven were closed, moved or became independent. The company also decided to move its headquarter from Eindhoven to Amsterdam because of its closer proximity to Schiphol Airport and the central location in one of the most important financial districts of the Netherlands.

This move evoked much resistance among the Eindhoven municipality, due to loss of employment and municipal income, and the damage to the image of Eindhoven as one of the most important centres of industrial knowledge. The labour unions complained about the loss of jobs and the high costs in a period of heavy cutbacks. Nevertheless, the Executive Board persevered because they wanted to be located in what they perceived to be a Centre of Competence. Since 2001 about five hundred employees are accommodated in the Breitner Tower at the South Axis in Amsterdam. The tower's twenty-three storeys are equipped with high-tech facilities for the control of access, lighting, ventilation and energy saving. This example shows that the added value of a move can be very different for different stakeholders.

Source: Hoendervanger et al., 2012.

Box 2.2   LEGO Look and Feel

The "LEGO look and feel" concept is an example of changing the appearances in a corporate setting. The concept involves the interior decoration and layout of both the main foyer and common spaces in administrative areas with a modern design utilising LEGO products as design objects and thereby putting focus on LEGO's brand for both visitors and staff. The LEGO Service Centre provides projects as part of this concept for a fixed price to LEGO's internal clients and customers.

Source: Jensen and Katchamart, 2012.

The eight interventions concerning changing the physical environment mentioned in our ten interviews are listed in Table 2.2, together with the value each intervention was related to (according to the interviewees) and the KPIs used to measure the value.

Table 2.2 Interventions concerning changing the physical environment mentioned in interviews

FM intervention	Prioritised value	KPIs
Ongoing focus on space utilisation	Core business objectives	
m² reduction most important	Cost reduction	Most tangible and measurable Less m², less movements, lower costs by less building adaptations
Less m², e.g. by New Ways of Working; costs and length of rental contracts. Space management is key Cost reduction by FM is more difficult: slower, more painful, less improvement (e.g. less cleaning, less catering)	Cost reduction	€/m², €/workplace Not €/fte or headcount
Reduce m²	Cost reduction	Cost/m², cost/fte; depends on product (accommodation, IT, catering)
Standardisation of m²/work place and m²/fte Important decision: when to open or close an office	Efficient use of space and other facilities	m²/work place and m²/fte
Changing the size of work desks to create more workplaces, more intense communication and less paper mess	Innovation	
Disposal of real estate to gain cash (depends on book value, local market etc.) Sale and lease back No 'cheese slicer' approach: disposal of one floor is not interesting; focus is on whole locations/buildings	Profit (Ebit); improving cash position	Impact of lower real estate costs on profit Footprints per region Policy depends of region
Choice of materials, e.g. cradle-to-cradle floor and wall finishing and furniture Being able to recycle	Sustainability	Often not really measured (due to costs of measuring)

Changing facilities services

The facilities services (FS) are the operational FM activities. In the European standard on taxonomy for FM the FS are divided in demand related to Space and Infrastructure (S&I) and demand related to People and Organisation (P&O), with both categories sub-divided in standardised facility products as shown in Table 2.3.

Table 2.3 only includes the operational products at the top level. Each of the standardised facility products are sub-divided in up to two more levels. The last FS product in each group – Primary activity specific and Organisation specific – is intended to be a basis for benchmarking across different organisations and sectors. Aspects that are specific to one organisation or sector should be separated from the other FS products. For instance, in a sector using process energy this part of the energy consumption should be separated when benchmarking electricity consumption for ordinary building use. We will not deal further with these FS products.

The taxonomy is based on a lifecycle perspective of the FS products. This is of particular importance for the first product called Space (Accommodation). This covers both the acquisition and the development of buildings (in the taxonomy these are divided into the sub-products Building initial performance, Asset replacement and refurbishment, Enhancement of initial performance, and Portfolio development) as well as the daily operation of buildings (in the taxonomy, divided into the sub-products Property administration, Maintenance and operation, and Utilities). The Space sub-products related to the acquisition and development of buildings are very CREM-related and are covered by the above-mentioned interventions related to changing the physical environment. The other Space sub-products are more related to changing facilities services. Typical examples of such Space related interventions are:

Table 2.3 FM taxonomy with standardised facility products (CEN, 2011)

Demand related to	Standardised facility product
Space and Infrastructure (S&I)	Space (Accommodation)
	Outdoors
	Cleaning
	Workplace
	Primary activity specific
People and Organisation (P&O)	HSSE (Health, Safety, Security and Environment)
	Hospitality
	ICT (Information and Communication Technology)
	Logistics
	Business Support (Management Support)
	Organisation specific

- Changing the property administration from no charging for the use of space to introducing internal rent to increase the transparency of cost and optimise space utilisation
- Changing the maintenance regime from being mostly reactive to being more proactive by focusing on preventive maintenance to improve the conditions of facilities
- Changing the monitoring of the indoor climate to improve temperature regulation and air quality
- Changing the monitoring and management of energy to reduce energy consumption.

Typical examples of interventions related to the other Space and Infrastructure related products Outdoors, Cleaning and Workplace are:

- Changing the gardening of green areas to organic gardening without any use of pesticides to reduce the negative impact on environment
- Changing the cleaning regime from invisible night cleaning to morning cleaning to make the activities visible and establish possible contact between cleaners and users
- Changing the workplaces with more flexible furniture to increase ergonomic quality and adaptability to individual work styles.

The product HSSE, related to People and Organisation, covers Health, Safety, Security and Environment. Typical examples of interventions are:

- Changing the facilities and providing the users with the opportunity to engage in sport and fitness activities in the corporate building
- Changing the monitoring of corporate facilities by installing video surveillance to increase safety and security
- Changing the environmental management to engage the users more in reducing the negative environmental impact in relation to energy and waste.

The facility product Hospitality covers, among other things, catering, reception and meeting rooms. Typical examples of interventions are:

- Changing the catering in canteens to become more ecological and/or increase the selection of different types of food, to accommodate a more international and diverse workforce and to cope with the need for healthy food
- Changing the system for booking meeting rooms to an online system with improved information for users and higher utilisation of the facilities.

An example of research on hospitality in a FM context is presented in Box 2.3.

Box 2.3   Hospitality in hospital facilities

A recent Dutch research project involving three hospitals in the Netherlands explored what aspects of a hospital stay are related most to hospitality according to the patients. The study adopts a definition of hospitality which refers to a host that provides security, psychological and physical comfort for a guest who is away from home, the coming together of a provider and receiver, and a blend of tangible and intangible factors. Hospitality serves as a means for the host and the guest to protect both from hostility. The study also builds on a definition of hospitality as a contemporaneous human exchange, which is voluntarily entered into and is designed to enhance the mutual wellbeing of the parties concerned through the provision of accommodation and food or drink. Third, it adopts the four dimensions of hospitality: spatial, temporal, behavioural, and physical. Many of the tangible factors (accommodation, food and drink, cleanliness) are taken care of by FM, whereas the attitude and behaviour dimensions refer to both FM and medical staff.

The study showed that taking adequate time, listening, involvement and quality of care were associated most frequently with hospitality. Within the behavioural dimension patients valued 'being taken care of' highest. They want to be reassured and put at ease. Other often-mentioned aspects were friendliness, reception, respect, adequate information and empathy. Aspects of space and facilities were also mentioned, but less often, mainly referring to the availability of coffee in waiting areas and adequate space, especially for people in wheelchairs or who use walkers. The results show some particularity of hospitality in a hospital environment compared with similar research from the hospitality industry (hotels etc.). Hospital patients belong to a special type of guest, classified as 'those in need'.

Source: Groen, 2014 (reviewed in Jensen and Van der Voordt, 2015).

Typical examples of interventions related to the other People and Organisation related products, ICT (Information and Communication Technology), Logistics and Business Support, are:

- Changing the ICT services by introducing new software solutions, new mobile solutions and/or improved IT end user services with increased responsiveness from support staff and training of super users
- Changing the handling of post to become more digital by introducing an electronic document handling system, scanning technology and scanning and archiving services.

Table 2.4 Interventions concerning changing the facilities services mentioned in interviews

FM intervention	Prioritised value	KPIs
Waste handling Use of materials Handling of chemicals	Corporate Social Responsibility (CSR)	Energy consumption Water consumption Emission of CO_2 Number of apprentices Number of disabled staff
Meeting facilities that are ready to use	Create time: release the core business staff from having to spend time or effort on support-related tasks which the FM organisation is better suited to carry out	
Good food and coffee in the canteen	Create wellbeing: make good working conditions with attractive work environment and internal services	User satisfaction
Delivery of high quality to retain staff Leadership Social innovation	Engagement/commitment to the company	Audits (surveys) by HRM (e.g. is work environment inspiring?)
Prompting to take the staircase Healthy food	Health	Sick leave Health complaints
Digitalisation of all documents → improved efficiency; new ways of working → better working climate → improved productivity; ditto indoor air quality and comfort	Improving core business/ productivity	Measuring activities: by whom, when; measuring workplace occupancy Many issues not really measured, mainly discussed e.g. new ways of working → better private life
Giving choices, for instance in catering	User centrality and service orientation	User surveys Exit polls Mystery visits
Friday morning brunch	User satisfaction	People's Opinion (external measurements)

The eight interventions concerning changing the facilities services mentioned in our ten interviews are listed in Table 2.4, together with the value each intervention was related to and the KPIs used to measure the value.

Changing the interface with core business

When organisations reach a certain size and complexity FM and CREM are typically established as separate functions. The interface between the core business and FM/CREM is defined specifically in each organisation and is not static. If the FM/CREM function is successful, in many cases it will increase its area of responsibility. This is often part of a centralisation of the responsibility from several parts of the core business organisation to the FM/CREM function. Examples of changes in this interface are presented in Box 2.4.

Box 2.4 Changes in responsibilities regarding the interface with core business

Chapter 2 of our first book included a number of cases from companies participating in a Nordic workgroup on "Highlighting the added values for the core business provided by Facilities Management". Two cases from the Danish pharmaceutical company Lundbeck show a change in responsibility between the core business and the inhouse FM unit. The first case concerned technical production support, where the staff responsible for maintenance of machinery etc. were transferred from various production departments to the FM unit because the equipment became more and more complicated and demanded a higher number of technical disciplines.

The second case concerned the management of internal moving. Previously the departments that needed to move staff had to contact the different service units responsible for ordering handymen to move furniture, rearrange phone connections, reorganise IT equipment and connections and update signage. Instead the FM unit offered a new streamlined moving process, where they provided a one-stop ordering process on the company intranet and took care of the coordination of all the services involved in an internal move of staff. This meant that the coordination of the necessary tasks was transferred from the core business department to the FM unit. This is a typical service management product innovation, which relieves the customers from trivial tasks and implements a smooth service delivery with increased speed, thereby making the organisation more flexible.

Source: Jensen and Malmstrøm, 2012.

Changes in communication and information channels and procedures are typical areas where FM and CREM can change the interface between the core business and the support function and make the business more proactive rather than reactive. For instance, the cleaning staff or the security staff, who get around to most rooms in a building during their work, could be asked to report any faults like non-functioning lighting, cracked windows, bad smells from drains, water leaks, running water in toilets and dripping taps. This makes it possible for the support staff to make the necessary repairs without waiting for ordinary users to report the faults.

Changing the supply chain

In most cases FM is organised as a mixture of an inhouse FM function and a number of external providers of FS, which constitutes a FM supply chain. The situation is to some degree similar for CREM, but the CREM supply chain is more project-related and mostly consists of consultants, designers and contractors. Changes in the supply chain are primarily changes in the delivery process, but they often also have consequences for the incentives for different parties and the management of the mutual relationships between the parties.

The number of external providers varies a lot depending on the type of company and the sourcing strategies. Outsourcing in FM has over the last decades been constantly increasing in most countries and is a common way to achieve cost reductions in FM. Outsourcing of single services to different providers working in parallel for the same company has been a dominant pattern, but recently outsourcing of bundled services has become more common. Integrated FM contracts (I-FM), where one external provider is responsible for most facilities services as well as management, is not unusual today, particularly among multinational companies. A well-known example of an integrated contract including FM and CREM is DBFMO, in which Design, Build, Finance, Maintain and Operate are the bundled responsibility of a private supplier or included in a Public Private Partnership (PPP). The benefits of bundled and integrated contracts are often listed as synergies between different services, more professional management of the service provision and less need for inhouse management resources. For I-FM this is typically supplemented by the existence of one point of contact between the customer company and the provider. Even though the general trend is towards more outsourcing in most countries, there are also many examples of insourcing of formerly outsourced services. Besides, outsourcing may be differentiated with respect to strategic, tactical and operational activities. Box 2.5 shows examples of changes in the supply chain involving insourcing as well as outsourcing.

The two interventions concerning changing the supply chain mentioned in our ten interviews are listed in Table 2.5, together with the value each intervention was related to and the KPIs used to measure the value.

Table 2.5 Interventions concerning changing the supply chain mentioned in interviews

FM intervention	Prioritised value	KPIs
Outsourcing all operational services	Release management resources Line of business can focus on their core tasks	Follow-up on FM suppliers from Sourcing and FM
Long-term relationship/ partnership with key FM suppliers and a "Management by Exceptions" concept: reduce the need for detailed control on adaily business	Satisfaction with service providers	Key figures for price, quality and user satisfaction Monthly, quarterly and yearly follow-up at different management levels

---

Box 2.5   Changing responsibilities in the supply chain

The companies participating in the Nordic workgroup to "Highlight the added values for the core business provided by Facilities Management" showed various examples of a change in responsibility between the internal FM function and external providers. In a case from the Danish pharmaceutical company Lundbeck about standard spare parts, the purchasing was transferred from each individual provider to the FM unit. In another case from the same company about calibrating scales, tasks were transferred to the FM unit from the vendors of each piece of technical equipment involving scales. The motivation in both cases was to increase quality and reduce cost, but the consequence was also a change in the supply chain resulting in a certain degree of insourcing.

Two other cases from the Swedish-based service provider Coor Service Management also provide examples of changes in responsibility, but here the responsibility was transferred from inhouse to the external service provider. In a case on the provision of work wear the responsibility was taken over by the service provider, who reorganised the work process. As a consequence the internal employees were relieved to do other work for the core business. In a case on deliveries for hospital wards the responsibility for ordering and receiving deliveries was transferred to the service provider, who employed dedicated and trained staff to do the work. The nurses and other internal staff were relieved to do the healthcare core business work. Transfer of responsibility from inhouse is an important aspect of adding value for service providers and an essential element in outsourcing. However, neither of these two cases were part of an outsourcing process involving tendering and transfer of staff, but rather part of an ongoing collaboration between providers and clients.

Source: Jensen and Malmstrøm, 2012.

Changing the internal processes

What we deal with here is mostly increasing the efficiency of operational processes within a specific organisation without necessarily changing either the product or the supply chain. The organisation can be inhouse or an external provider. Within management theory and practice there are a number of concepts aimed at increasing productivity and process efficiency, for instance Total Quality Management, Business Process Re-engineering, Benchmarking and Lean Management. Typical elements in such concepts are eliminating waste, implementing new technological solutions and optimising the work flow. Many companies conduct projects using such concepts, and the FM function is often included in the project. Many provider companies also work systematically to develop process innovations, and this is also the case for some of the larger inhouse organisation. Usually reorganising FM in Shared Service Centres (SSC) will change the organisation of support services and result in integration and centralisation, changes in the tasks and responsibilities of FM staff, and multitasking.

One of the main arguments from service providers for the benefits of delivering bundled or integrated services is the possibility to create synergy between services. The same is obviously also possible for an inhouse organisation. An example is to change part-time jobs to full-time jobs, for instance by combining cleaning in the early morning with catering later in the day. Thereby the number of service staff and costs can be reduced, flexibility increased, and work perhaps enriched. An example of multiskilling of FM staff is presented in Box 2.6.

Box 2.6   Multiskilling of FM staff in Copenhagen Airport

A book showcasing thirty-six cases of FM best practice in the Nordic countries presented a case from Copenhagen Airport, where the organisation worked systematically to multiskill their service staff. An important activity in the airport during winter time is snow clearing on runways etc., which is difficult to predict and needs a lot of manpower immediately for short periods of time. Therefore, they have trained many of the service staff to be qualified to operate the snow clearing equipment as a task alongside their usual job in periods when it is needed. They argue that multiskilling both creates flexibility for the airport and provides the staff with higher value on the labour market.

Source: Jensen et al., 2008.

The two interventions concerning changing the internal processes mentioned in our ten interviews are listed in Table 2.6, together with the value each intervention was related to and the KPIs used to measure the value.

Table 2.6 Interventions concerning changing the internal processes mentioned in interviews

FM intervention	*Prioritised value*	*KPIs*
Expression of being sustainable Improving image of FM department, e.g. by improving recognisability of FM staff by providing appropriate clothing	Identity	Monitoring image of FM department, internally (employee monitoring) and externally (customer monitoring)
Best practice sharing	Make processes smarter	Cost measures Response times

Strategic advice and planning

Strategic advice and planning are essential elements in the strategic and tactical activities of FM and CREM. The FM taxonomy (CEN, 2011) includes some products at a strategic and tactical level, including a number of central (horizontal) functions with sub-products mentioned in parentheses:

- Sustainability (Life Cycle Planning/Engineering)
- Quality (Standards and Guidelines)
- Risk (Risk Policy)
- Identity (Innovation).

In Part II there are specific chapters concerning Sustainability (15) and Risk (12), and we refer to these for more information. Concerning identity we refer to Chapter 6 on Image and Chapter 7 on Culture. The areas of strategic advice and planning can cover many other aspects, which will often change over time according to what is of strategic importance for the company.

A typical area of strategic advice from FM to top management concerns the development of a long-term strategy for the corporate property portfolio. This requires a profound and up-to-date understanding of the overall corporate strategy to identify the future demand for property, and close dialogue with the evaluation of options, scenarios and proposals concerning the future supply of property. According to O'Mara (1999), corporate real estate and facilities have two main purposes. One is to support the work processes (see the chapters on Satisfaction, Health and Safety, Productivity, Flexibility and Innovation in Part II) and the other is to symbolically present the corporation (see the chapters on Image, Culture, Sustainability and CSR in Part II). The property strategy should reflect this. Research within CREM has resulted in a number of frameworks to support the development of strategies for property portfolios. The publication by De Jonge et al. (2009) includes a comparison of such frameworks and presents a generic framework for "Designing an Accommodation Strategy" (DAS frame).

Another typical area is investment planning and feasibility studies, which concerns decision support on choosing between alternative options for fulfilling a need for change in the capacity of space or similar. For instance, this might conern whether the company should extend existing facilities, relocate, build a new property, sell or buy property, rent or rent out space, etc. In these kinds of decisions it is often necessary to evaluate both the initial investment costs and the consequences for the ongoing operational expenses. In such evaluations it can be useful to make calculations of Life Cycle Costing (LCC) – also called Whole Life Costing (Jensen, 2008).

The nine interventions concerning strategic advice and planning mentioned in our ten interviews are listed in Table 2.7, together with the value each intervention was related to and the KPIs used to measure the value.

Table 2.7 Interventions concerning strategic advice and planning mentioned in interviews

FM intervention	Prioritised value	KPIs
Close collaboration between tenant and real estate administration about adapting offices to changes in tenant's business activities.	Coherent strategy between core business and FM	
Knowledge about how to create wellbeing, safety and comfort for staff	Productivity of core business	
Guidelines for staff when they move into a new tenancy Use of "cartoons" with drawings of rebuilding projects	Communication	
Challenge the customer	Improvements and innovation	Number of proposed improvements and cost reductions Innovation pools
Priority matrix	Transparency of cost and priorities	Cost
Site master plans/Space Management	Scalability	Space utilisation
Being transparent in annual reports (very important for shareholders)	Transparency of real estate	Compliance Current obligations (e.g. rent contracts) Future liabilities
Interviews with staff about trends and expected growth or consolidation; if main trends are known, then further investigation in depth	Forecasting of future needs	
Selling of corporate campus to generate capital	Right balance between owned buildings and rented buildings (in connection with capital needed for the organisation)	

Conclusion

In this chapter we have presented a number of examples of different types of FM/CREM interventions. The examples are not meant to be exhaustive. Many more interventions are possible to improve the performance of the organisation, to attain the organisational objectives and – when better performance fits with the objectives – to add value to the organisation. In order to get the "best value for money" it is recommended to conduct a SWOT analysis of both the real estate portfolio and buildings-related facilities and services and the organisation, in order to explore interventions that might result in quick wins and to explore which interventions should be prioritised.

References

Brand, S. (1997) *How buildings learn; what happens after they're built*. London: Phoenix Illustrated.

CEN (2011) *Facility Management – Part 4: Taxonomy of Facility Management – Classification and Structures*. European Standard EN 15221-4. European Committee for Standardization.

De Jonge, H., Arkesteijn, M.H., Den Heijer, A.C., Vande Putte, H.J.M., De Vries, J.C. and Van der Zwart, J. (2009) *Corporate Real Estate Management – Designing and Accommodation Strategy (DAS Frame)*. Teaching Compendium, Delft University of Technology.

Groen, B. (2014) 'Contribution of facility management to hospital(ity) issues'. In Alexander, K. (Ed.) *Proceedings of EFMC 2014*, Berlin.

Hoendervanger, J.G., Van de Voordt, T. and Wijnja, J. (2012) *Huisvestingsmanagement. Van strategie tot exploitatie*. Groningen: Noordhoff Uitgevers.

Jensen, P.A. (2008) *Facilities Management for Practitioners and Students*. Centre for Facilities Management – Realdania Research, DTU Management Engineering.

Jensen, P.A. and Katchamart, A. (2012) 'Value Adding Management: A Concept and a Case'. Chapter 10 in Jensen, P.A., Van der Voordt, T. and Coenen, C. (Eds.) *The Added Value of Facilities Management – Concepts, Findings and Perspectives*. Centre for Facilities Management – Realdania Research, DTU Management Engineering, and Polyteknisk Forlag, pp. 164–176.

Jensen, P.A. and Malmström, O.E. (2012) 'The Start of a Nordic Focus on the Added Value of FM'. Chapter 2 in Jensen, P.A., Van der Voordt, T. and Coenen, C. (Eds.) *The Added Value of Facilities Management – Concepts, Findings and Perspectives*. Centre for Facilities Management – Realdania Research, DTU Management Engineering, and Polyteknisk Forlag, pp. 31–43.

Jensen, P.A., Nielsen, K. and Nielsen, S.B. (2008) *Facilities Management Best Practice in the Nordic Countries – 36 cases*. Centre for Facilities Management – Realdania Research, DTU Management Engineering. Technical University of Denmark.

Jensen, P.A. and Van der Voordt, T. (2015) *How Can FM Create Value to Organisations – A critical review of papers from EuroFM Research Symposia 2013–2015*. Research report. A EuroFM Publication, April.

Jensen, P.A., Van der Voordt, T. and Coenen, C. (Eds.) (2012) *The Added Value of Facilities Management – Concepts, Findings and Perspectives*. Centre for Facilities Management – Realdania Research, DTU Management Engineering, and Polyteknisk Forlag, May.

O'Mara, M. (1999) *Strategy and Place – Managing Corporate Real Estate and Facilities for Competitive Advantage*, New York: The Free Press.

Van der Voordt, D.J.M. and Jensen, P.A. (2014) 'Adding Value by FM: Exploration of management practice in the Netherlands and Denmark'. Paper in Alexander, K. (Ed.) 'Promoting Innovation in FM'. Research Papers: Advancing knowledge in FM. *International Journal of Facilities Management, EuroFM Journal*. March.

3 Value Adding Management

Per Anker Jensen and Theo van der Voordt

Introduction

The purpose of this chapter is to explain what the second part of the generalised Value Adding Management model, called "Management" or "Implementation", is in a FM and CREM context. It concerns the notion of Value Adding Management and related concepts.

The first section introduces the concept of Value Adding Management and the second section presents the concept and research findings on strategic alignment between FM/CREM and core business. The following two sections concern stakeholder management and relationship management, respectively. Stakeholder management and relationship management are closely related. We distinguish between stakeholder management as related to management of multiple stakeholders, while relationship management concerns a dyadic relationship between two parties. The chapter finishes with a presentation of some results from our recent research on the current practice of value adding management.

Value Adding Management as a concept

"Value Adding Management" (VAM) and related terms are widely used in the business and management literature. A Google search for "Value Adding Management" returns 11.6 million results, and a search for "Value Add Management" returns 33.0 million results. The combinations of "Value Adding Management" and "Facilities Management" or "Facility Management" return 2.5 and 2.1 million results, respectively.

In the literature related to manufacturing, "Value Adding Management" or VAM is often used in a similar way to Lean Management with a focus on eliminating non-value-adding or "waste" activities. However, VAM is also seen as part of an overriding strategy, where the corporate mission is *what* and VAM is *how* (Anonymous, 2014). This is a parallel to our generalised Added Value process model, but there is no mention of *why*, except indirectly by including "value adding" in the term. The industrial consultant Carlo Scodanibbio even refers to VAM as the philosophy of the second industrial revolution and the guiding light for the year 2000 industries (Scodanibbio, 2014).

The research group on the Added Value of FM finished the conclusion section of our first joint article by stating that one of the next steps in our collaborative research would be to develop practice guidelines for value adding management (Jensen et al., 2012). Our first book included a chapter with the title "Value Adding Management: A concept and a case" (Jensen and Katchamart, 2012). They distinguish VAM from other forms of management in relation to effectiveness and efficiency, as shown in Figure 3.1.

If there is a lack of management focus it is likely that both efficiency and effectiveness will be low, which is the situation shown in Figure 3.1 in the bottom left corner called Laissez Faire Management. The situation where the management focus is on optimising efficiency is shown in the bottom right corner, and is called Industrial Management. This is equivalent to traditional management methods in manufacturing based on Tayloristic scientific management tools like Motion and Time Methods (MTM). Modern concepts like lean or agile management are typical examples of management with a dominating focus on efficiency. The opposite situation, where the management focus is on optimising effectiveness, is shown in the top left corner and is called Preparedness Management. A fire brigade is an extreme example of this situation, where one has an organisation standing by in case of the occurrence of a certain undesirable event, but any management concept where high effectiveness has priority whatever the cost is in this category. VAM is placed in the top right corner where both effectiveness and efficiency have high priority.

In the following we will use "Value Adding Management" or VAM strictly as the management and implementation of FM/CREM interventions with the aim of creating added value to the core business.

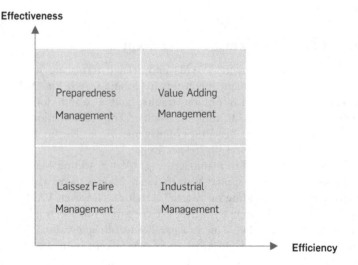

Figure 3.1 VAM compared with other forms of management (Jensen and Katchamart, 2012)

Strategic alignment

An overall aspect of VAM is strategic alignment between FM/CREM and core business. Alignment, in an active sense, implies moving in the same direction, supporting a common purpose, being synchronised in timing and direction, being appropriate for the purpose and, in a passive sense, the absence of conflict (Then et al., 2014).

Regarding CREM, Van der Voordt (2014) developed the scheme in Figure 3.2, connecting the terms alignment and added value to show that corporate real estate only adds value when it supports the organisational objectives.

Figure 3.2 shows that alignment of the accommodation and building related facilities and services requires a thorough understanding of the organisational strategy and its structure, culture, primary processes and so on. When the FM and/or CREM department develops its mission, vision and strategy, this should be done in connection with the mission, vision and strategy of the organisation. FM/CREM interventions should not only be checked according to their impact on FM/CREM performance and organisational performance, but also according to their impact on attaining organisational goals. A better performance does not by definition deliver added value. For instance, if an FM intervention results in a higher ranking on "green buildings" but the organisation was fully satisfied with

Figure 3.2 Connections between alignment and adding value (Van der Voordt, 2014)

the original ranking, this higher ranking does not add any value to the organisation.

Kaplan and Norton, who introduced the Balanced Scorecard, have published a book with the title "Alignment" (Kaplan and Norton, 2006). Their focus is on how the different business units and service units in an organisation can be aligned to the corporate strategy to create synergy. When the enterprise aligns these disparate activities it creates additional sources of value – besides the customer-derived value – which they call enterprise-derived value. Chapter 5 in their book concerns aligning support functions. They recommend a systematic set of processes for support units to create value through alignment:

1   Align support unit strategies with corporate strategies and with business unit strategies by determining the set of strategic services to be offered
2   Align their internal organisation so that they can execute the strategy and develop strategic plans
3   Close the loop by assessing the performance of their functional initiatives.

Arkesteijn and Heywood compared fourteen CRE alignment models that were developed in the period from 1987 to 2013 (De Jonge et al., 2009; Heywood, 2011; Arkesteijn and Heywood 2013). This analysis traced four building blocks and twelve components that are necessary to model CRE alignment – see Table 3.1. Though all studied models included the four main topics, not every model includes all twelve components, but all include at least seven of them.

Table 3.1 Building blocks and components to model the alignment of CRE to organisational strategies (based on Arkesteijn and Heywood, 2013)

	Building Blocks			
	1 Understanding the corporate strategy	*2 Understanding real estate performance*	*3 Making the real estate strategy*	*4 Implementing the real estate strategy*
Components	Business drivers and forces	Audit of existing real estate	CRE strategy (formation)	Implementing the real estate intervention
	Internal strategic drivers	Assess the effect of CREM actions	Strategy integration (alignment)	Implementing the required CREM practices
	Strategic triggers	Real estate market data/information	Integration with other corporate functions	
	Corporate strategy (formation)			

NB. This categorisation differs slightly from a previous publication (Heywood, 2011) as subsequent work has tested and refined the original work resulting in different components and names.

In the following we will summarise three recent studies from FM and CREM research on strategic alignment: Then et al. (2014), Beckers et al. (2015), and Appel-Meulenbroek and Haynes (2014). All these studies involve contributors to this book.

Then et al. (2014) present a conceptual model for the alignment of Real Estate and FM to business needs, inspired by Kaplan and Norton (2006) and the results of an international survey based on the model. The model consists of four alignment variables between four categories:

1   Supply and Demand Alignment: Between Business Need (space, operational functions, revenue etc.) and Facility Solution (real estate)
2   FM Service Alignment: Between Facility Solution and FM Services (service levels, processes and practices)
3   FM Resources Alignment: Between FM Services and FM Resources (people, systems, budget)
4   Organisational Alignment: Between FM Resources and Business needs (policy, organisation structure, culture etc.).

A number of criteria were defined for each alignment variable, and the importance of these criteria were evaluated on a Likert scale from 1 (low) to 5 (high) in an international questionnaire survey, obtaining ninety-seven responses from Brazil, Asia, Europe, USA and Australia/New Zealand. The framework with the average results is shown in Figure 3.3.

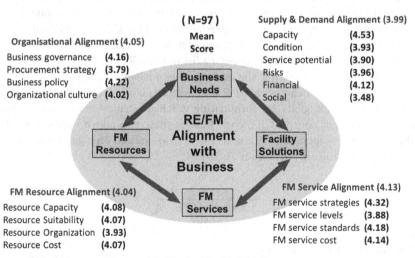

Likert Scale: 1 (low) to 5 (high)

Respondents – Senior RE/FM/Property professionals in USA, UK & Europe, Brazil, Asia through email questionnaire survey or direct contact.

Figure 3.3 Business-RE/FM alignment model with average survey results (Then et al., 2014)

The results of the validation of the framework showed that all four alignment variables achieved average scores around 4.0, with 5.0 as the possible maximum and with FM Service alignment reaching the highest score of 4.13. The criterion with the highest average score was capacity with 4.53, and the lowest was social with 3.48. Both are part of Supply and Demand Alignment, which had six criteria, while the other three each categories had four criteria. One of the main differences in agreement was between the importance of strategy versus cost in the alignment between Facility Solutions and FM Services.

Beckers et al. (2015) carried out an empirical study in aligning Corporate Real Estate (CRE) with the corporate strategies of higher education institutions in the Netherlands. The focus of their study was on the differences in alignment between the espoused CRE strategy and CRE strategy-in-use in relation to both corporate strategy and CRE operating decisions. As such they assessed four types of alignment, as shown in the framework in Figure 3.4. Compared with the alignment model of Then et al. (2014) shown in Figure 3.3, the framework of Beckers et al. (2015) focuses on Supply and Demand Alignment. The empirical research was based on a mixture of qualitative methods, combining document studies and interviews at thirteen of the largest out of a total of thirty-nine Dutch universities of applied science. The sample represents a market share of 75% of all students, and responsibility for 145 buildings with a total of approximately 1.5 million square metres gross floor area.

The results showed that the alignment of espoused strategies had a number of gaps, which were seen to be related to a limited understanding of the possible added value of CRE for the organisation and the end users. CRE strategies-in-use, on the other hand, seemed to be aligned quite well with the corporate goals, which suggests that the CRE managers are aware that CRE can be more than just a support function or a necessity, and of how CRE can add value to serve the corporate interest. The analysis applied a list of ten CRE strategies found in the literature, mostly resembling the added values included in Chapter 4 in this book. The study concludes that CRE managers mainly steer CRE alignment towards corporate goals by formulating CRE strategies that support user activities and that control CRE cost. Besides these two added values of CRE, aligning CRE operating decisions was also related to supporting the organisational culture (Beckers et al. 2015).

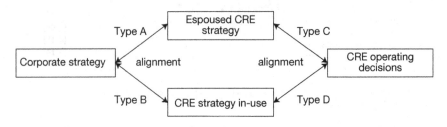

Figure 3.4 Corporate Real Estate alignment framework (Beckers et al., 2015)

Appel-Meulenbroek and Haynes (2014) present a normative paper with an overview of steps and tools for the corporate real estate strategy alignment process. The paper is based on a concept of strategic thinking and several existing models of added value of CRE, applying a basic distinction between exchange value strategies and use value strategies. The focus is mostly on the physical aspects of real estate, including structural, installation and location aspects, but organisational design is also included. The result is a checklist with twelve tool-supported steps that may guide CRE practitioners on their quest for CRE alignment. The twelve steps are divided into three phases starting with situational analysis (five steps), followed by strategy formulation (two steps) and finishing with planning the implementation (five steps).

According to Appel-Meulenbroek and Haynes (2014), however, the alignment of CREM does not end there. The actions need to be evaluated to learn from the process through Post-Occupancy Evaluations (POEs). With regard to exchange value, performance measurement is usually efficiency driven. This means it is largely within the control of the CRE department, but it will also predominantly take place at the operational level. Unfortunately, this is often not the place where added value is proven, but rather the place where costs are justified. In order to be able to talk about adding use value, performance measurement should focus on effectiveness as well. This requires the interconnection between CRE and corporate strategy and therefore portrays full strategic alignment. However, perhaps it is not as important to get the alignment exactly right (as things will change anyway) as it is to adopt a strategic mindset. Strategic management is a continuous process during which tools will have to be reused again and again, until CREM obtains a strategic mindset and CRE strategy alignment becomes evident and natural for general management as well.

Both Beckers et al. (2015) and Appel-Meulenbroek and Haynes (2014) point out that the difference between the time horizon of the real estate strategy and the corporate strategy is a major problem in achieving strategic alignment. Beckers et al. (2015) mentions that the studied strategic corporate plans of Dutch universities of applied science mostly utilise a planning horizon of four years, whereas the planning horizon of CRE decisions can span fifty years or even longer. Appel-Meulenbroek and Haynes (2014) suggest a possible solution to help give the CRE manager the flexibility and agility to respond to changing business demand, which is to shift from owned real estate to a model of real estate on demand. This approach would require working with third parties to provide their real estate needs. However, to ensure that any time delay is minimised as much as possible, the on-demand real estate would have to be fully integrated into the evaluation of the total real estate portfolio. In addition, the CREM tools need to be used as early in the alignment process as possible. Only then can CREM reach its full potential for adding value.

Stakeholder management

The existing frameworks of added value presented in Chapter 1 include various groups of stakeholders. The concept of stakeholders in FM was dealt with in detail in our first book, in Chapter 5 about "FM as a Value Network: Exploring Relationships amongst FM stakeholders" (Coenen et al., 2012a). The chapter presents the stakeholders in FM as shown in Figure 3.5. It shows that there are many relevant stakeholders of FM inside and outside the organisation, but the most relevant in relation to value – the key stakeholders – are the Client, the Customer and the End User on the demand side of FM. This is in line with the European standard on FM terms and definitions (CEN, 2006). An example of research based on this concept is presented in Box 3.1.

Another way to make a distinction between different types of stakeholders is shown in a scheme that was developed by Den Heijer (2011) in her PhD thesis on Managing the University Campus. According to Den Heijer, CREM has to integrate the interests and preconditions of the policy makers, such as the Executive Board and the middle management, the requirements of the end users, the cost effectiveness of interventions, i.e. the responsibilities of financial controllers, and the technical conditions, i.e. the responsibility of the technical managers. She links these four main stakeholders to strategic value, functional value, financial value and energy value respectively; see Figure 3.6. The same scheme can be used to zoom in to a lower scale such as a department or a business unit, or even a particular place such as an entrance hall, or to zoom out to a higher scale such as a real estate portfolio or an urban area.

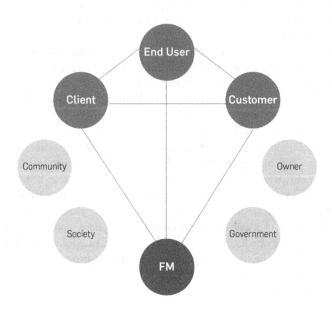

Figure 3.5 Key Stakeholders of Value in FM (Coenen et al., 2012a)

Box 3.1 Example of different appraisals of added value by different stakeholders

Kok et al. (2013) have applied the FM Value Network in an empirical study about "Gaps in Perceived Quality of Facility Services between Stakeholders in the Built Environment". Like the study by Beckers et al. (2015) on CREM alignment presented in section 3.3, the study by Kok et al. (2013) concerns Dutch universities of applied science, but the two studies are not otherwise related. The research methodology was quantitative with an online questionnaire survey covering data from eighteen out of thirty-nine institutions. The study included responses from members of the Boards of Directors (17) as clients, education managers (211) as customers, teachers (1,755) as end users and facility managers (76). The average evaluation of the quality was 5.04 on a seven-point Likert-scale ranging from 1 (very poor) to 7 (very good).

The study produced the unexpected result that the members of the Boards of Directors and the facility managers had very similar and well above average perceptions of the quality of facility services, while education managers and lecturers also had very similar evaluations but these were well below average. This indicates that the quality of FM is well aligned with the expectations of the board of directors but misaligned with the needs of the education managers and the lecturers. As possible reasons the authors suggest that the facilities managers do not manage their relationships with end users effectively, or else that the perception gaps are a result of service standardisation, which serves the purpose of the Board and FM but fails to satisfy users' needs. It may also indicate a flawed expectation management, resulting in unrealistic or too high expectations of end users. These results are interesting, but it is a limitation that the study does not include responses from students.

Source: Kok et al., 2013
(reviewed in Jensen and Van der Voordt, 2015).

Jensen (2011) investigated the organisational relationship between FM and core business at the strategic and operational level with inspiration from theory on governance and forms of coordination. The basic forms of coordination of business activities between different actors or units are, according to this theory, via a market based on price or by hierarchy in an organisation. A number of other forms of coordination, such as coalition and negotiation, have been identified as well (Grandori, 1997).

The conclusions in relation to FM are that for decision-making related to strategic FM concerning common corporate capacity and infrastructure, it is important to create a close collaboration and alignment between the FM organisation and the core business to achieve the necessary business orientation.

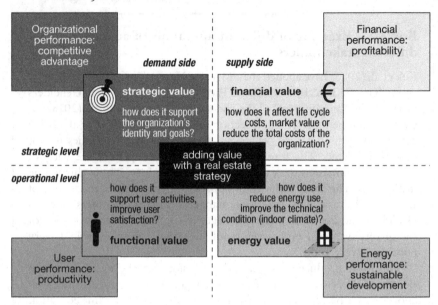

Figure 3.6 Four types of stakeholders in relation to different types of added value (Den Heijer, 2011)

Such collaboration could take the form of a coalition managed by a forum of representatives from FM and the different parts of the company. In contrast, for FM provisions with a differentiation in relation to various internal users, decentralised decision-making seems to be the obvious solution (Jensen, 2011). That is particularly the case where the quality of the provision is easily defined and understood by both parties. In those cases price seems to be the best form of coordination and a service orientation is essential. Examples of this could be cleaning, catering, internal removals, hiring of conference rooms and procuring of standard products. For more complex provisions with the need for dialogue about specific customisation more centralised decision-making may be needed, involving negotiation between managers at some level. Space management issues, like rebuilding projects and workplace design, could be typical examples (Jensen and Katchamart, 2012).

Therefore, it is essential that the relationship management in VAM is differentiated on the three levels shown in Table 3.2. Business orientation means that considerations for the whole corporation are in focus. This calls for joint decision-making involving all main stakeholders among the corporate management, which can take the form of a coalition at the strategic level. Customer orientation means that the specific needs of each business unit or department are in focus. This calls for bilateral negotiation and decision-making at the tactical level. Service orientation means that individual users' needs are in focus at the operational level. In this case the services are either provided based on price per order, for instance catering and transportation, or based on a service

Table 3.2 VAM relationship differentiation (Jensen and Katchamart, 2014)

Level	Demand side	Relationship focus	Coordination form
Strategic	Client	Business orientation	Coalition
Tactical	Customer	Customer orientation	Negotiation
Operational	End user	Service orientation	Price per order/Service charge

charge such as part of internal rent or similar, for instance cleaning and security. An example of collaboration between FM and top management reflecting these principles is presented in Box 3.2.

Kaya et al. (2004) claim that FM units often have problems demonstrating their strategic value. FM units should increase their integration within the organisation, for instance through increased collaboration with other organisational units, and increasingly contribute to the business outcomes. This goal can be achieved by completing roles as translators, processors or demonstrators. As a translator the FM unit should enable strategic alignment by translating the business needs to a strategy that defines the various project activities (interventions) to be implemented. As a processor the FM unit should implement and operate the projects (interventions) contributing to the business needs, and develop relationships with end users. As a demonstrator the FM unit should demonstrate the project's contribution to the business strategy, both in financial and non-financial terms, with a breakdown into different sub-categories, i.e. at corporate, business unit and individual level. The three roles are seen as part of a cyclical sequence.

Box 3.2 Example of collaboration between FM and top management

Jensen and Katchamart (2012) analysed the Danish construction toy company LEGO, where the FM department – a part of the LEGO Service Centre (LSC) – has managed FM in a way that resembles VAM for some years. The management of LSC participates in an annual meeting with LEGO's top management – the leadership team – to evaluate performance and discuss development plans. In order to align strategic management decisions between top management and FM on a continuous basis, LEGO has established an LSC Facility Committee with a main focus on three aspects: projects, capacity and competency. Meetings are held every six weeks and decisions are made jointly by the members of the Facility Committee. The LSC's service levels are negotiated and decided bilaterally with the management of each business unit as customers. The LSC also measures their performance based on satisfaction surveys at regular intervals. These surveys are differentiated in relation to the client, customers, end users and FM staff.

Source: Jensen and Katchamart, 2012

Relationship management

The different aspects of VAM and the studies described above mostly refer to inhouse FM/CREM functions. Many of the principles will be similar for external providers, but their possibility of managing relationships with all groups of stakeholders will often be more limited or more formalised. For external providers the relationship between the customer representative and the provider is of particular importance.

This relationship has recently been the focus of research conducted by a research group headed by our co-editor of the first book, Professor Christian Coenen, ZHAW. The basis for relationship management research in FM is primarily Business to Business (B2B) Marketing research, where there is a strong focus on buyer-seller relationships. For FM this is translated into customer-provider relationships. Even though the provider of FM services can be internal or external, the main focus is on the relationships between contractual partners consisting of a customer organisation and an external service provider. Coenen et al. (2012b) presented an extensive literature review as a basis for developing this research area. They identified various dimensions of relationship value and different relationship benefits as value drivers. The difference between value, quality and satisfaction was investigated as well. Relationship value is considered as an antecedent of relationship quality and also as an antecedent of relationship satisfaction. Relationship quality is seen as consisting of three components: satisfaction, trust and commitment.

A conference paper by Coenen and Schäfer-Cui (2013) investigated the concept of relationship value in FM further, both from a theoretical perspective and empirically. A general theoretical framework of relationship value with nine value dimensions and related drivers was presented. The following seven dimensions were related to benefits: FM (Service) Quality, Service Delivery, Supplier Know-how, Core Business Support, Trouble-shooting Support, Personal Interaction, and Sustainability. The remaining two dimensions – Direct FM Costs (Price) and Process Costs – were related to costs. This framework was the basis for an online questionnaire survey distributed by email to targeted FM customers in different European countries, resulting in sixty responses. The results showed, not surprisingly, that all benefits dimensions had a positive correlation with the overall relationship value, with Supplier Know-how and Service Delivery being most important. However, more surprisingly, both cost dimensions also had a positive correlation with the overall relationship value. The authors explain this by the fact that services are intangible, and customers need to contribute time and effort to discussions with FM providers to make sure that the services are performed correctly and on time. They also conclude that this finding supports the idea of value co-creation between customers and providers.

Research on relationship management in FM shows new ways to develop knowledge on VAM in FM by introducing terminology from marketing psychology, and new ways of combining qualitative and quantitative methodologies. It would be interesting to extend the present research focus on dyadic relationships between customers and providers to the broader FM Value Network involving relationships between multiple stakeholders.

Value Adding Management in practice

As mentioned in Chapter 1, we have recently explored how people cope with the added value of FM/CREM in practice, and if and how they incorporate this concept in their daily practice (Van der Voordt and Jensen, 2014). We conducted ten interviews with experienced senior facility managers, corporate real estate managers, consultants and service providers – five in Denmark and five in the Netherlands. Criteria for selection were a senior level of practical experience, a mix of FM and CREM professionals, and a mix of inhouse FM, service providers and consultants. All interviewees were from the private sector. The final sample represents various sectors such as biotechnology, technical services, maintenance management, FM service providers and consultancy. Years of experience range from twelve to thirty-four years.

Almost all interviewees use the term Added Value (AV) in daily practice, in various settings:

- Internally in inhouse FM organisations, between FM organisations and corporate management, and within provider companies
- Externally between clients and providers (in contract negotiations and ongoing collaboration), clients and consultants, and clients and deliverers of IT systems and equipment.

The AV concept is used both to demonstrate the added value of one's own function or department, and to discuss the added value of FM/CREM interventions. Related terms are Value Creation, Value Increase, Appreciation, Total Value Add (TVA) and Economic Value Added (EVA). In the 1990s AV was mainly linked to Economic Value Add and Shareholder Value, whereas nowadays the concept has a wider scope depending on who you talk to, e.g. a CEO, an operational manager, a supplier or an end user. One of the advantages of applying the AV concept is that the dialogue is moved away from the contractual agreement and the SLAs. According to one respondent: "It makes the customer feel that you are interested in his business and not just in submitting the next bill. It makes it possible to raise the level of the whole FM provision." It helps to speak the language that top managers understand.

Downsides of the AV concept are that AV is perceived differently by different people, and it is therefore difficult to make concrete and operational and to document. AV concerns things that cannot always be measured in economic terms. It is very important to understand which value is most important for the client or customer and what he or she really needs (often more than simply solving the current problem). In addition to sound data, storytelling can also be used to convince clients of the added value of FM and CREM provisions and proposed interventions.

Most practitioners perceive AV as the trade-off between benefits and costs, and focus on value for money and making the core business more effective. AV is connected to the term "value", which has both an economic meaning and meanings related to feelings and other subjective and qualitative aspects such as comfort, making complex things simpler and easier to manage, and high speed delivery. Various interviewees made a distinction between what they called hard economic aspects and more soft aspects related to health, safety, environment and quality.

Benefits are mainly linked to clients, customers and end users, but also to shareholders and – less often – to society as a whole. All respondents include different types of added values, without a clear classification into – for instance – user value versus customer value, or economic value versus environmental value. Practitioners mainly focus on the impact of FM and CREM on the core business and organisational performance, and this is also essential in provider companies' sales arguments.

Though the term AV is not always used explicitly, practice is always concerned with balancing the benefits of e.g. flexibility of short-term contracts, speed of delivery or better quality, and the costs of extra investments or higher running costs. AV depends very much on the client's perception. One of the service providers makes a distinction between the value they provide as part of their standard package at the start of a new contract, and the value they create during the contract. The latter changes a lot depending on what is important for the customer over time.

The focus on particular types of value depends on the involved stakeholders. According to one of the CREM interviewees:

- Shareholders focus almost exclusively on a high Return On Investment (ROI) and low risk, costs and reliabilities
- The board of management usually connects added value to their strategic vision and policy and focuses on maximum turnover (volume of business), minimum costs, and a high Ebit (Earnings before interest and taxation)
- Heads of regional units have to cope with both top management needs (profit), regional customers and employee requirements. They try to find a balance between cost reduction and benefits such as the attraction and retention of talented staff
- Site managers focus more on operational issues and employee satisfaction.

There is also a difference in VAM at strategic, tactical and operational levels. According to one of the CRE managers, adding value at a strategic level means developing site master plans and implementing the real estate strategy. Its focus is on long-term decisions and the avoidance of complaints. AV at a tactical level involves, for instance, speed of delivery and doing what is being asked. Issues an the operational level include cost reduction, employee satisfaction and customer satisfaction. Although AV is mostly addressed at the strategic level, it is of relevance at all levels and for everybody in the FM organisation. It should be part of the organisational culture. However, according to one respondent FM is not really a strategic issue in most organisations, and CEOs are not really interested in FM. Talking about AV at an operational level can even be counterproductive because "operational managers don't have a clue about what AV actually means". Focus points in FM also depend on the context. When the economy is booming avoiding dissatisfaction and disruption might be key issues, whereas in times of economic recession cost reduction will be at the core of strategy. The size of the company is a factor as well; in small firms FM is mainly operational.

One of the interviewees pointed to Maslow's pyramid of needs as a starting point for management of value. In his own words: "FM does not create value by supporting the lower levels in the pyramid. They are taken for granted and you will get criticism if they are not fulfilled, but you will not receive any appreciation if they are fulfilled. That is just doing the work that is necessary. To be appreciated you need to deliver something that is beyond basic expectations."

Besides KPIs there are a number of other ways to visualise or document added value. Providers often prepare performance reviews for their customers at fixed intervals. Other examples are business cases for specific initiatives and reports on finished projects. Added value is also included in communication with stakeholders in less formal ways, as part of ongoing dialogue and storytelling. Management of expectations is an important aspect of adding value. One of the providers was attempting to create an annual added value report on key accounts, but they had not yet managed to find the right way to meet their customers' expectations. Their experience with using the Balanced Scorecard approach was that the economic and people perspectives were quite easy to document, while the customer and process perspectives were much more difficult to measure. It also depends a lot on what triggers the specific customer and user.

The interview survey also included a number of questions about the prioritised added values. This will be presented in the next chapter.

References

Anonymous (2014) *Value Adding Management*. Book chapter – no author and book title indicated. http://www.productivity.in/knowledgebase/Material%20Management/Value%20adding%20management.pdf (accessed 3 September 2014).

Appel-Meulenbroek, R. and Haynes, B. (2014) 'An Overview of Steps and Tools for the Corporate Real Estate Strategy Alignment Process', *Journal of Corporate Real Estate*, 4 (1), pp. 44–61.

Arkesteijn, M.H. and Heywood, C., (2013) *Enhancing the Alignment Process between CRE and Organisational strategy*, Amsterdam: CoreNet Global EMEA summit.

Beckers, R., Dewulf, G. and Van der Voordt, T. (2015) 'Aligning Corporate Real Estate with the Corporate Strategies of Higher Education Institutions'. *Facilities*, 33 (13/14), pp. 775–793.

CEN (2006) *Facility Management – Part 1: Terms and definitions*. European Standard EN 15221-1. European Committee for Standardization.

Coenen, C, Alexander, K. and Kok, H. (2012a) 'FM as a Value Network: Exploring Relationships amongst FM Stakeholders'. Chapter 5 in Jensen, P.A., Van der Voordt,

T. and Coenen, C. (Eds.) *The Added Value of Facilities Management – Concepts, Findings and Perspectives*. Centre for Facilities Management – Realdania Research, DTU Management Engineering, and Polyteknisk Forlag, pp. 75–91.

Coenen, C., Von Felten, D. and Waldburger, D. (2012b) 'Beyond Financial Performance: Capturing Relationship Value in FM'. Chapter 7 in Jensen, P.A., Van der Voordt, T. and Coenen, C. (Eds.) *The Added Value of Facilities Management – Concepts, Findings and Perspectives*. Centre for Facilities Management – Realdania Research, DTU Management Engineering, and Polyteknisk Forlag, pp. 105–122.

Coenen, C. and Schäfer-Cui, Y.Y. (2013) 'Relationship Value in FM: A Customer Perspective'. Paper in Alexander, K. (Ed.) 'FM for a Sustainable Future'. Conference Papers. 12th EuroFM Research Symposium, Prague, Czech Republic, May. *International Journal of Facilities Management, EuroFM Journal*.

De Jonge, H., Arkesteijn, M.H., Den Heijer, A.C., Vande Putte, H.J.M., De Vries, J.C. and Van der Zwart, J. (2009) *Corporate Real Estate Management: Designing an Accommodation Strategy*. Delft: Delft University of Technology.

Den Heijer, A. (2011) *Managing the university campus*. PhD thesis. Delft: Eburon.

Grandori, A. (1997) 'Governance Structures, Coordination Mechanisms and Cognitive Models'. *The Journal of Management and Governance*, 1 (1), pp. 29–47.

Heywood, C. (2011) 'Approaches to aligning corporate real estate and organisational strategy'. In Appel-Meulenbroek, R. and Janssen, I. (Eds.) *European Real Estate Society Conference*. Eindhoven, the Netherlands: European Real Estate Society.

Jensen, P.A. (2011) 'Organisation of Facilities Management in relation to Core Business'. *Journal of Facilities Management*, 9 (2), pp. 78–95.

Jensen, P.A. and Katchamart, A. (2012) 'Value Adding Management: A Concept and a Case'. Chapter 10 in: Jensen, P.A., Van der Voordt, T. and Coenen, C. (Eds.) *The Added Value of Facilities Management – Concepts, Findings and Perspectives*. Centre for Facilities Management – Realdania Research, DTU Management Engineering, and Polyteknisk Forlag, pp. 164–176.

Jensen, P.A. and Van der Voordt, T. (2015) *How Can FM Create Value to Organisations – A critical review of papers from EuroFM Research Symposia 2013–2015*. Research Report. A EuroFM Publication, April.

Jensen, P.A., Van der Voordt, T., Coenen, C., Von Felten, D., Lindholm, A.-L., Nielsen, S.B., Riratanaphong, C. and Pfenninger, M. (2012) 'In Search for the Added Value of FM: What we know and what we need to learn', *Facilities* 30 (5/6), pp. 199–217.

Kaplan, R.S. and Norton, D.P. (2006) *Alignment: Using the Balanced Scorecard to Create Corporate Synergies*. Boston, MA: Harvard Business School Press.

Kaya, S., Heywood, C.A., Arge, K., Brawn, G. and Alexander, K. (2004) 'Raising facilities management's profile in organisations: Developing a world-class framework', *Journal of Facilities Management*, 3 (1), pp. 65–82.

Kok, H., Mobach, M. and Omta, O. (2013) 'Gaps in Perceived Quality of Facility Services between Stakeholders in the Built Environment'. Paper in Alexander, K. (Ed.) 'FM for a Sustainable Future. Conference Papers'. 12th EuroFM Research Symposium, Prague, Czech Republic, 22–14 May. *International Journal of Facilities Management, EuroFM Journal*.

Then, D.S.-s., Tan, T.-h., Santovito, R.S. and Jensen, P.A. (2014) 'Attributes of Alignment of Real Estate and Facilities Management to Business Needs – An international comparative analysis', *Journal of Corporate Real Estate* 16 (2), pp. 80–96.

Scodanibbio, C. (2014) *Training programme – Value Adding Management*. www. scodanibbio.com (accessed 3 September 2014).

Van der Voordt, T. (2014) *Adding Value by Corporate and Public Real Estate*. Position paper, Faculty of Architecture, Delft University of Technology.

Van der Voordt, D.J.M. and Jensen, P.A. (2014) 'Adding Value by FM: Exploration of management practice in the Netherlands and Denmark'. Paper in Alexander, K. (Ed.) 'Promoting Innovation in FM'. Research papers: Advancing knowledge in FM. *International Journal of Facilities Management, EuroFM Journal*, March.

4 Value parameters

Theo van der Voordt and Per Anker Jensen

Introduction

The purpose of this chapter is to explain what the third part of the generalised Value Adding Management model, called "Added Value" or "Outcome", is in a FM and CREM context. The chapter starts with an overview of the different added value parameters that have been included in various studies on the added value of FM and CREM. Next we present the results from our recent research on how FM and CREM practitioners in Denmark and the Netherlands prioritise added value parameters, and how that relates to the FM Value Map developed by Jensen (2010). Based on the literature and our recent research we have made a selection of twelve added value parameters, which we present in the third section. The final section shows how added value parameter can be measured according to the literature, and how it is measured in practice.

Added value parameters in the literature

In the last decade a number of interesting studies have been conducted to improve our understanding of different types of added value and the prioritisation of different values, in different sectors such as offices, higher education and healthcare (Lindholm, 2008; De Vries et al., 2008; Den Heijer, 2011; Prevosth and Van der Voordt, 2011; Riratanaphong, 2014; Van der Voordt and Jensen, 2014; Van der Zwart, 2014).

Table 4.1 presents an overview by Riratanaphong and Van der Voordt (2015) of the value parameters that were discussed in various studies and which have been classified in the six categories of performance measurement mentioned by Bradley (2002). With the division of the Organisational development category into five sub-groups the table provides ten different value parameters with slightly different names. Remarkably, the list of parameters by De Vries et al. (2008) is lacking from this list. Van der Zwart (2014) also compared a number of different lists of added values including Nourse and Roulac (1993), De Jonge (1996) and Den Heijer (2011). He excluded Van Meel et al. (2010) and Jensen et al. (2012), but added Scheffer et al. (2006) and Niemeijer (2013). He also refers to Lavy et al. (2010). This illustrates that so far researchers working on the added

Table 4.1 Different value parameters classified into the six categories

Bradley (2002)	Nourse and Roulac (1993)	De Jonge (1996)	Lindholm & Gibler (2005); Lindholm (2008)	Van Meel et al. (2010)	Den Heijer (2011)	Van der Zwart and Van der Voordt (2013)	Jensen et al. (2012)
1. Stakeholder perception (employee satisfaction)	Promoting HRM objectives	–	Increasing employee satisfaction	Attracting and retaining talented staff	Supporting user activities Increasing user satisfaction Improving quality of place	Increasing user satisfaction	Satisfaction
2. Financial health	Capturing real estate value creation of business	Increasing of value	Increasing the value of assets	–	Increasing real estate value	Improving finance position	–
3. Organisational development	Flexibility	Increasing of flexibility	Increasing flexibility	Increasing flexibility	Increasing flexibility	Improving flexibility	Adaptation
	Facilitating managerial process and knowledge work	Changing culture		Encouraging interaction Supporting cultural change	Supporting culture Stimulating collaboration	Improving culture	Culture
	Promoting marketing message Promoting sales and selling process	PR and marketing	Promoting marketing and sales	Expressing the brand	Supporting image	Supporting image	–
	Facilitating and controlling production, operation and, service delivery	Risk control	–	–	Controlling risk	Controlling risk	Reliability
	–	–	Increasing innovation	Stimulating creativity	Stimulating innovation	Increasing innovation	–

Table 4.1 continued

	Bradley (2002)	Nourse and Roulac (1993)	De Jonge (1996)	Lindholm & Gibler (2005); Lindholm (2008)	Van Meel et al. (2010)	Den Heijer (2011)	Van der Zwart and Van der Voordt (2013)	Jensen et al. (2012)
4. Productivity	–		Increasing productivity	Increasing productivity	Enhancing productivity	Supporting user activities	Improving productivity	Productivity
5. Environmental responsibility	–		–	–	Reducing environmental impact	Reducing the footprint	–	Environmental
6. Cost efficiency	Occupancy cost minimisation		Cost reduction	Reducing costs	Reducing costs	Decreasing costs	Reducing costs	Cost

value of FM and CREM use both similar and dissimilar value parameters and have not yet attained consensus about a standardised list of value parameters.

In our first book (Jensen et al., 2012, pp. 270–274) we made a comparison of the four models presented in Chapter 1 of the current book. One difference is that we used a more recent version of the model by Anna-Liisa Sarasoja, which includes Supporting environmental sustainability (Lindholm and Aaltonen, 2011 – also included in our first book as Figure 17.1, p. 271). The different value parameters have been categorised under the four headings People, Process, Economy, and Surroundings. The results are shown in Table 4.2.

The parameters related to People are quite similar in models A and D. All models include (employee) satisfaction. Model B defines "Culture" as including "Image", but these factors are separated as different parameters in models A and D. Model C only includes "Increase employee satisfaction" under People, but this model is the only model to include "Promote marketing and sale" under Economy, which can be seen as an economical expression of "Image", understood as brand. All four models include at least three parameters for Process with

Table 4.2 Comparison of added value parameters in four models (Jensen et al., 2012)

	A. De Vries et al., 2008	B. Jensen et al., 2008	C. Lindholm and Aaltonen, 2011	D. Den Heijer, 2011
Core business				
People	Image Culture Satisfaction	Satisfaction Culture	Increase employee satisfaction	Increasing user satisfaction Supporting image Supporting culture
Process	Production Flexibility Innovation	Productivity Reliability Adaptability	Increase innovation Increase productivity Increase flexibility	Increasing flexibility Supporting user activities Improving quality of place Stimulating innovation Stimulating collaboration
Economy	Cost Possibility to finance Risk control	Cost	Increase value of assets Promote marketing and sale Reduce cost	Controlling risk Increasing real estate value Decreasing cost
Surroundings		Economic Social Spatial Environmental	Supporting environmental sustainability	Reducing the footprint

many overlaps; the differences can partly be seen as different degrees of sub-dividing. In relation to Economy, model B (the FM Value Map) only includes the parameter "Cost", while the three other more CREM-based models include parameters for "Value of real estate", "Value of assets" or "Possibility to finance". The parameter "Controlling risk" in model D is defined as related to financial goals, but it is also strongly related to the Process parameter "Reliability" in model B. In model A "Risk control" is included as well, partly related to reducing financial risks, but also to improving health and safety. Model B was the first model to include parameters related to Surroundings, including the "Environmental" parameter, but the other more recent CREM-based models, C and D, also include a parameter for "Environmental sustainability".

Added value parameters in practice

In Chapters 2 and 3 we presented some results from a recent study, where we explored how practitioners in Denmark and the Netherlands cope with the added value of FM/CREM (Van der Voordt and Jensen, 2014). In Chapter 2 we presented the kinds of interventions that may be applied in practice to attain added value by FM/CREM and what KPIs are used to measure the outcomes (see Tables 2.1, 2.2 and 2.4 to 2.7). In Chapter 3 we discussed how practitioners manage the added value of FM and CREM. In the following we present results from the same study about how the practitioners prioritise added value parameters.

The interviewees were asked, as an open question: "What are your top five main values to be included in management of accommodations, facilities and services?" The responses per respondent are presented in Table 4.3. The abbreviations between brackets refer to the list of impact parameters in Table 4.4, which depicts the frequencies of the values mentioned in Table 4.3, categorised according to the impact parameters from the FM Value Map (Jensen, 2010). The responses are divided into Danish and Dutch interviewees and in total. The results show that values related to Satisfaction and Cost are most frequently prioritised, but with a striking difference between the interviewees from Denmark and the Netherlands. Satisfaction is seen as more important than Cost in Denmark, while Cost is seen as more important than Satisfaction in the Netherlands. Productivity is also important to all respondents, but more so in Denmark. Values in relation to Adaptation and Environmental are mentioned in both countries, while Culture only is represented in the Netherlands. The remaining four impact factors – Reliability, Economic, Social and Spatial – are not represented in the response to the open question.

The differences in the frequencies of prioritised values by Danish and Dutch respondents might be caused by the different contexts, but also by the selection of respondents with more Dutch representatives from CREM than in the Danish sample.

The open question on prioritised values was followed by a more closed question, based on a list of possible added values found in the literature and

Table 4.3 Main values from open question (impact parameters from Table 4.2 in brackets)

ID	1	2	3	4	5
DK1	Transparency of cost and priorities (C)	Scalability (A)	Release management resources (P)	User satisfaction (S)	Satisfaction with service provider (S)
DK2	Core Business objectives (A)	Innovation (C)	Coherent strategy between Core Business and FM (A)	Productivity of core business (P)	Communication (S)
DK3	Create time (P)	Create wellbeing (S)			
DK4	Satisfaction of outsourced staff (S)	Make processes smarter (C) (P)	Improvements and innovation (P)	User centricity and service orientation (S)	Corporate social responsibility (E)
DK5	Increase energy conscience and CO_2 emissions (E)	Ease of operation (P) (C)	Deliver better service with less or the same cost (S) (C)	Satisfaction (S)	
NL1	Profit (Ebit); improving cash position (C)	Cost reduction (C)	Transparency of real estate data for shareholders (C)		
NL2	Cost reduction (C)	Affordability (C)			
NL3	Sustainability (E)	Cost reduction (C)	Identity (Cu)	Satisfaction (S)	
NL4	Cost reduction (C)	Improving core business/ productivity (P)	Health (S)		
NL5	Efficient use of space (C)	Forecasting future m² needs (A)	Balance between owned buildings, rented buildings and sale and lease back (C)	Forecasting of future capital need (C)	Engagement (Cu)

Table 4.4 Frequency of the main values in Table 4.1 related to impact parameters from the FM Value Map

Impact parameter	Abbr.	Denmark 24 (100%)	Netherlands 17 (100%)	Total 41 (100%)
Satisfaction	S	8 (33%)	2 (12%)	10 (24%)
Cost	C	5 (21%)	10 (59%)	15 (37%)
Productivity	P	6 (25%)	1 (6%)	7 (17%)
Reliability	R			
Adaptation	A	3 (13%)	1 (6%)	4 (10%)
Culture	Cu		2 (12%)	2 (5%)
Economic	Ec			
Social	So			
Spatial	Sp			
Environmental	E	2 (8%)	1 (6%)	3 (7%)

asking how these values are related to the prioritised main values. The list was based on the impact parameters from the FM Value Map (see Figure 1.1) and was divided into impacts on core business (Satisfaction, Cost, Productivity, Reliability, Adaptability and Culture) and impacts on surroundings (Economic, Social, Spatial and Environmental). The findings are presented in Table 4.4. In response to this more closed question all possible outcomes were discussed with the respondents. In the following each of the impact parameters will be commented on based on the interview results. The presentation of each parameter starts with the definition used in the FM Value Map.

Impact on core business

- *Satisfaction* concerns the impact of FM/CREM on satisfaction of customers, staff/end users and owners. One of the respondents mentioned that customer satisfaction has been the most important, but user satisfaction has become increasingly important too. Satisfaction is a subjective parameter and is often measured quantitatively by surveys or more qualitatively, for instance by mystery visits. Survey results are often benchmarked across organisations. It should be noted that values concerning wellbeing and health in Table 4.3 are categorised as part of satisfaction.
- *Cost* covers operational cost, staff turnover and capital investments. Cost reduction is obviously an important measure, but transparency is also mentioned by interviewees from inhouse FM/CREM. An interviewee from a provider mentions that cost receives the most attention when there is a problem. Cost impacts are obviously often measured and also benchmarked, both in € and m² per person, per full time equivalent (fte) or per workplace, occupancy level, total costs of ownership per m², or in terms of affordability, e.g. the ratio between facility costs and total costs of running a business.

- *Productivity* is related to efficiency, low staff absence and effectiveness. Impact on core business productivity can be difficult to measure, but a typical method for providers is to measure the number of proposals for improvements and innovations. Often productivity impact is not measured directly but addressed more qualitatively in discussions, business cases and performance reviews. Impact on productivity is rarely benchmarked.
- *Reliability* is associated with business continuity, security and safety. The respondents' views on reliability varied a lot. One view is that reliability is at the lowest level of the Maslow pyramid of needs and therefore is not a motivation factor, which can add value. Another view is that business continuity has become increasingly important. For one of the interviewees it has top priority, e.g. regarding fire safety and data security. An interviewee in a biotech company mentions that downtime is important to control, and that compliance to legal requirements has top priority. Reliability is mostly measured in terms of response time and business continuity and is not often benchmarked.
- *Adaptation* is linked to foresight, flexibility and responsiveness. Adaptation is mostly considered at a high management level in relation to capital investments and contract negotiations. An inhouse CREM interviewee mentions that technical flexibility and flexibility in renting are becoming more important. Adaptation is rarely measured or benchmarked.
- *Culture* concerns organisational identity, corporate image and corporate brand. For some companies branding is important, but not for others. Some view culture as related to the image of FM and not as a corporate concern. An interviewee from the Netherlands mentions monitoring the image of FM internally (employee monitoring) and externally (customer monitoring), and remarks that external image is often more important than internal image. Engagement, i.e. a sense of belonging and being committed to the company, was mentioned once as well.

Impact on surroundings

- *Economic* covers income, commerce and tax. Some of the interviewees asked for a more clear definition. Others regarded the economic impact of FM on society to be mostly indirect. However, one interviewee claimed that economic impact is his company's reason for being. There were no examples of measuring and benchmarking economic impact.
- *Social* is related to employment, education and integration. Social impact can be important in relation to location of new facilities, and it is important for some service providers in terms of integration. As examples of measures of social impact an interviewee from a provider mentioned the number of apprentices and the number of disabled staff.
- *Spatial* is associated with architectural expression, landscaping and townscaping. Spatial impact is mostly important for inhouse FM/CREM organisations and specialist consultants and is rarely important for service

providers. An inhouse facilities manager mentioned that they participate in working groups with the local municipality concerning transportation and infrastructure. There were no examples of measuring and benchmarking spatial impact.

- *Environmental* concerns resource consumption, pollution and environmental sustainability. There are clear indications that sustainability has become increasingly important, but it is still not given high priority in many companies. The environmental impact is typically measured and benchmarked quantitatively in terms of energy consumption, but in some cases it was also documented qualitatively in terms of choice of environmental suitable materials and treatment of chemicals.

Summary

To summarise, it is clear that the most prioritised values are costs and satisfaction, followed by productivity. These findings were confirmed in research on prioritised values in the healthcare sector (Van der Voordt et al., 2012; Van der Voordt, 2015). Remarkably, four out of ten outcome parameters – reliability and economic, social and spatial impact on the surroundings – were not spontaneously mentioned at all in response to the open question about prioritised values. These issues came only to the fore when we asked for comments on the list of possible added values that was shown after the open questions. Not all values shown on the list – in particular possible impacts on the surroundings – did immediately ring a bell, and some raised different interpretations or misunderstanding. Sustainability was mainly perceived as a building characteristic. Most respondents made no clear distinction between impacts on the core business and impacts on the surroundings, and focused more on a distinction between interventions regarding buildings and building-related facilities and services versus choices regarding the location and the surroundings. Because practitioners use different terms, various responses could not be allocated clearly to one particular value.

Selection of added value parameters

Based on the parameters in Tables 4.1 to 4.4 we have decided in this book to focus on the twelve value parameters listed in Table 4.5. As in Table 4.2 they are organised under four headings, but the Process heading has been changed to Process and Product. The main differences between Table 4.5 and Tables 4.1 and 4.2 are:

- "Health and Safety" is added as a separate parameter, which is in line with Table 4.3
- "Adaptability" has been chosen as parameter instead of "flexibility"; we see flexibility as a building characteristic and as such as an input parameter and adaptability as an outcome

- "Innovation" has been renamed "Innovation and Creativity", which is in line with Van Meel et al. (2008) in Table 4.1, and also a request from the authors of Chapter 11 on this parameter
- "Risk" has been chosen as a parameter, which also covers "Reliability"
- "Sustainability" has been chosen as a parameter, due to its wider scope than "Reducing the footprint"
- "Corporate Social Responsibility" (CSR) has been added and can be seen as covering the Economic and Social parameters in relation to surroundings in model B in Table 4.2 (the FM Value Map by Jensen et al, 2008, Figure 1.1)
- From model A in Table 4.2 (De Vries et al., 2008, Figure 1.3) the parameter "Possibility to Finance" has not been selected, but it is seen as related to "Value of Asset"
- From model B in Table 4.2 the Spatial parameter in relation to surroundings is not included as such, but it can be seen as part of the Image parameter
- From model C in Table 4.2 (Lindholm and Aaltonen, 2011, Figure 1.2) the parameter "Promoting Marketing and Sale" has not been selected, but it is seen as related to Image
- From model D in Table 4.2 (Den Heijer, 2011, Figure 1.4) the parameters "Supporting user activities", "Improving quality of place", and "Stimulating collaboration" have not been chosen:
  - Supporting user activities is seen as included in Productivity
  - Improving quality of place is seen as included in Satisfaction
  - Stimulating collaboration is seen as included in Innovation and Creativity.

Figure 4.1 shows the Value Adding Management model from Chapter 1 with the six types of interventions presented in Chapter 2, the different aspects of value adding management presented in Chapter 3 and the twelve added value parameters presented in Chapter 4. The interventions are expected to have

Table 4.5 Added value parameters used in this book

Group	Parameter
People	Satisfaction
	Image
	Culture
	Health and Safety
Process and Product	Productivity
	Adaptability
	Innovation and Creativity
	Risk
Economy	Cost
	Value of Assets
Societal	Sustainability
	Corporate Social Responsibility

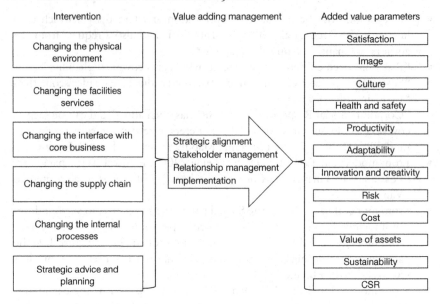

Figure 4.1 Value Adding Management model with types of interventions and added value parameters

potential impacts in different degrees and likelihoods on the different added value parameters. These relationships will be explored in Part II of this book.

Ways to measure added value

In order to be able to define the added value of an intervention by FM/CREM it is important to measure the outcomes of any intervention, ex post and preferably also ex ante, as input to a business case. Clear performance indicators make it possible to assess how well people or facilities perform. The outcomes can provide the inspiration to achieve higher levels of effectiveness, efficiency, quality and competitiveness. As such, performance measurement is an important aid for making judgments and decisions. Performance measurement can help managers to answer five important questions: 1) where have we been; 2) where are we now; 3) where do we want to go; 4) how are we going to get there; and 5) how will we know that we got there (Lebas, 1995).

Sinclair and Zairi (1995) provided a list of seven topics to emphasise the importance and need for performance measurement, stating that performance measurement:

- Enhances improvement
- Can ensure that managers adopt a long-term perspective
- Makes communication more precise ('say it in numbers')
- Helps an organisation to allocate scarce resources to the most attractive improvement activities

- Is central to the operation of an effective and efficient planning, control or evaluation system
- Can affect the motivation of individuals by setting challenging but achievable targets, and can encourage the right organisational behaviour
- Can support management initiatives including Total Quality Management and managing change.

Parker (2000) mentioned both similar and additional reasons, such as:

- Identify success
- Identify whether the organisation meets customer requirements
- Understand their processes (to confirm what they know or to reveal what they do not know)
- Identify where problems, bottlenecks and waste exist and where improvements are necessary
- Ensure that decisions are based on facts, not supposition, emotion or intuition
- Show if the planned improvements actually happened.

Apart from the need to operationalise the various value parameters in SMART (Specific, Measurable, Assignable, Realistic and Time-related) performance indicators, performance measurement requires companies to be precise about the performance of *what*, e.g. people, facilities or services.

Performance measurement in practice

A comparison of performance measurement in three cases (Riratanaphong and Van der Voordt, 2015) showed that the Balanced Scorecard (BSC) from Kaplan and Norton (1992) still seems to be a leading conceptual framework. According to the BSC organisational performance should be evaluated from four perspectives: 1) Financial: profitability, revenue, sales growth; 2) Customer: customer retention, customer satisfaction, market research; 3) Internal business processes: processes to meet or exceed customer expectations; and 4) Learning and growth: how to grow and meet new challenges. Tables 4.6a to 4.6c show a list of possible ways to measure performance of facilities according to Bradley (2002), classified according to six performance criteria (presented in Table 4.1 as well) and linked to the four perspectives of the BSC concept. Accommodation usage is allocated to organisational development but also relates to cost effectiveness. The list is compared with performance measurement in practice, i.e. to the three cases that were discussed by Riratanaphong and Van der Voordt (2015), and to the KPIs that came to the fore in our ten interviews with practitioners in Denmark and the Netherlands (Van der Voordt and Jensen, 2014).

Apart from clear performance indicators, it is also important to be able to define the causes of high or low performance, and to understand which changes are needed to improve what kind of performance. De Vries et al. (2008) concluded

Table 4.6a Performance indicators and ways to measure according to Bradley (2002) and what was found in practice

Performance criteria	Bradley (2002)	Case 1	Case 2	Case 3	10 interviews
1. Stakeholder perception (customer perspective)					
Employee satisfaction with work environment	Quality of indoor environment (lighting, air conditioning, temperature, noise) Provision of safe environment Location success factors (access to employees, amount of local amenities) Ratio of office space to common areas Provision of amenities Amount of workplace reforms and space modifications	Employee satisfaction with regard to lighting, temperature, noise disturbance, odour or dust disturbance, safe environment, availability of self-protection, and equipment in case of accident	Employee satisfaction with regard to diversity of the spaces, opportunities to work outside the office, atmosphere, interaction and knowledge exchange, and IT	Employee satisfaction with regard to air (dust, odours, ventilation), temperature, adequate space, lighting, noise level, appearance of the workplace	User surveys Exit polls Mystery visits Audits (surveys) by HRM Health complaints Sick leave
Employee satisfaction with FM/CREM services	Employee satisfaction with professional skills Employee satisfaction with information sharing	Employee satisfaction with regard to IT-related support services and the management of facilities	Employee satisfaction with regard to IT-related support services and the management of facilities	Employee satisfaction with regard to IT-related support services and the management of facilities	Follow-up on FM suppliers from sourcing and FM Key figures for price, quality and user satisfaction Monitoring image of FM department, internally and externally Response times Number of proposed improvements and cost reductions Innovation pools

Customer satisfaction with facilities	Survey rating (e.g. customer/tenant survey of the facilities, building, property management and CRE services) Number of complaints Average call frequency and cost per square foot of help desk Location success factors (proximity to transportation, access to customers, distance to other sites and businesses)	User satisfaction about the building	Rank in customer survey Number of complaints*	Customer satisfaction survey*	Customer satisfaction surveys

* not directly related to real estate; NA = not applied, i.e. not measured or no data available

Table 4.6b Performance indicators and ways to measure according to Bradley (2002) and what was found in practice

Performance criteria	Bradley (2002)	Case 1	Case 2	Case 3	10 interviews
2. Financial health (financial perspective)					
Value of property, plant and equipment	Business return on real estate assets Real estate return on investment Real estate return on equity Sales or revenue per square foot (metre) Space (square feet or metres) per unit of revenue Return on property management	Income from commercially rented area Return on asset	NA	NA	Impact of real estate costs on profit
3. Organisational development (internal business process perspective)					
Quality of facilities	Physical condition of facilities Suitability of premises and functional environment Number of building quality audits	Work done according to the development of building management and IT standard	Risk management and business control (strategic, operational, compliance and financial risks)*	Risk Inventory and Evaluation (RI&E) including the physical condition of facilities	NA
Accommodation usage	m² per employee Effective utilisation of space e.g. amount of teamwork space, vacancy rates, time wasted with interruptions due to open space layout	NA	NA	m² per desk	Occupancy rate Space utilisation
CRE unit quality	Time used in project versus time budgeted for the project Money spent on project versus money budgeted on the project Amount of advice given to other business units	Delivering rentable area to other government agencies Percentage of allocating commercial area	% reduction in process cycle time* Number of engineering changes* Capacity utilization* Order response time* Process capability*	Design process descriptions and optimising business processes*	NA

* not directly related to real estate; NA = not applied, i.e. not measured or no data available

Table 4.6c Performance indicators and ways to measure according to Bradley (2002) and what was found in practice

4. Productivity (learning and growth perspective)

Employee productivity	Productivity (% of perceived productivity support from working environment) Absentee rates by buildings	Health and wellbeing in the workplace Perceived productivity support	Health and wellbeing in the workplace through workplace innovation (WPI) Perceived productivity support	Health and wellbeing through workplace design Perceived productivity support	Perceived productivity (user survey)
Strategic involvement	CRE involved in corporate strategic planning CRE integrated with HR strategies CRE actively involved in firm-wide initiatives such as special asset use, consolidations, shared services	Master plan of the IT system Management of the information system IT solution in HRM	Implementation of WPI Smart IT solutions for the introduction of WPI	Implementation of flex workplaces	NA

5. Environmental responsibility (internal business process perspective)

Resource use	Energy consumption Number of energy audits	Introduction of green building Construction materials and equipment meet local content	Green products* Energy efficiency improvement Collection and recycling of company's products* Amount of recycled materials in company's products*	Introduction of sustainable approach to the new building EU Energy label	Footprints per building and per region Energy consumption Water consumption Emission of CO_2
Waste	Contaminated sites management Amount of garbage	NA	NA	NA	NA

* not directly related to real estate; NA = not applied, i.e. not measured or no data available

Table 4.6d Performance indicators and ways to measure according to Bradley (2002) and what was found in practice

6. Cost efficiency (financial perspective)

Occupancy costs	Total occupancy cost per employee Occupancy cost as a % of total operating expense Occupancy cost as a % of operating revenue by building or business unit	Taxes (property and land)	Office rent /m^2/ month	Depreciation expense	Cost/m^2 Cost/workplace Cost/fte m^2/workplace m^2/fte m^2 reduction
Operating costs (building and FM)	Total operating expenditures versus budget, including: general administration; capital expenditures; moves, additions, rearrangements; facility/properties services; other business services (mail and copy centres, risk, and/or security) Facility management costs (environment, working conditions, quality)	Operating costs - Facility costs (buildings and equipment) - Overhead costs (employees and committee)* - Fees and services*	Utility (electricity and water) cost/unit Parking cost/month Overhead cost*	Operating costs* - Salary costs* - Social charges* - Personnel costs of third party*	Number of movements Cost reduction by less building adaptations

* not directly related to real estate; NA = not applied, i.e. not measured or no data available

that cause-effect relationships are difficult to prove due to the impact of many interrelated input factors, and the way that interventions are implemented. This also comes to the fore in Part II of this book. The twelve selected value parameters will be assessed on what we know and what we still need to know, and what Key Performance Indicators could be applied to measure different added values.

References

Bradley, S. (2002) 'What's working? Briefing and evaluating workplace performance improvement', *Journal of Corporate Real Estate,* 4, pp. 150–159.

De Jonge, H. (1996) *De toegevoegde waarde van concernhuisvesting.* NSC-conference, Amsterdam.

De Vries, J.C., De Jonge, H. and Van Der Voordt, D.J.M. (2008) 'Impact of real estate interventions on organisational performance', *Journal of Corporate Real Estate,* 10 (3), pp. 208–223.

Den Heijer, A. (2011) *Managing the University Campus: Information to support real estate decisions.* Doctoral Dissertation, Delft University of Technology.

Jensen, P.A. (2010) 'The Facilities Management Value Map: A conceptual framework', *Facilities,* 28 (3/4), pp. 175–188.

Jensen, P.A., Van der Voordt, T., Coenen, C. and Sarasoja, A.-L. (2012) 'Comparisons and Lessons Learned'. Chapter 17 in Jensen, P.A., Van der Voordt, T. and Coenen, C. (Eds.) *The Added Value of Facilities Management – Concepts, Findings and Perspectives.* Centre for Facilities Management – Realdania Research, DTU Management Engineering, and Polyteknisk Forlag, pp. 268–291.

Kaplan, R.S. and Norton, D.P. (1992) 'The balanced scorecard – measures that drive performance', *Harvard Business Review,* 70 (1), pp. 71–79.

Lavy, S., Garcia, J.A. and Dixit, M.K. (2010) 'Establishment of KPIs for facility management performance measurement: Review of the literature', *Facilities,* 28 (9/10), pp. 440–462.

Lebas, M.J. (1995) 'Performance measurement and performance management', *International Journal of Production Economics,* 41, pp. 23–35.

Lindholm, A.L. (2008) *Identifying and measuring the success of corporate real estate management.* PhD thesis, Helsinki University of Technology.

Lindholm, A.L. and Aaltonen, A. (2011) '*Green FM as an adding value element for the core business',* CFM's Nordic Conference 22–23 August. Centre for Facilities Management – Realdania Research, DTU Management Engineering. Technical University of Denmark. Revised printed version published as Sarasoja A.-L. and Aaltonen, A. (2012): 'Green FM as a way to create value'. Chapter 12 in Jensen, P.A., Van der Voordt, T. and Coenen, C. (Eds.) *The Added Value of Facilities Management – Concepts, Findings and Perspectives.* Centre for Facilities Management – Realdania Research, DTU Management Engineering, and Polyteknisk Forlag, pp. 195–204.

Lindholm, A.L. and Gibler, K.M. (2005) 'Measuring the added value of corporate real estate management beyond cost minimization', *12th Annual European Read Estate Society Conference,* Dublin.

Niemeijer, C.E. (2013) *De toegevoegde waarde van architectuur.* PhD thesis. Delft University of Technology, Faculty of Architecture.

Nourse, H. and Roulac, S. (1993) 'Linking Real Estate Decisions to Corporate Strategy', *Journal of Real Estate Research,* 8, pp. 475–494.

Parker, C. (2000) 'Performance measurement', *Work study,* 49, pp. 63–66.

Prevosth, J. and Van der Voordt, T. (2011) *De toegevoegde waarde van FM. Begrippen, maatregelen, en prioriteiten in de zorgsector.* Naarden: Vereniging Facility Management Nederland.

Riratanaphong, C. (2014) *Performance measurement of workplace change in two different cultural contexts.* Doctoral dissertation, Delft University of Technology.

Riratanaphong, C. and Van der Voordt, T. (2015) 'Measuring the Added Value of Workplace Change. Comparison between Theory and Practice'. *Facilities,* 33 (11/12), pp. 773–792.

Scheffer, J.L., Singer, B.P. and Van Meerwijk, M.C.C. (2006) 'Enhancing the contribution of corporate real estate to corporate strategy', *Journal of Corporate Real Estate,* 8 (4), pp. 188–197.

Sinclair, D. and Zairi, M. (1995) 'Effective process management through performance management', *Business Process Re-engineering and Management Journal,* 1, pp. 75–88.

Van der Voordt, D.J.M. and Jensen, P.A. (2014) 'Adding Value by FM: Exploration of management practice in the Netherlands and Denmark'. Paper in Alexander, K. (Ed.) (2014) *Promoting Innovation in FM.* Research Papers. Advancing knowledge in FM. *International Journal of Facilities Management, EuroFM Journal,* March.

Van der Voordt, T., Prevosth, J. and Van der Zwart, J. (2012) *Adding Value by FM and CREM in Dutch Hospitals.* Chapter 13 in Jensen, P.A., Van der Voordt, T. and Coenen, C. (Eds.) *The Added Value of Facilities Management: Concepts, Findings and Perspectives.* Lyngby, Denmark: Centre for Facilities Management and Polyteknisk Forlag

Van der Voordt, T. (2015) 'Value adding management of corporate real estate – parameters and priorities'. Conference paper, in Arslanli, K.Y.A. (Ed.) *Proceedings of European Real Estate Society 22th Annual Conference,* 24–27 June, Istanbul, Turkey, pp. 187–198.

Van der Zwart, J., and Van der Voordt, T. (2013) 'Value adding management of hospital real estate. Balancing different stakeholders' perspectives', *(E)Hospital,* 15 (3), pp. 13, 15–17.

Van der Zwart, J. (2014) *Building for a better hospital. Value-adding management and design of healthcare real estate.* PhD thesis, Delft University of Technology, Faculty of Architecture.

Van Meel, J., Martens, Y. and Van Ree, H.J. (2010) *Planning office spaces: A practical guide for managers and designers,* London: Laurence King Publishing Ltd.

Part II
Value parameters

5 Satisfaction

*Theo van der Voordt, Sandra Brunia and
Rianne Appel-Meulenbroek*

Introduction

Although work environments are mostly designed with a focus on supporting
the primary processes – enhancing productivity, stimulation of collaboration and
social interaction, and cost reduction or cost-effectiveness – employee satisfaction
is often one of the objectives as well (Van der Voordt, Ikiz-Koppejan and
Gosselink, 2012). Satisfied employees can be an objective in itself, a means to
avoid trouble, or a means to support labour productivity, based on the assumption
that a happy worker is a productive worker. For the same reasons satisfaction is
often a highly prioritised value in FM and CREM (Van der Voordt et al., 2012;
Van der Voordt and Jensen, 2014). Although the relationship between employee
satisfaction and labour productivity is not always statistically significant and can
be interfered with by variables such as intrinsic motivation and competition with
colleagues, various studies have confirmed that both constructs are correlated
(Zelenski et al., 2008; Fassoulis and Alexopoulos, 2015). Besides, in this
knowledge age the human capital of an organisation, i.e. highly skilled employees,
is one of its most important resources for production and as such leads to its
success and performance.

In this chapter we first present a theoretical framework of the possible impact
of FM and CREM on employee satisfaction with the physical environment and
building-related facilities and services. Other types of satisfaction, such as
satisfaction with how the facilities are being managed, job satisfaction, and the
satisfaction of other stakeholders such as clients and customers, will be left out
for reasons of limited space. Then we present an overview of the available
empirical evidence that confirms the impact of facilities on employee satisfaction.
This section will also discuss which aspects employees find most important. The
chapter ends with a discussion of ways to measure and manage employee
satisfaction – including a proposed shortlist of KPIs – and future perspectives,
i.e. input for a future research agenda.

State of the art

Theoretical framework of influencing factors

Many researchers have tried to conceptualise and visualise possible relationships between the built environment and building-related facilities and employee satisfaction. As input to the debate on the experience and use of open offices, Sundstrøm (1986) developed a conceptual model that includes both physical factors (e.g. sound, temperature, status, privacy) and non-physical factors (e.g. supervision, salary, autonomy, content of the job) on labour satisfaction. Van der Voordt and De Been (2010) developed a conceptual model that is used in connection with the so-called Work Environment Diagnostic Instrument (WODI) of the Centre for People and Buildings in Delft, the Netherlands. Their model visualises the assumed impact of facilities, the organisation, personal characteristics and work processes on employee satisfaction and labour productivity, and also takes into account the impact of the external context.

Building on various conceptual frameworks, Riratanaphong and Van der Voordt (2012) developed a theoretical framework that incorporates the influence of both product and process oriented variables; see Figure 5.1. This conceptual framework visualises the assumed impact of people, business processes, preconditions and the external context on workplace design. Besides, the model shows the impact of these variables and the implementation process of workplace change on the appraisal of workplace change. This model is based on both former frameworks and insights from studies on drivers to change and Pre- and Post-Occupancy Evaluations of organisations that introduced New Ways of Working (Van der Voordt, 2003; Van der Voordt et al., 2012; Brunia and Beijer, 2014; Appel-Meulenbroek et al., 2014; Riratanaphong, 2014). Figure 5.1 shows the complexity of real life with many factors that may influence employee satisfaction. This complexity makes it rather difficult to trace clear cause-effect relationships between facilities and end-user satisfaction.

Empirical data about employee satisfaction

Empirical research about employee satisfaction with (new) work environments is often based on a single case study or a limited number of cases (e.g. Van Wagenberg, 1996; Van der Voordt, 2004; Mallory Hill et al., 2005; Van der Voordt and Van der Klooster, 2008; Blok et al., 2009; Gorgievsky et al., 2010). In this chapter we will focus on two large datasets from the Center for People and Buildings knowledge centre, Delft, the Netherlands (Brunia and Beijer, 2014) and Leesman Office, London, UK (Van Susante, 2014) respectively. Both datasets have been collected by standardised questionnaires (Maarleveld et al., 2012; Appel-Meulenbroek et al., 2014).

Since its foundation in 2001 the Center for People and Buildings (CfPB) conducted numerous Pre- and Post-Occupancy Evaluations using a standardised questionnaire to measure the employees' (dis)satisfaction with different office

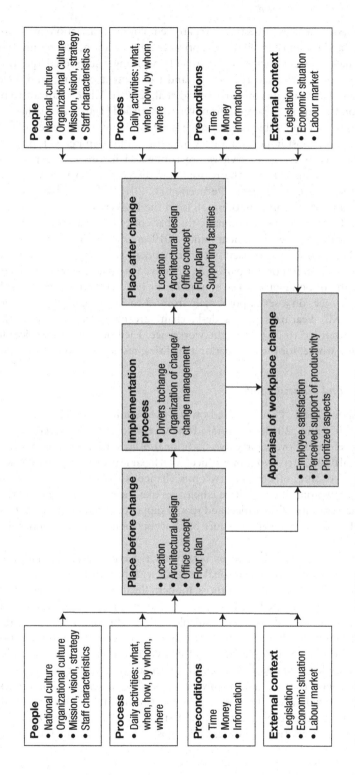

Figure 5.1 Conceptual model of Riratanaphong and Van der Voordt (2012)

environments. The key questions regard employee satisfaction with the architecture and layout of the building, privacy, concentration, communication, indoor climate, various facilities and remote working. Additional questions regard satisfaction with the organisation and the content of work, and which items are found most important (Maarleveld et al., 2009). All items are answered on a five-point Likert scale ranging from very dissatisfied (1) to very satisfied (5), and the option 'not applicable'. Importance is measured by asking the employees to mark the three most important aspects in a list of nineteen items. If possible, additional in-depth group interviews are conducted in order to explain the results and to get insight into the reasons for (dis)satisfaction with particular aspects of the work environment (De Been et al., 2015). The database includes both public and private organisations, all from the Netherlands.

Leesman Office is a private company that has collected data about employee satisfaction with office environments since 2010, mainly in Western European countries. For each feature or facility the respondent is asked to state if he/she considers it to be an important part of an effective workplace. If it is considered important, the respondent is asked to rate his or her satisfaction with it in the current workspace, on a seven-point scale from -3 to 3. There is also an option "Not Provided". Features such as desk, chair, greenery, art, and air quality control are regarded to be part of the workspace. Facilities refer to services for the office as a whole, for instance parking facilities, restaurants and cleanliness.

Satisfaction and dissatisfaction

Table 5.1 shows the CfPB data based on 105 case studies that were conducted between 2007 and 2013. In total 14,980 employees responded to the questionnaire, spread over thirty-five traditional offices, fourteen combi-offices with personal desks, fifty-two combi-offices with flexible use of shared workplaces, two open 'landscape' offices, and two miscellaneous spaces. It shows that on average, over two out of every three employees like their organisation and their work, the accessibility of the office, and spatial support of communication. Least appreciated topics include the support of concentration, storage facilities, and the indoor climate.

Figure 5.2 shows the differences between satisfaction levels in traditional offices and combi-offices with personal desks or with hot desks.

In comparison with traditional offices, modern flexible 'non-territorial' offices are much more appreciated with respect to the architectural appearance, its atmosphere and interior design, and available facilities (Brunia and Beijer, 2014). This might be due to the fact that these offices have been built more recently, with more attention to design issues and facilities. However, new flexible offices also evoke high levels of dissatisfaction, in particular with respect to privacy and support of concentration, user involvement in the implementation process, comfort of the workplace itself, storage opportunities and acoustics. Similar findings came to the fore in a CfPB case study including a Pre-Occupancy Evaluation of four existing buildings and a Post-Occupancy Evaluation of two

Table 5.1 Percentage of satisfied employees per item in 105 cases (Brunia and Beijer, 2014)

Aspects of the work environment	Satisfaction average 2007–2013*	Lowest %	Highest %	Range
Organization	66%	26%	98%	72%
Content and complexity of work	80%	40%	100%	60%
Sharing own ideas about work environment	43%	7%	78%	71%
Accessibility of the building	78%	19%	99%	80%
Appearance of building's exterior	55%	7%	96%	89%
Number and diversity of places	44%	10%	81%	71%
Position of the spaces (lay-out)	53%	20%	86%	66%
Openness work environment	52%	19%	89%	70%
Comfort of the workspaces	56%	17%	82%	65%
Appearance and ambiance interior design	49%	4%	92%	88%
Privacy	37%	8%	89%	81%
Opportunities to concentrate	38%	9%	86%	77%
Opportunities for communication	70%	35%	92%	57%
Archive facilities	36%	2%	71%	69%
IT facilities	53%	17%	95%	78%
Provided facility services	53%	23%	87%	64%
Indoor climate	34%	0%	78%	78%
Light	58%	23%	88%	65%
Acoustics	42%	14%	79%	65%
Opportunities for remote working	49%	5%	90%	85%

*Based on 14,980 respondents from 105 case studies (period 2007–2013)

new locations belonging to the same organisation. Whereas most items were rated more positively than before, privacy, concentration opportunities, acoustics and storage space got lower satisfaction scores (De Been, 2014).

Figure 5.3 shows the Leesman data about employee satisfaction with various features in the work environment. The data are based on responses from 47,913 respondents working for 115 organisations in more than 370 locations in five different Western European countries, representing all kinds of office types (Van Susante, 2014). The figure shows that satisfaction is specifically low for climate aspects (temperature control, air quality) and privacy aspects (noise levels, lack of quiet rooms), whereas temperature and noise are ranked high according to the percentage of employees who find these aspects important. The aspects that respondents marked most frequently as being satisfactory are desks, chairs, and several types of equipment.

The Leesman dataset also includes data about satisfaction with fifteen facilities. Figure 5.4 shows that on average tea/coffee, security and mail services are highly appreciated, whereas hospitality services, leisure facilities and atriums and communal areas evoke high levels of dissatisfaction (Van Susante, 2014).

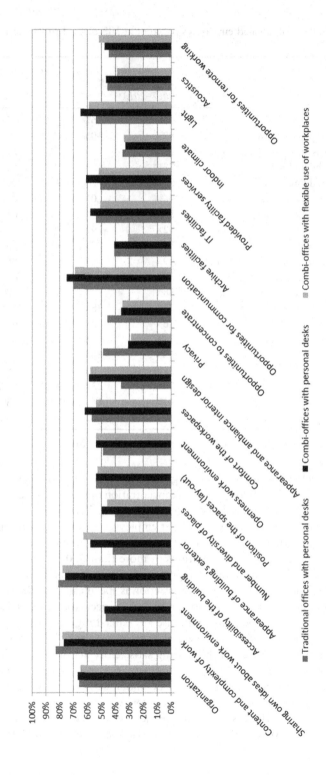

Figure 5.2 Percentage of satisfied employees per item per office type (Brunia and Beijer, 2014)

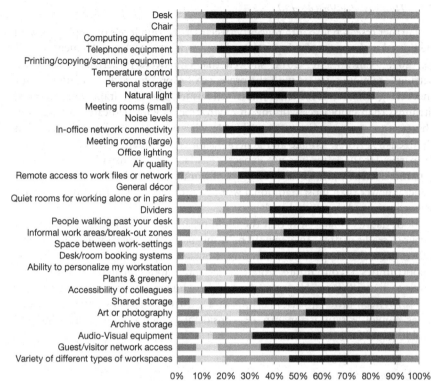

Figure 5.3 Percentage of employees satisfied with each feature, based on 47,913 responses from employees working at 115 different organisations, ranked according to the percentage of respondents that marked the topic as "important" (Van Susante, 2014)

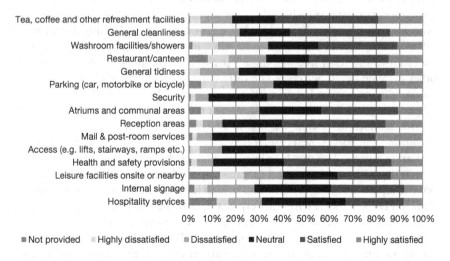

Figure 5.4 Percentage of employees satisfied with each facility, based on 47,913 responses by employees from 115 different organisations (Van Susante, 2014)

Further research into differences between satisfaction levels of employees with dedicated desks versus sharing activity-based workplaces (Appel-Meulenbroek et al., 2015; Van Susante, 2014) showed that employees in activity-based work settings are more satisfied with the presence of rooms for seclusion (alone or for meetings), climate, décor, cleanliness and leisure (e.g. restaurant, refreshments, atriums) than employees working in environments with dedicated seating. On the contrary, workers in modern flexible work settings are less satisfied with desk/chair, privacy, storage and general facilities (e.g. mail, security, reception). In particular the latter was not expected, as facilities are usually given a lot of attention when implementing new ways of working. Flexible work settings are in general newer than most traditional work environments, and this might explain why climate and décor are highly scored.

Both datasets make clear that the building and building-related facilities and services really matter. In general the atmosphere and interior design of new offices and the support of communication are appreciated, whereas privacy, opportunities to concentrate, storage space and indoor climate often evoke high levels of dissatisfaction. However, there is also a huge range in average satisfaction levels between different cases (Brunia et al., 2016). This shows that in various cases employee satisfaction can be considerably increased by improving the building and facilities.

Most important aspects

Regarding the level of importance, the CfPB dataset showed that on average the comfort of the workplace, accessibility of the building and support of concentration are most often allocated to the top three most important aspects. The same holds true for indoor climate and support of communication, whereas acoustics, lighting and storage opportunities are surprisingly rarely mentioned in the top three most important aspects (Brunia and Beijer, 2014). The top five most important aspects include the same topics in traditional offices and combi-offices with personal desks and hot desks. However, the level of importance may differ greatly between different cases. For instance, in one case privacy was allocated to the top three by only 29 per cent in the respondents, and at the other extreme by 70 per cent. The difference regarding the importance of the indoor climate is even larger, with a range of people between 2 per cent and 67 per cent who allocated this topic in their top three most important aspects.

In a number of cases it was found that aspects with a high percentage of dissatisfied employees were also often mentioned as one of the three most important aspects. This may have been because employees are more aware of these aspects due to a dissatisfier-effect (negative aspects lead to dissatisfaction, whereas positive aspects do not necessarily lead to high levels of satisfaction; see for instance Kim and De Dear, 2012). Another explanation may be that employees expect that a high ranking will result in measures to improve the environment.

However, this pattern is not confirmed in all cases. Figure 5.5 shows that taking all 105 cases together, no clear pattern came to the fore regarding the

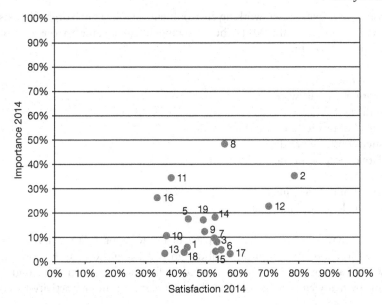

Figure 5.5 Relationship between responses regarding satisfaction and level of importance, based on 105 cases (Brunia and Beijer, 2014)

relationship between the percentages of employees ranking a particular aspect as one of the three most important aspects and the percentages of satisfied employees about these topics. Most important topics include both highly appreciated items and much less appreciated items.

In the Leesman questionnaire the respondents were just asked which aspects they found important and not which aspects they found most important. However, the percentage of employees that perceives a particular feature as important can be used as an indication of the most important aspects. The data show that in addition to an appropriate desk, chair and computer equipment, tea/coffee and other refreshments, cleanliness, washrooms/showers and the restaurant/canteen are seen as important by more than 80% of the respondents, and as such are apparently the very basics of an office. Surprisingly, modern facility services like leisure and hospitality are only seen as important by 50% or less of the office population (Van Susante, 2014).

Focusing on activity-based working environments in four Dutch organisations (Appel-Meulenbroek et al., 2011) the following features came to the fore in terms of their importance:

- Ergonomics
- Quality of IT hardware
- Comfort
- Control of privacy and social interaction
- Dimensions of work desk
- Relative location of the workstation
- Ambiance
- Control of indoor climate (actual/perceived)
- Use of colours and materials.

Overall we may conclude that in particular the appropriate support of communication and concentration, well-designed desks and chairs, the comfort of the work place, indoor climate, and sound IT facilities are perceived as highly important by many employees, in general and in particular in new activity-based offices.

Benefits and costs

Table 5.2 shows a number of physical interventions that could be taken to increase employee satisfaction, but which always should be counterbalanced against costs.

Table 5.2 Typical physical interventions with benefits and costs

Intervention	Management	Benefits	Sacrifices
More closed spaces	Change of spatial layout	Better opportunities for concentration, more privacy	Risk of social isolation. More space needed, so extra costs
Lower density	Change of spatial layout	Larger distance between people, fewer feelings of crowdedness, more privacy	More space needed, so extra costs
More (personal and team) storage space	Providing extra cupboards and other storage facilities	Increased personal control, easier access to personal hard copies	More space needed, so extra costs
Advanced climate design	Replacement of technical installations	Better indoor climate	Extra costs

Intervention	Management	Benefits	Sacrifices
Acoustic measures	Installing of acoustic ceiling and sound absorbing materials	Less noise, less distraction, less fatigue	Extra costs
Improvement of interior design	Redesigning of the interior, including furniture and finishing	Nice atmosphere, improved ergonomics, better comfort, improved health and wellbeing	Extra costs
More collaborative (meeting) spaces	Change of spatial layout	Better opportunities for communication and knowledge sharing	More space needed, so extra costs Possible increase of noise/disturbance

How to measure and manage

Key Performance Indicators

The aspects that the respondents to the surveys of the Centre for People and Buildings and Leesman Office found most important can be used to conduct a quick scan of employee satisfaction with the work environment, and as such removes the need for CREM/FM to ask employees to fill in long questionnaires. It is proposed to at least measure satisfaction according to those aspects that many employees find important in achieving high levels of satisfaction. Also, it is important to incorporate important dissatisfiers such as temperature control, air quality, noise levels and seclusion rooms that enable employees to work alone or in pairs (Van Susante, 2014). In order to avoid dissatisfaction, managers have to take care of a high level of performance in these features and facilities.

Based on the analyses, we propose that employers should conduct a (preferably annual) employee survey that at least measures the employee satisfaction with the following issues:

1 Opportunities to communicate
2 Opportunities to concentrate
3 Meeting rooms
4 Seclusion rooms
5 Personal storage
6 Indoor climate
7 Noise levels
8 Chairs, desks and other office equipment
9 Office leisure (e.g. tea/coffee, washroom/shower and restaurant/canteen)
10 General cleanliness
11 IT facilities.

Measuring methods

The simplest way to measure employee satisfaction is to ask them how satisfied or dissatisfied they are with their (work) environment or particular facilities and services. Also ask them why, in order to be able to explain the findings and to explore cause-effect relationships. Nowadays online surveys are quite common, with questions such as:

- "How satisfied are you with ... a number of facilities" on a 5-point scale, ranging from 1 = very dissatisfied to 5 = very satisfied (Maarleveld et al., 2009; Appel-Meulenbroek et al., 2015) or on a 10-point scale, like marks in educational settings (Brunia and Beijer, 2014)
- In case of a questionnaire after a change: "Taking everything into account, would you prefer to stay or to go back to the former situation? Stay, go back, don't mind?" (Vos and Dewulf, 1999)
- "What are the most important aspects?", either unspecified or with a focus on "being able to work in a pleasant way". Responses may vary from a request to mark the top three in a list of topics (e.g. Brunia and Beijer, 2014), mark the relevance of topics with yes/no, or mark a list of various topics according to how well they are supported on a 6-point scale, ranging from -3 = not supported at all to +3 = very well supported by the work space (Van Susante, 2014).

Other measurement methods could be:

- Observation of the actual use of places as an indicator of which places are preferred and which ones are disliked and therefore are not well-occupied (provided people have a choice) (Appel-Meulenbroek et al., 2011, 2015)
- Conducting open or semi-structured interviews (De Been et al., 2015)
- Walk-throughs with employees or other stakeholders, asking them to think aloud which places, facilities and services they like or dislike and why (e.g. Hansen et al., 2010)
- Using narratives by asking people to report about a normal or particular day at the office (or in another environment), what they like and dislike (with a focus on facilities), and why (Smith, 2000; Mitchell and Egudo, 2003)
- Showing pictures of different environments or facilities and asking respondents to rate them on a five-point scale, ranging from 1 = not pleasant at all to 5 = excellent, or asking them to mark their most preferred option out of a sample of two or more solutions (Van Oel and Van den Berkhof, 2013).

To get a better understanding of cause-effect relationships, both inductive reasoning (exploring new ideas based on empirical data) and deductive reasoning (starting with hypotheses and testing them) may be applied. To analyse quantitative data, many statistical techniques are available, such as partial correlation analysis, multiple regression analysis and factor analysis. More information about these methods can be found in methodological and statistical textbooks.

For CREM/FM to manage employee satisfaction, it might not be feasible to use very long lists of workplace features and facilities each time. A quick scan could be used to focus on a limited list of items that are perceived as the most important by employees (as discussed above), or on specific measures to find out whether certain interventions have been successful in increasing satisfaction (and which costs have been incurred). Another option was to focus on groups of features and facilities that came to the fore in a factor analysis of the satisfaction scores of 43,800 respondents in the Leesman database. This factor analysis provided only seven unique factors for features and three factors for facilities (Appel-Meulenbroek et al, 2015; Van Susante, 2014). These are discussed below.

Groups of features:

1   Seclusion rooms (meeting rooms, quiet rooms for working alone or in pairs, desk/room booking systems, informal work areas/break-out zones, accessibility of colleagues)
2   Privacy (people walking past your desk, noise levels, dividers, space between work-settings, ability to personalise the workstation)
3   Indoor climate (temperature control, air quality, lighting, natural light)
4   Office décor (art/photography, plants/greenery, general décor, variety of different types of workspace)
5   ICT and equipment (remote access, in-office network connectivity, guest/visitor network access, audio-visual equipment, printing/copying/scanning equipment)
6   Desk and chair (desk, chair, computing equipment, telephone equipment)
7   Storage (shared, archive, personal).

Groups of facilities:

1   General facilities (building access, health/safety provisions, mail and post room, security, internal signage, reception, hospitality, parking)
2   Office leisure (restaurant, atriums/communal areas, refreshment facilities, leisure facilities onsite or nearby, washroom/showers)
3   Cleanliness (general cleanliness, general tidiness).

Within these groups many aspects correlate, and thus a survey with ten main questions could shed light on these groups of items. The analysis also made clear that features and facilities are separate items when they are used to manage employee satisfaction, and thus both deserve managerial attention to increase added value.

Perspectives

Although a large number of case studies and surveys have been conducted into employee satisfaction with the physical environment and building-related services, many knowledge gaps remain to be filled. Besides, current data have to be transformed into appropriate design and management guidelines and decision support tools. Possible research questions for future research include:

- How can the differences in satisfaction between employees within the same case and between different cases be explained?
- What theories might be helpful or should be developed to understand different satisfaction levels? Can we build on motivation theory and insights regarding age, gender, personal characteristics and different organisational cultures to explore a "workplace satisfaction theory"?
- How can complex cause-effect relationships be disentangled? Is it possible to define input-throughput-output/outcome relationships in a reliable and valid way?
- (How) can cause-effect relationships be quantified?
- Is it possible to predict the impact of interventions in the work environment on the percentage of satisfied employees and the costs and sacrifices?
- What kinds of synergy or conflicts occur between measures to increase employee satisfaction and measures to increase other values, such as productivity and cost reduction?
- What managerial decisions will result in an optimal balance between benefits and costs, for instance regarding communication and concentration and the preferences and needs of employees and organisations as a whole?
- What is the impact of employee satisfaction with physical assets on job satisfaction? Which physical, social, behavioural and other aspects of work environments have an influence here?

In order to answer these questions, further research is needed with different methodological approaches, including standardised questionnaires, interviews, narratives, in-depth research into topics that cause high levels of dissatisfaction (concentration, privacy, filing, indoor climate), preferably in an interdisciplinary way and preferably including Pre- and Post-Occupancy Evaluations of workplace change. Future studies should also include respondents from different countries in more evenly spread groups.

References

Appel-Meulenbroek, R., Groenen, P. and Janssen, I. (2011) 'An end user's perspective on activity-based office concepts', *Journal of Corporate Real Estate*, 13 (2), pp. 122–135.

Appel-Meulenbroek, R., Kemperman, A., Liebregts, M. and Oldman, T. (2014) 'Designing the modern work environment to support important activities: An analysis of different preferences in 5 European countries', *Proceedings of the 21st ERES conference*, June 25–28, Bucharest, Romania.

Appel-Meulenbroek, R., Kemperman, A., Van Susante, P. and Hoendervanger, J.G. (2015) 'Differences in employee satisfaction in new versus traditional work environments', *European Facility Management Conference*, Glasgow.

Appel-Meulenbroek, R., Kemperman, A., Kleijn, M. and Hendriks, M. (2015) 'To use or not to use; which type of property should you choose? Predicting the use of activity based offices', *Journal of Property Investment and Finance*, 33 (4), pp. 320–336.

Barber, C. (2001) 'The 21st-Century Workplace', in Kaczmarczyk, S. et al. (Eds.) *People and the Workplace*, Washington, DC: GSA Office of Governmentwide Policy.

Blok, M., De Korte, E., Groenesteijn, L., Formanoy, M. and Vink, P. (2009) 'The effects of a Task-facilitating Working Environment on office space use, communication, concentration, collaboration, privacy and distraction', *Proceedings of the 17th World Congress on Ergonomics*, Beijing, China.

Brill, M. and Weidemann, S. (2001) *Disproving widespread myths about workplace design*, Buffalo: Kimball International.

Brunia, S. and Beijer, M. (2014) *Trends in werkplekbeleving. Een analyse van de CfPB-indicatoren 2007-2014*. Delft: CfPB.

Brunia, S., de Been, I. and Van der Voordt, T. (2016) 'Accommodating New Ways of Working: Lessons from best practices and worst cases', *Journal of Corporate Real Estate*, 18 (1), pp 30–47.

De Been, I. (2014) *Vergelijking van de WODI nameting in 2014 met de voormeting in 2013 bij HTM*. Internal report. Delft: Centre for People and Buildings.

De Been, I., Beijer, M. and Den Hollander, D. (2015) 'How to cope with dilemmas in activity based work environments: Results from user-centred research', *European Facility Management Conference*, Glasgow, 2–3 June.

Fleming, D. (2005) 'The application of a behavioral approach to building evaluation', *Facilities*, 23 (9/10), pp. 393–415.

Fassoulis, K. and Alexopoulos, N. (2015) 'The workplace as a factor of job satisfaction and productivity', *Journal of Facilities Management*, 13 (4), pp. 332–349.

Gorgievski, M.J., Van der Voordt, T.J.M., Van Herpen, S.G.A. and Van Akkeren, S. (2010) 'After the fire. New ways of working in an academic setting', *Facilities*, 28 (3/4), pp. 206–224.

Hansen, G.K., Blakstad, S.H., Knudsen, W. and Olsson, N. (2010) 'Usability Walkthroughs', *CIB W111 Research Report Usability of Workplaces*, Phase 3, 31–44. Rotterdam.

Kim, J. and De Dear, R. (2012) 'Nonlinear relationships between individual (IEQ) factors and overall workspace satisfaction', *Building and Environment*, 49 (0), pp. 33–40.

Maarleveld, M., Volker, L. and Van der Voordt, T.J.M. (2009) 'Measuring employee satisfaction in new offices – the WODI toolkit', *Journal of Facilities Management*, 7 (3), pp. 181–197.

Mallory-Hill, S., Van der Voordt, T.J.M. and Van Dortmont, A. (2005) 'Evaluation of innovative workplace design in the Netherlands', in Preiser, W.F.E. and Vischer, J.C. (Eds.) *Assessing Building Performance*. Oxon, UK: Elsevier, pp. 160–169 and pp. 227–228.

Mitchell, M. and Egudo. M. (2003) *A Review of Narrative Methodology*, Edinburgh, Australia: DSTO Systems Sciences Laboratory.

Riratanaphong, C. (2014) *Performance measurement of workplace change in two different cultural contexts*. PhD thesis. Delft University of Technology.

Riratanaphong, C. and Van der Voordt, T. (2011) 'Satisfaction and productivity after workplace change', *European Facility Management Conference*, Vienna, 23–25 May.

Roelofsen, P. (2002) 'The impact of office environments on employee performance: The design of the workplace as a strategy for productivity enhancement', *Journal of Facilities Management*, 1 (3), pp. 247–264.

Smith C.P. (2000) 'Content analysis and narrative analysis', in: Reis H.T. and Judd, C.M. (Eds.) *Handbook of research methods in social and personality psychology*, New York: Cambridge University Press.

Sundstrøm, E.D. (1986) *Workplaces: The psychology of the physical environment in offices and factories.* Cambridge: Cambridge University Press.

Van der Voordt, D.J.M. (2003) *Costs and benefits of innovative workplace design,* Delft: Centre for People and Buildings.

Van der Voordt, D.J.M., Ikiz-Koppejan, Y.M.D. and Gosselink, A. (2012) 'Evidence-Based Decision-Making on Office Accommodation: Accommodation Choice Model', in Mallory-Hill, S., Preiser, W.F.E. and Watson, C. (Eds.) *Enhancing Building Performance.* Chichester, West Sussex, UK: Wiley-Blackwell, pp. 213–222.

Van der Voordt, D.J.M. and van der Klooster, W. (2008) 'Post-Occupancy Evaluation of a New Office in an Educational Setting', in *Proceedings of CIB Conference in Facilities Management*, Heriot Watt University, Edinburgh, June.

Van der Voordt, T. (2004) 'Productivity and employee satisfaction in flexible offices', *Journal of Corporate Real Estate*, 6 (2), pp. 133–148.

Van der Voordt, T. and De Been, I. (2010) 'Werkomgeving: breinbreker of tevredenheidsgenerator?' in Bakker, I. (Ed.), *De Breinwerker.* Naarden: Uitgeverij FMN, pp. 67–86.

Van der Voordt, T., De Been, I. and Maarleveld, M. (2012) 'Post-Occupancy Evaluation of Facilities Change', in: Finch, E. (Ed.) *Facilities Change Management*, Chichester, West Sussex: Wiley-Blackwell,.

Van der Voordt, T. and Jensen, P.A. (2014) 'Adding value by FM: Exploration of management practice in the Netherlands and Denmark'. *European Facility Management Conference*, Berlin, 4–6 June.

Van der Voordt, T., Prevosth, J., and Van der Zwart, J. (2012) 'Adding Value by FM and CREM in Dutch Hospitals', in Jensen, P.A., van der Voordt, T. and Coenen, C. (Eds.) *The Added Value of Facilities Management: Concepts, Findings and Perspectives.* Lyngby, Denmark: Centre for Facilities Management and Polyteknisk Forlag.

Van Oel, C. and Van den Berkhof, F.W. (2013) 'Consumer preferences in the design of airport passenger areas', *Journal of Environmental Psychology*, 36, pp. 280–290.

Van Susante, P. (2014) *Differences in employee satisfaction in new versus traditional work environments.* MSc thesis, Eindhoven University of Technology.

Van Wagenberg, A.F. (1996) 'Redesign and evaluation of experimental Dutch office layouts', *Proceedings of World Workplace 1996*, Salt Lake City, pp. 715–725.

Vos, P.G.J.C. and Dewulf, G.P.R.M. (1999) *Searching for data – A method to evaluate the effects of working in an innovative office.* Delft: Delft University Press.

Wagner, A., Gossauer, E., Moosmann, C., Gropp, T., and Leonhart, R. (2007) 'Thermal comfort and workplace occupant satisfaction. Results of field studies in German low energy office buildings', *Energy and Buildings*, 39 (7), pp. 758–769.

Zelenski, J.M., Murphy, S.A. and Jenkins, D.A. (2008) 'The Happy-Productive Worker Thesis Revisited', *Journal of Happiness Studies*, 9, pp. 521–537.

Interview 1: Liselotte Panduro, ISS, Denmark

Liselotte Panduro has nine years of experience in FM. Her educational background includes a Higher Diploma in accounting and organisation and an International MBA in e-business. Since 2009 she has been a Director in ISS with responsibility for large contracts in Denmark. Before that she worked for three years in a real estate administration company. Earlier in her career she worked in business development in large telecom and finance corporations. She is a member of the board of the Danish Facilities Management Association.

Added Value in general

Q. What does the term "Added Value" mean to you?

A. *It depends very much on the client's perception. I distinguish between the value that ISS provides as part of our standard package when we start a new contract, and the value that ISS creates during the contract. The latter changes a lot depending on what is important for the customer over time.*

Q Are there other related terms that you prefer to use rather than "Added Value"?

A. *Value creation is also used, as well as benefits and the purpose, which is very much to make a difference to the customer.*

Q. How do you see the relation between "Added Value" and cost reduction?

A *There is a strong coherence. Cost reduction is a part of added value.*

Q. Is "Added Value" mostly treated at a strategic, tactical or operational level?

A. *Conceptually I handle added value at the strategic level but with involvement at the tactical level, which consists of my key account managers. But basically the whole organisation is involved. It is important that every person in the organisation sees the purpose of their work – that they are part of building a cathedral and not just cutting stones. Storytelling is an important way to create understanding and motivation.*

Q. In which context or dialogue is "Added Value" mostly considered, in your experience?

A. *I use it frequently in ongoing collaboration with clients.*

Q. How do you see the relation between innovation and "Added Value" in FM?

A. *Innovation is a part of adding value. ISS distinguishes between continuous improvement and innovation. Inspiration for continuous improvements, for instance, comes from ISS's centres of excellence, key account knowledge sharing groups and account specific customer inputs and user groups. Innovation is much more segment specific and inspiration comes, for instance, from IFMA, ISS Global innovation projects, the ISS Innovation Board, ISS Segment knowledge and joint innovation panels on accounts.*

Benefits and limitations of "Added Value"

Q. What are the benefits of considering and talking about "Added Value"?

A. *The benefits are that it makes you focus on the impact of FM on the client organisation and gives a more constructive dialogue than focusing on cost only. Added value is important to make visible that we make a difference. We strive to create partnerships, where we are proactive and develop together with the customer.*

Q. What are the limitations of considering and talking about "Added Value"?

A. *The limitation is that it is difficult to document. My experience with using the Balanced Scorecard is that the economic and people perspectives are quite easy to document, while the customer and process perspectives are much more difficult to measure. It also depends a lot on what triggers the specific customer and user. Management of expectations is an important aspect of adding value.*

Management of Added Value

Q. What are your top five of main values to be included in management of accommodations, facilities and services? Could you mention a few examples of concrete FM interventions to attain these added values and KPIs to measure them?

A. *See table below.*

Prioritised values in FM	Concrete FM interventions	KPIs
1. The satisfaction of outsourced staff		HSE aspects in general Staff satisfaction index Absence due to sickness Staff turnover Occupational accidents
2. Make processes smarter	Best practice sharing	Cost measures Response times
3. Improvements and innovation	Challenge the customer	Number of proposed improvements and cost reductions Innovation pools
4. User centricity and service orientation	Giving choices, for instance in catering	User surveys Exit polls Mystery visits
5. Corporate Social Responsibility (CSR)	Waste handling Use of materials Handling of chemicals	Energy consumption Water consumption Emission of CO_2 Number of apprentices Number of disabled workers

Q. Do you benchmark your data with data from other organisations?

A. *We do not at the moment benchmark in a structured way, but it is coming.*

Q. What other methods do you use to document "Added Value"?

A. *We produce yearly performance reviews on key accounts.*

Q. How is "Added Value" included in your communication with your stakeholder?

A. *Besides the yearly performance reviews on key accounts we use storytelling. We have also attempted to make an annual added value report on key accounts, but we have not yet managed to find the right way to meet the customers' expectations.*

Last comment

Q. Are there other topics that you find important in connection to the concept of "Added Value"?

A. *A reason why it is important to focus on added value is that it concerns the impact of FM, which requires the development of profession specific FM knowledge. Such knowledge is not transferable from other industries and professions, while management knowledge about developing the internal processes in FM is more easily transferable. This also means that an increased professionalisation of FM requires an increased specialisation of FM in relation to different segments and industries. An important part of my job is to develop strategies in relation to specific segments, for instance the pharmaceutical industry.*

6 Image

Theo van der Voordt

Introduction

In this very competitive society organisations have to make significant efforts to attract and retain customers and talented staff. For this reason organisations try to distinguish themselves from other organisations by low prices, better products and services, a good price/quality ratio, innovations, being service oriented, and operating in a socially responsible way. In addition to company logos, a company's accommodation, facilities and services can also be used as a means to evoke a positive image and communicate brand values (Van Kempen, 2008; Zoetemelk, 2010; Appel-Meulenbroek et al., 2010; Khanna et al. 2013). This may be seen as supporting a corporate's image *by* FM/CREM. Real estate and other facilities can have image effects that are hard to imitate or substitute with other means. It is assumed that this might reduce communicating costs in advertising campaigns (Heywood and Kenley, 2008). An example of evoking a positive image by facilities and real estate is the way modern hospitals express their hospitality by providing a pleasant entrance area, a reception desk with friendly and welcoming staff, voluntary hosts that help visitors and patients to find their way, and the availability of a coffee bar, flower stand and professional restaurant. The Rotterdam Eye Hospital delivers valet parking as a means to reduce the stress of their patients and accompanying visitors and to give them a feeling of being welcome and comfortable (Van der Zwart et al., 2009) – a good example of customer intimacy (Treacy and Wiersema, 1995).

Another connection between image and FM/CREM regards the image *of* FM and CREM. How do clients, customers and end users perceive the content of FM and CREM, the position of FM/CREM in the overall structure of the organisation, and its products and services? What do they find important? How satisfied or dissatisfied are they with the FM/CREM processes, and the input, output and outcomes? Regarding the image *of* FM and CREM and how to manage and measure it, we refer to the literature; see, for instance, Coenen et al. (2010, 2013); Von Felten et al. (2014); Redlein and Zobl (2013, 2014). For the image of CREM see, for instance, Omar (2015) and Omar and Heywood (2014).

This chapter will focus on how to improve the image of an organisation *by* FM and CREM. It starts with a conceptual analysis of corporate image in connection with other concepts such as corporate identity and corporate branding, and examines how to support a corporation's image by FM and CREM. Then it presents an overview of possible interventions to enhance a positive image and how the impact of FM and CREM on a corporation's image can be managed and measured. The chapter ends with suggestions for the future research agenda.

State of the art

The image of an organisation may be defined as the mental representation of how customers, end users and the public perceive the organisation and what it stands for. It represents the picture of an organisation as perceived by various target groups (Van Riel, 1995). According to Worcester (1997), corporate image is the net result of the interaction of all the experiences, impressions, beliefs, feelings and knowledge people have about a company. Corporate image is related to corporate identity, which can be defined as the way in which an organisation's identity is revealed through behaviour and communications, as well as through symbolism to internal and external audiences (Van Riel and Balmer, 1997). Corporate identity refers to what the organisation is, what it does, how it does it, and where it wants to go (Markwick and Fill, 1995). It represents a sense of individuality or uniqueness that can help to improve a company's competitive position (Balmer and Gray, 2000; Balmer, 2001; Hatch and Schultz, 2003). Figure 6.1 shows the assumed relationships between image and identity and the possible impact of corporate real estate and other influencing factors (Van Loon et al., 2006).

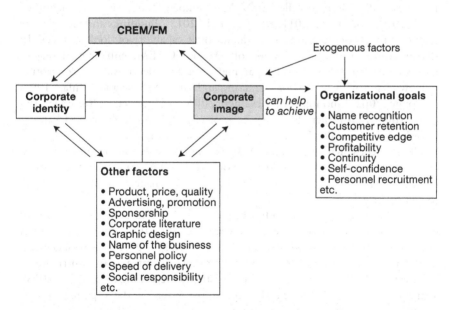

Figure 6.1 Relationship between corporate image, corporate identity and influencing factors (adapted from Van Loon et al. 2006)

Establishing the desired corporate identity entails positioning the entire company, also called corporate branding (Balmer, 1995). Branding is directed to the overall perception that internal and external stakeholders have of an organisation and the image they have in mind of the corporate identity (Balmer, 2001). Corporate branding is believed to create, communicate and deliver value to customers in a manner that benefits the organisational performance and supports competitive advantage (Balmer and Gray, 2000). Corporate branding can be part of a marketing strategy by using the vision, culture and image of a company as a unique selling proposition (Hatch and Schultz, 2003). According to Harris and de Chernatonay (2001), a real estate strategy that is aligned with branding objectives can add value to the core business by attracting and retaining customers and talented staff, and as such it contributes to distinctiveness and competitive advantage.

Corporate identity is propagated through all the facets of a company with which people come into contact. For this reason branding requires the engagement of everyone in the organisation, from CEO to HR, IT, RE and production departments (Khanna et al., 2013). Human resources managers can incorporate cultural matters and corporate values in training and development programmes in order to encourage employees to express cultural values in their behaviour. Production departments are responsible for designing products and delivering services that are coherent with the brand values of the company. Although marketing/PR and staff are shown to be more important than real estate, the possible contribution of FM and CREM in communicating brand values to internal and external stakeholders is recognised as well (Krumm and De Vries, 2003; HermanMiller, 2007; Van Kempen, 2008; Appel-Meulenbroek et al., 2010; Zoetemelk, 2011; Khanna et al., 2013). Buildings and facilities are strong channels of communication due to their capability of 'look and feel'. In the literature on the added value of FM and CREM, enhancing image is mentioned as an important value parameter and included in various conceptual frameworks, usually as part of marketing and sales (Nourse and Roulac, 1995; De Jonge, 2002; Lindholm and Leväinen, 2006; Lindhom et al. 2006; De Vries et al., 2008, Den Heijer, 2011; Jensen et al., 2012; Van der Zwart and Van der Voordt, 2013). See also Figures 1.1 to 1.4 in Chapter 1.

More recently, Khanna et al. (2013) elaborated a conceptual framework that visualises the alignment of a corporation's real estate strategy with corporate identity and brand values in order to evoke a positive image, thereby contributing to competitive advantage; see Figure 6.2.

According to Figure 6.2, branding by FM and CREM starts at an organisational level by defining vision, culture and strategy regarding the corporate identity. A next step is to define brand values and to explore how portfolio management, location strategy, building strategy and workplace strategy can contribute to express these brand values clearly and consistently. According to Wheeler (2013), appropriate brands should clearly express the vison of the organisation, be meaningful, authentic, different, sustainable, coherent and flexible, and express commitment. Based on an extensive survey in the USA, Gensler (2013) concluded

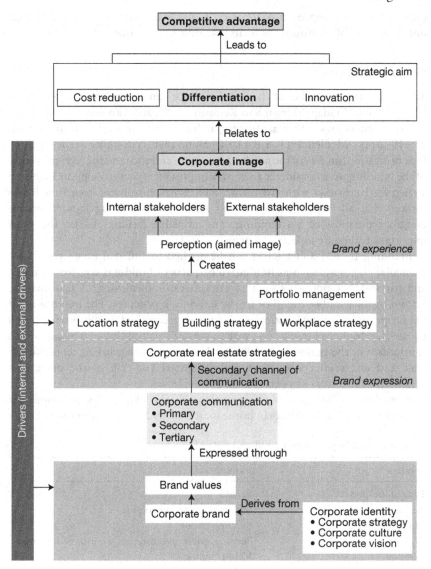

Figure 6.2 Conceptual framework to visualise the relationships between image, identity, branding and CREM strategy (Khanna et al., 2013, adapted from Balmer and Gray, 2000)

that quality outranks any other attribute contributing to brand loyalty, followed by trust, familiarity and price. Brand strength is influenced by the extent to which the interpretations of the brand are congruent. If the perceived corporate image is similar to the conceived corporate identity by the organisation, it strengthens the corporate identity and supports competitive advantage. Brand values are not static, due to changing visions and core business strategies. For instance, in a growth phase a company may focus on transparency and innovation, whereas during a financial downturn the focus may shift to low price.

Benefits and costs

Table 6.1 shows a number of characteristics that can contribute to branding by FM and CREM (adapted from Van Kempen, 2008, and Van den Assem, 2015). Based on interviews with practitioners, architects, and real estate brokers (total N = 19) Appel-Meulenbroek et al. (2010) found that accessibility of the location, type of location, quality of the finishing, the main entrance and the recognisability of the building were considered to be very important aspects to support branding strategies. Interviews with two banks and eighty customers from four banks showed that all respondents perceived accessibility, building façade, interior design, reputation of surroundings and available facilities to be the most important real estate aspects (Zoetemelk, 2011).

Branding by FM and CREM can be applied to full real estate portfolios, particular buildings such as headquarters, parts of buildings e.g. front offices, and particular components such as workplaces or furnishing. A firm may be known by its neighbourhood or by the building it occupies. The importance of the image effects of the location has long been recognised, and is often referred to as the 'right address'. An analysis of 787 office lease transactions in Amsterdam in the period from 1996 to 2007 showed that urban economies of scale and image effects together can explain 64 per cent in the office rent variations in this city (Koppels et al., 2009).

Regarding building level, corporations might choose to accommodate themselves in iconic buildings in prime locations, in order to communicate their

Table 6.1 Branding options by FM and CREM

Location	Building
Type of location	Façade
Reputation	Building height and volume
Landscape architecture	Building layout
Facilities	Forms, colours and materials, exterior and interior
Visibility	Main entrance
Accessibility by car	Architectural style of interior and exterior
Accessibility by public transport	Horizontal zoning
	Vertical zoning
	Accessibility
	Appropriateness for its function
	Thermal comfort
	Lighting
	Security
	Catering services
	Visibility of sustainability measures

corporate brand values to their employees, clients and other stakeholders. By choosing a particular form, an organisation transmits messages about its values and aspirations – for example, by opting for the representation of progress, power and success associated with a skyscraper, or by architecture referring to industrial design and a living machine of innovation. The physical image of the building may advertise and attract attention to a firm's goods and services. The increased interest of companies to buy or lease sustainable office buildings can partly be explained from this perspective. It provides organisations with a means to communicate their corporate social responsibility (Thyssen, 2010). As Lindholm et al. (2006) argued, the work place strategy also provides possibilities to communicate brand values. For example, the office and facility layout might reflect the organisational emphasis on team work, transparency, innovation or any other set of corporate brand values. According to Ward and Holtman (2000) the "narrative office" brings brand values alive, acts as a receptacle for corporate memory, and gives employees constant visual stimuli to promote a service ethos. While the location, building, and portfolio strategy affect both the perception of the corporate identity by internal and external stakeholders, the workplace strategy mainly aims at expressing brand values to internal stakeholders (Van Loon et al., 2006).

A preliminary draft of the framework of Khanna et al. (2013) shown in Figure 6.2 has been used to investigate the use of corporate real estate to communicate the key values of seven multinationals. Most interviewees said they believe that branding by real estate supports their marketing policy and results in competitive advantage. As one of the interviewees said: "Real estate adds value by supporting the brand loyalty and brand awareness among employees. The move of the headquarters to a more exciting location attracted two hundred talented workers and evoked a positive image among employees, customers and investors. The WOW factor adds to job satisfaction and top employer brand." However, another respondent stated that the company does not see marketing as one of the added values of real estate: "our money is made by consultancy hours, most of our consultants don't stay in the office very often but visit clients, so our headquarters is just a facility to accommodate staff and other employees."

Overall the findings of Khanna et al. (2013) confirmed that brand values are incorporated in a company's location strategy, building strategy, workplace strategy and portfolio management, but in different ways and with different focus points. The most commonly used brand values are sustainability, reliability, transparency, innovation and people orientation.

Sustainability

At location level the value of being green and sustainable can be expressed through the proximity to public transport (particularly a train station) and centralisation vs. decentralisation. Regarding the building strategy, sustainable brand value can be communicated by implementing energy management programmes, BREEAM or LEED certified headquarters, video conferencing facilities, and optimising the footprint by desk sharing to achieve CO_2 reduction targets. An interesting application is the use of core business sustainable products like sustainable lighting and carpeting to maximise energy efficiency, e.g. by Philips in its Amsterdam headquarters. Regarding workplace strategy, organisations focus on flexible/smart/alternative workplace concepts to showcase their sustainable brand value, which also provides flexibility and reduces occupancy and facilities costs. The use of cradle to cradle materials is another option that is incorporated in workplace strategies to showcase the sustainable brand value. On a portfolio level the sustainable brand value can be communicated through optimisation of the real estate portfolio, adaptive reuse of redundant office space, long lease spans, and BREEAM or LEED certified buildings.

Reliability

Measures to communicate the brand value reliability are closely linked to continuity (e.g. lifespan use of existing buildings) and smart and efficient use of capital and other resources (e.g. adaptive reuse of vacant buildings). Ways to show the investors and customers that the organisation uses capital intelligently may include the involvement of renowned real estate consultants to optimise the real estate portfolio, and making socially responsible decisions such as adaptive reuse instead of new building to reduce the problem of high vacancy in office stock.

Transparency/openness

Transparency can be expressed by the use of glass, open voids or atriums, and by omitting walls to enhance visual connections; see the example in Figure 6.3. The workplace strategy can be used to support and express the organisation's culture (see Chapter 7 on Culture) and to reflect the organisational structure.

Innovation

Companies considering innovation as one of their brand values closely link innovation to the technological developments in their core businesses. The IT department is strongly involved in the translation of this brand value into real estate by the application of new technological developments. In the location strategy, innovation is usually depicted by choosing the location in regions where talented labour is concentrated, e.g. in Central Business Districts. In building strategy, innovation can be expressed by the use of biometrics, i.e. technology to make buildings smart, for instance by scanning human physical characteristics for security and to control physical access. Innovative work place strategies focus on flexible working with non-assigned "hot" desks.

Figure 6.3 ING House, Amsterdam: example of a building that represents modernity and transparency

People orientation

This brand value is mainly focused on internal stakeholders as the targeted audience. The involvement of employees in designing their workplaces can support a sense of belonging and is an interesting way to share the corporation's vision and core values with the internal stakeholders. Other measures include the incorporation of employee values in location choices, the supply of employee services, and the application of new office concepts to improve social interaction. In one of the interviewed companies the "connecting people" brand value had been translated to a location in the city centre, with a high exposure to the outside world and customers, visible from the highway and served by two train stations.

Other values

In order to strengthen the distinctiveness and reputation of a company, various organisations apply guidelines for workplace design worldwide at all levels of real estate. Large organisations often choose to be accommodated in a high rise building that is clearly visible in the skyline of the city, or use a huge and well-designed reception desk with the company logo to evoke an atmosphere of spaciousness and richness. In the Post NL headquarters in the Netherlands a huge typographical art work with metre-high letters on a glass panel, with different coloured foil, shows the various promises of this company to the customers. The large size of the conference hall expresses the magnitude of the target group of Post NL. The colour red in the interior refers to the traditional use of red in post offices and post boxes. Branding by real estate is also applied here through the sorting centres that are considered to be a window to external stakeholders, and the post shops, where the company gets in touch with its customers.

Some companies express 'trust' by choosing a location in a safe area, espousing transparency, and providing facilities that support employees' wellbeing to make them feel comfortable. Vodafone connects trust to new ways of working with a high level of employee autonomy in deciding when, how and where to work, and expressing the value of 'speed' in training modules in the work environment. In the retail and leisure sector large banners are used to show sub-brands like ETOS or Gall and Gall, and as such to illustrate the large product range of the company. At the Unilever headquarters in Rotterdam, Netherlands, the company's own products are shown all over the building. At the Nike headquarters the building is full of cues to sports shoes and sport facilities. Young web designers and successful firms such as Google like to express themselves as being modern, progressive, young and different, and like to be accommodated in 'funky' offices (Van der Voordt et al., 2003; Van Meel, 2015). A particular location or building type can also evoke a negative image. For instance, one of the municipal organisations in Rotterdam, Netherlands, operates from the ground floor of a high quality building at an expensive location. This has led to negative reactions from the citizens of Rotterdam, because they see this localisation as a waste of public money (Van den Assem, 2015).

How to measure and manage

Key Performance Indicators

There are various ways to measure the image and identity of an organisation and possible gaps between brand identity and corporate image, such as surveys, (semi-structured) interviews, group discussions, and heuristic analyses of historical sources (Van Riel and Balmer, 1997; Roy and Banerjee, 2014). The results may be used as KPIs to benchmark the image with the image of other organisations or of the image before and after one or more interventions. We suggest conducting a survey among customers, employees and other stakeholders and asking them:

- To express their perception of Corporate Identity, Corporate Values and Corporate Brands, for instance by marking the most appropriate items in a list of possible characteristics such as modern, progressive, innovative, sustainable, reliable, transparent, socially responsible, people oriented (Khanna et al., 2013), or to mark on a five-point Likert scale to what extent a particular value is associated with this particular company
- To mention which words the respondents find to be most connected with the organisation (open question).

- To mark on a five-point scale extent the extent to which a number of location and building characteristics contribute to particular brand values (Van den Assem, 2015).

Other more indirect way to measure a companies' image are, for instance:

- Counting the number of shares and likes on Facebook and other social media (Wheeler, 2013)
- Analysing social news, blogs, wikis, internet fora etc. (Wheeler, 2013).

Step-by-step plan

In order to manage the enhancement of a positive image and to express brand values by corporate real estate, Van Loon et al. (2006) presented a step-by-step plan to translate brand values into accommodation choices. Khanna et al. (2013) presented another step-by-step plan to develop a well-considered and well-argued strategy for 'branding by real estate'. In combination, this leads to the process proposed below.

Step 1: Internal analysis – outline the corporate brand values

- (Re)define the general mission, vision and strategy of the organisation
- Identify the basic organisational culture characteristics
- List the elements of brand essence
- Summarise the corporate brand values.

The use of pictures can help to explore the key values of an organisation and to think "out of the box", to enable latent needs and values to come to the fore. For instance, the analysis might use pictures of building façades, interior designs, cars, strip figures, dogs or sports. Involved stakeholders may be invited to freely associate and link preferred pictures to the company, and to explain why this picture represents the organisation more clearly than the other pictures. In a workshop that was organised by Van Loon a participant linked his organisation to Suske and Wiske, two well-known Flemish comic strip figures: "because they represent a well collaborating team, based on equality, and always search for innovative solutions". Another participant associated the company with the comic strips of Maarten Toonder, i.e. with Oliver B. Bommel and Tom Pouce, because they demonstrate a combination of conservative reliability, richness and being smart (Van Loon et al., 2006).

Step 2: External analysis

Conduct a SWOT analysis of the organisation, taking into account the current market, the role of competitors, the needs and interests of the customers, expected trends and future scenarios.

Step 3: Define CREM and FM strategies to communicate brand values

- Define the strategic directions and the desired supply for well-integrated brand values
- Decide which potential properties are important for communicating the listed corporate brand values
- Define the target audience to whom the brand values need to be transmitted
- Outline the strategic means of translating brand values in CRE and FM.

Step 4: Create a tactical action plan

- Compare the brand communication of the current supply with the desired brand expression
- Identify actions to translate corporate brand values in FM and CREM strategies.

Step 5: Evaluate and communicate the success of branding by FM/CREM

- Define the success factors of the translation of brand values in CREM and FM strategies
- Evaluate the employer brand rank after implementation
- Conduct a brand values awareness campaign among internal and external stakeholders
- Communicate the results internally and externally.

The different steps could be prepared by the FM and CREM department and discussed in workshops with general managers, facility managers, real estate managers, HRM representatives, end users and customers, if necessary with the support of a consultant, a designer or a communications expert. The outcomes can be used as input to a programme of requirements for new buildings or adaptations of existing buildings.

Perspectives

So far the literature on corporate image has mainly concerned definitions and conceptual analyses of 'image' in connection with other concepts such as corporate identity, brands and branding, drivers to brand, branding strategies and practical examples of branding on different scale levels. FM/CREM-related empirical research on branding is often focused on the perceived relevance of particular location and building characteristics. Empirical data about the *actual* impact of branding by FM and CREM on corporate image and corporate performance is still barely available. One of the reasons for this might be that measuring the actual impact is rather complex, costly and time consuming. Nevertheless it would be challenging to collect more empirical data about the image of organisations, and discover the extent to which corporate real estate, facilities and services contribute to this image.

Another challenge is to explore the trade-off between the benefits of typical interventions and their costs and sacrifices by conducting customer and employee surveys and calculating the expected and actual impacts on profit and costs. A third issue for further research is to conduct field research among different types of companies to generate a list of profiles of corporate identities within types of companies, related brand values, and how its location, buildings and facilities actually influence a company's image and performance. Each profile might then be matched to guidelines for branding by FM and CREM. Other issues for further research may involve finding a way to determine the actual influence of CRE on branding in relation to other relevant company resources, and studying whether the direct or indirect influence is the most important (Appel-Meulenbroek et al., 2010).

References

Appel-Meulenbroek, R., Havermans, D., Janssen, I., and Van Kempen, A. (2010) 'Corporate branding: An exploration of the influence of CRE', *Journal of Corporate Real Estate*, 12 (1), pp. 47–59.

Balmer, J.M.T. (1995) 'Corporate branding and connoisseurship', *Journal of General Management*, 21 (1), p. 2246.

Balmer, J.M.T. (2001) 'Corporate Identity, Corporate Branding, and Corporate Marketing: Seeing through the fog', *European Journal of Marketing*, 35 (3/4), pp. 248–291.

Balmer, J.M.T. and Gray, E.R. (2000) 'Corporate Identity and corporate communication: Creating a competitive advantage', *Industrial and Commercial Training*, 32 (7), pp. 256–261.

Coenen, C., Von Felten, D. and Schmid, M. (2010) 'Reputation and public awareness of facilities management – a quantitative survey', *Journal of Facilities Management*, 8 (4), pp. 256–268.

Coenen, C., Waldburger, D. and Von Felten, D. (2013) 'FM Servicebarometer: Monitoring customer perception of service performance', *Journal of Facilities Management*, 11 (3), pp. 266–278.

De Jonge, H. (2002) 'De ontwikkeling van Corporate Real Estate Management', *Real Estate Magazine*, 22, pp. 8–12.

De Vries, J.C., De Jonge, H. and Van der Voordt, T. (2008) 'Impact of Real Estate on Organisational Performance', *Journal of Corporate Real Estate*, 10 (3), pp. 208–223.

Den Heijer, A. (2011) *Managing the University Campus*. Delft: Eburon.

Gensler (2013) *Brand engagement survey: The emotional power of brands*. Available at http://www.gensler.com/design-thinking/research/2013-brand-engagement-survey.

Harris, F. and de Chernatonay, L. (2001) 'Corporate branding and corporate brand performance', *European Journal of Marketing*, 35 (3/4), pp. 441–456.

Hatch, J.M. and Schultz, M. (2003) 'Bringing the corporation into corporate branding', *European Journal of Marketing*, 37 (7/8), pp. 1041–1064.

HermanMiller (2007) *Three-Dimensional Branding. Using space as a medium for the message*. Available at http://www.gensler.com/design-thinking/research/2013-brand-engagement-survey.

Heywood, C. and Kenley, R. (2008) 'The sustainable competitive advantage model for Corporate Real Estate', *Journal of Corporate Real Estate*, 10 (2), pp. 85–109.

Jensen, P.A., Van der Voordt, T. and Coenen, C. (Eds.) (2012) *The Added Value of Facilities Management: Concepts, Findings and Perspectives*, Lyngby, Denmark: Centre for Facilities Management and Polyteknisk Forlag.

Khanna, C., Van der Voordt, D.J.M. and Koppels, P. (2013) 'Real Estate mirrors Brands. Conceptual framework and practical applications', *Journal of Corporate Real Estate*, 15 (3/4), pp. 213–203.

Koppels, P.W., Remøy, H., Weterings, A. and De Jonge, H. (2009) 'The added value of image: A hedonic office rent analysis', *16th Annual European Real Estate Society Conference*, Stockholm, Sweden.

Krumm, P.J. and De Vries, J.C. (2003) 'Value Creation through management of real estate', *Journal of Property Investment and Finance*, 21, pp. 61–72.

Lindholm, A.-L., Gibler, K.M. and Levaïnen, K.I. (2006) 'Modelling the value adding attributes of real estate to the wealth maximization of the firm', *Journal of Real Estate Research*, 28 (4), pp. 443–475.

Lindholm, A.-L. and Leväinen K. I. (2006) 'A framework for identifying and measuring value added by corporate real estate', *Journal of Corporate Real Estate*, 8 (1), pp. 38–46.

Markwick, N. and Fill, C. (1995) 'Towards a framework for managing corporate identity', *European Journal of Marketing*, 31 (5/6), pp. 396–409.

Nourse, H.O. and Roulac, S.E. (1995) 'Linking Real Sstate Decisions to Corporate Strategy', *Journal of Real Estate Research*, 8 (4), pp. 475–494.

Omar, A.J. (2015) *Positioning Corporate Real Estate Management (CREM) Using A Branding Approach*. PhD thesis, Melbourne School of Graduate Research, Australia.

Omar, A.J. and Heywood, C.A. (2014) 'Defining a corporate real estate management's (CREM) brand', *Journal of Corporate Real Estate*, 16 (1), pp. 60–76.

Redlein, A. and Zobl, M. (2013) 'Facilities Management in Austria 2012 – Value Add?' in Alexander, K. (Ed.) 'FM for a Sustainable Future', *EuroFM Journal*. Conference papers EFMC 2013, Prague.

Redlein, A. and Zobl, M. (2014) 'Facility Management in West- and Eastern Europe', in Alexander, K. (Ed.) 'Promoting innovation in FM', *EuroFM Journal*. Research papers EFMC 2014, Berlin.

Roy, D. and Banerjee, S. (2014) 'Identification and measurement of brand identity and image gap: A quantitative approach', *Journal of Product and Brand Management*, 23 (3). pp. 207–219.

Thyssen, J. (2011) *Corporate Social Responsible Real Estate Management – Understanding how a corporate real estate object interacts with society*. MSc Thesis, Delft University of Technology.

Treacy, M. and Wiersema, F. (1995) *The discipline of market leaders*, Massachusetts, USA: Addison Wesley.

Van den Assem, J. (2015) *Corporate branding and CREM – Alignment between Corporate Identity and CRE-Strategy*. MSc thesis, Eindhoven University of Technology.

Van der Voordt, T., Van Meel, J., Smulders, F. and Teurlings, S. (2003) 'Corporate culture and design. Theoretical reflections on case-studies in the web design industry', *Environments by Design*, 4 (2), pp. 23–43.

Van der Zwart, J., Arkesteijn, M.H. and Van der Voordt, T. (2009) 'Ways to study corporate real estate management in healthcare: An analytical framework'. Conference Proceedings *HaCIRIC 2009*, Health and Care Infrastructure Research and Innovation Centre, Brighton, UK April 2–3.

Van der Zwart, J. and Van der Voordt, T. (2013) 'Value adding management of hospital real estate. Balancing different stakeholders' perspectives', *(E)Hospital*, 15 (3), pp. 13, 15–17.

Van Kempen, A. (2008) *Corporate Branding en Real Estate: Hoe bedrijfshuisvesting de bedrijfsidentiteit kan uitdragen*. MSc thesis, Eindhoven University of Technology.

Van Loon, S., Van der Voordt, D.J.M., and Van Liebergen, M. (2006) 'Form follows identity: het vertalen van corporate image naar huisvestingseisen', *Real Estate Magazine*, 9 (46), pp. 46–49.

Van Meel, J. (2015) *Workplaces today*, Copenhagen: Centre for Facilities Management.

Van Riel, C. (1995) *Principles of Corporate Communication*. London: Prentice-Hall.

Van Riel, C.B.M. and Balmer, J.M.T. (1997) 'Corporate Identity: The concept, its measurement and management', *European Journal of Marketing*, 31 (5/6), pp. 340–355.

Von Felten, D., Böhm, M., Coenen, C. and Meier, G. (2014) 'Identity and image of FM: Two sides of a coin to promote productivity in FM', in Alexander, K. (Ed.) 'Promoting innovation in FM', *EuroFM Journal*. Research papers EFMC 2014, Berlin.

Ward, V. and Holtman, C. (2000) 'The role of public and private spaces in knowledge management', paper presented at *Knowledge Management: Concepts and Controversies*, University of Warwick.

Wheeler, A. (2013) *Designing Brand Identity*, Hoboken, New Jersey: John Wiley and Sons.

Worcester, R.M. (1997) 'Managing the image of bank: The glue that binds', *International Journal of Bank Marketing*, 15 (5), pp. 146–52.

Zoetemelk, P.M.N. (2011) *Bedrijfsidentiteit versus Bedrijfsimago. De rol van bedrijfshuisvesting bij branding in de bankensector*. MSc thesis, Eindhoven University of Technology.

Interview 2: Carel Fritzsche, Stork Technical Services, the Netherlands

Carel Fritzsche has twenty-five years of experience in CREM. He studied Urban Planning in Tilburg and Economic Geography in Utrecht. He has worked at Tandem Computers Europe, Twynstra Gudde Management and Consultancy and Johnson Controls – Global Workplace Solutions. Currently he is the global head of CREM for Stork, a leading company and partner for improving asset integrity for clients operating in the oil and gas, chemical and process, and power industries. The company is active globally, with a strong (historic) foothold in the Netherlands.

Added value in general

Q. What does the term "Added Value" mean to you?

A. *I don't have a sharp definition, it is no hard science, but key is the trade-off between costs and benefits (affordability), mainly financially driven (shareholder value), but also regarding other value domains ('areas of attention'). Within our industry, besides Operational Excellence, Safety and Compliance thinking is also leading. Added Value (AV) is not a single topic but should also be related to the different stakeholders within the company. Shareholder Value is mainly connected to Return on Investment, low risks, costs and reliabilities. The Board of Management is mainly focused on strategic direction, global client-market development, Ebit (i.e. Earnings before interest and tax) and overall cash-position, and less on strategic vision and policy. Heads of regional units equally focus on financial performance (maximum turnover (volume of business) and minimum costs) as well as on the attraction and retention of talented staff. They have to cope with both top management needs (profit), regional customers and employees, e.g. retaining good staff. Site managers mainly focus on operational issues and employee satisfaction.*

Q. Are there other related terms that you prefer to use rather than "Added Value"?

A. *TVA = Total Value Add.*

Q. How do you see the relation between "Added Value" and cost reduction?

A. *Cost reduction is a value in itself and a topic to check every action/intervention against. Savings on CRE costs directly result in extra profit, but should also be viewed in relation to the investment needed (pay-back). Though cost reduction is very important we also discuss the benefits and constraints of a proposed direction, for example the impact of closing locations on client accessibility and satisfaction.*

Q. Is "Added Value" mostly treated at a strategic, tactical or operational level?

A. *Strategic, tactical and operational management occur on all geographic levels and all scale levels. CRE is partly centrally managed from the head office and partly by regional offices (strategic-tactical) and at a building level (tactical-operational). Policy and direction strongly focus at the global portfolio level (including maintaining the real estate database, international reports and KPI measurements). Projects are also driven by local culture and practice. FM worldwide is more strategic and FM locally is more operational.*

Q. In which context or dialogue is "Added Value" mostly considered, in your experience?

A. *Both internally and between inhouse CREM organisations and our clients, being both the Board of Management as well as the regional and local business leaders. AV is used to balance what is really required (which locations, how many m²) and the added value of this (related cost and investment versus the business benefits to be achieved). For instance, what is the AV of the CREM department? Know what you have and what you need and challenge business leaders for alternative solutions, sometimes based on proven experiences in other regions/locations. For example, to what extent can we benefit from doing activities around shut-downs at client-sites, to minimise the size of dedicated Stork-locations? The right balance between functionality (fit-for-purpose), flexibility (preferably not too long rent contracts) and cost-effectiveness is important. AV is not explicitly mentioned in global CREM reports.*

Q. How do you see the relation between innovation and "Added Value"?

A. *Without innovation companies don't survive, so innovation is a key condition to deliver AV.*

Benefits and limitations of "Added Value"

Q. What are the benefits of considering and talking about "Added Value"?

A. *Added Value shows what you provide and why. All actions/interventions and services are assessed by a business case, with various scenarios to be considered. The 'language' of CRE managers varies according to the type of real estate (e.g. offices versus industry), geographical location (practice and culture) and position in the S-curve of maturity. Anglo-Saxon countries normally have a strong hierarchical direction and focus on financial performance and shareholder value, while mainland-Europe tends to be less hierarchical and more compromise-driven. Although Stork is less hierarchically driven, it has a strong financial focus, including a constant focus on financial business performance per business unit. Also, we try to align our CRE language to the nature/responsibility of the person we are speaking or reporting to, e.g. financial to the CFO, image to Marketing and effectiveness to site managers.*

Q.   What are the limitations of considering and talking about "Added Value"?

A.   *Added Value is perceived differently by different people. It is very difficult to make AV concrete and operational. A concise and 'scientific' definition is impossible. For example, satisfaction can mean employee satisfaction, financial satisfaction, and also satisfaction with operational effectiveness. Various components are very subjective and difficult to measure quantitatively.*

Management of "Added Value"

Q.   What are your top five main values to be included in the management of accommodations, facilities and services? Could you mention a few examples of concrete FM interventions to attain these added values and KPIs to measure them?

A.   *See table below. Currently a main CREM topic is to get a complete and clear view of the CRE-portfolio and secure the real estate related key-data.*

Prioritised values	Concrete interventions	KPIs
1. Profit (Ebit) Improving cash position.	Disposal of real estate to gain cash (depends of book value, local market etc.) Sale and lease back. Footprint optimisation: focus is more on the total (regional) footprint (total locations/ buildings) and less on space reduction/disposal of only part of a building	Impact of lower RE costs on profit Footprints per region Policy depends on region
2. Cost reduction	Footprint consolidations, renegotiations of existing leases (better terms and conditions) and energy consumption	Footprint per region or building Cost per m² or per f.t.e. Total Cost of Ownership, absolute and ratios (TCO/ profit, RE costs/total business costs)
3. Transparency of real estate	Being transparent in annual reports (very important for shareholders)	Compliance Current obligations (e.g. rent contracts) Future liabilities Note: Impact of New IFRS legislation; cleaning balance sheet meant extra cash, nowadays rent contracts have to be visible

The focus and priority of CRE depends of the context. In a period of economic decline, cost reduction is key to maintain/improve profit. Given the type of business, safety and security are always important topics. The impact of FM on satisfaction for customers and staff/end users is particularly important at the local/building-level. Space flexibility and flexibility in lease contracts are becoming more important as well. Branding by Real Estate is no issue. Our logo and a positive perception of our products and services is more important. A functional issue is whether people can easily find Stork buildings on industrial business parks. Regarding sustainability Stork is no 'trend-setter'; we normally follow the market-trend. Given the primary process, Stork has a strong focus on environmental issues, e.g. avoiding soil pollution. Sustainability is mainly linked to durability, i.e. the continuity of business.

Q. Do you measure whether you attain the intended added values? If so, what Key Performance Indicators do you apply for each of the added values you mentioned before?

A. *See table opposite. The impact of CREM on staff productivity is not yet being measured.*

Q. Do you benchmark your data with data from other organisations?

A. *Not yet. We are working on internal benchmarking between (comparable) locations and external benchmarking on real estate market rents. Our current focus is on benchmarking our primary process, e.g. the costs of staff and process improvements around installation shut downs. Due to the variety in buildings (offices, production plants and a mix of office and industry/production halls) benchmarking is difficult.*

Last comment

Q. Are there other topics that you find important in connection to the concept of "Added Value"?

A. *Try to clearly define Added Value and different types of values. Include AV of ownership versus renting premises in connection to market-conforming real estate and more built-to-suit buildings, flexibility, and costs. Relate AV to the context (Anglo-Saxon countries versus Mainland Europe, USA, Asia) and the different real estate sector (offices, healthcare, education, industry). Interview more people in the industrial sector, not only from office-based companies. Learn from other disciplines such as business management, economics, etc. Make a clear distinction between the market value of CRE and the company value of CRE. Pay attention to AV from a historical perspective: how did the concept develop over the last decades?*

7 Culture

Theo van der Voordt and Juriaan van Meel

Introduction

When visiting a building, one gets a strong first impression of the people and the organisation that occupy it. For example: is there a grand entrée, or is it a modest welcoming area? Is the receptionist smiling and friendly, or bored and curt? Are you taken to a bland anonymous meeting room right behind the entrance, or are you allowed onto the work floor, having the meeting at the coffee table in the kitchenette? All these factors tell – consciously or subconsciously – a story about the people that inhabit the building. To put it in scholarly terms: buildings act as cultural 'artefacts' and 'symbols' that reflect the culture of their inhabitants, expressing particular norms and values about human relations, power, and the nature of work (Gagliardi, 1990; Schein, 2004).

One of the first management books that explained this notion was called *Corporate Culture: the Rites and Rituals of Corporate Life*, written by Deal and Kennedy (1982). The book was the starting point of what would become a long stream of popular management books that stressed the importance of culture for organisational performance. The general message of these publications is that a strong corporate culture is a key driver for the success or failure of organisations, because it guides people's behaviour and work styles, for example affecting how people treat customers, how employees work together, and how committed people are to the organisation they work for. Interestingly, Deal and Kennedy explicitly mentioned the physical work environment as an element of corporate culture. They wrote that "a company's investment in bricks and mortar inevitably says something about its culture". This is interesting because organisational culture is often seen as a rather elusive concept, consisting of abstract aspects such as norms, values, expectations, attitudes, social patterns, beliefs, customs, unwritten rules of behaviour and other intangibles. Despite this inherent vagueness in the concept of culture, most theories on culture agree that it is reflected in the shape of buildings and spaces in the sense that their design reflects in many ways the culture of the organisations that use them (Gagliardi, 1990; Van der Voordt et al., 2003; Schein, 2004; Hofstede et al., 2010). An interesting question is whether buildings and spaces not only *reflect* the culture of the occupiers, but also *affect* their values and beliefs; see Figure 7.1.

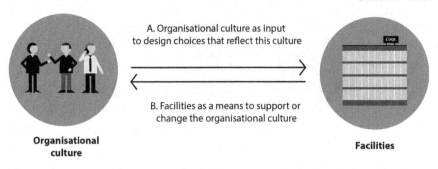

A. Organisational culture as input to design choices that reflect this culture

B. Facilities as a means to support or change the organisational culture

Organisational culture

Facilities

Figure 7.1 Facilities and corporate culture: a reciprocal relationship

In this chapter we will take a closer look at this possibly reciprocal relationship between culture and Corporate Real Estate Management (CREM) and Facility Management (FM), with an emphasis on the relationship between culture and the physical work environment. The concept of culture may refer to different scale levels, e.g. national culture, regional culture, industry culture, organisational culture and team culture. This chapter will focus on organisational culture, and to a lesser extent national culture. We will present a general model to explain what organisation culture is, and connect this model to facilities. Then we take a closer look at the cultural dimensions that are most relevant to CREM and FM practice. Furthermore, we will discuss the empirical data that are available concerning the relationship between culture and facilities. The chapter ends with a discussion of ways to measure the impact of FM and CREM on culture and possible KPIs, and a list of topics for further research. The content of this chapter is of a more tentative and explorative nature than the other chapters in this book, due to the lack of extensive research on the relation between CREM/FM and culture. This means that the ideas presented in this chapter should be seen as hypotheses rather than hard conclusions.

State of the art

Different layers of culture

There are many different definitions and models that try to describe what organisational culture is. Perhaps the most successful attempt to conceptualise organisational culture is Edgar Schein's much used 'layer model' (Schein, 2004); see Figure 7.2. This model is based on the idea that culture consists of several layers that build on top of each other. The model suggests that the deepest layer consists of a set of deeply embedded assumptions about fundamental issues such as human relations, human nature, time and power. These assumptions are often implicit and taken for granted by the people working in the organisation, but they have a strong impact on people's behaviour and the organisation's way of operating. The next layer consists of the espoused values of a company, i.e. the more formal or desired culture of an organisation. These are explicit principles,

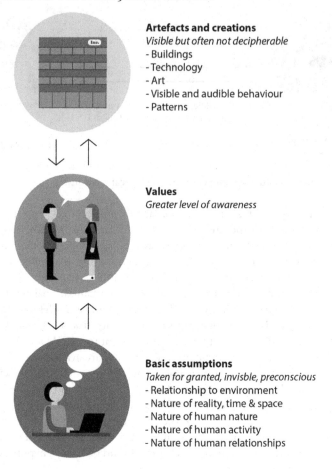

Figure 7.2 Several layers of culture (based on Schein, 2004)

norms and values that are intended to guide staff behaviour and norms, often expressed in official vision statements or behavioural rules – which are not necessarily the same as the 'real' values at the core of the culture. The third level includes the tangible, overt artefacts of an organisation. These include the organisation's dress code and the language people use, but also the physical setting and buildings of the organisation.

Culture and facilities

In Schein's model, the physical environment plays a fairly passive role. It is the part of the most outer layer which reflects and expresses the corporate culture, rather than influencing it. This type of relation (type A in Figure 7.1) is closely related to corporate image and branding – which is covered in chapter 6. Here, we want to look at the relation between culture and facilities from the other direction: to what extent do facilities not only reflect, but also affect culture (type B in Figure 7.1)? Can the physical environment be actively used to influence a company culture? Can it help to change processes and promote or discourage particular ways of working and thinking? In other words: is there an added value in using the physical environment as a tool for cultural change?

Practitioners in the field of facilities design – designers, consultants, managers – generally have a strong belief in the transformative powers of their actions. Workplace designers can frequently be caught making statements that their design has "generated a more inclusive culture where employees feel valued and recognized for their work" (Hassell, 2015), or that a new workplace model has "sparked cultural change". The authors of the popular management book *Change Your Space, Change Your Culture* even claim that redesign of workspace is "the fastest, easiest and most cost-effective way to shift culture" (Miller et al., 2014).

Scholars on the topic of corporate culture tend to deride such ideas. According to Schein and many other experts, culture is by definition deeply ingrained in people's ways of thinking and working. In their view, cultural change is a daunting task that demands profound changes in leadership, hiring and HR policies, working practices and almost any other aspect of management and organisation. In this view, a change in the physical environment will never suffice to change a company's culture. It can even be argued that such a change will fail or be counterproductive if it is not part of a wider change process. To use a common example: when moving people from cellular offices into open-plan offices, one should not expect that this intervention alone will suddenly create a collaborative culture. In fact, when such culture is lacking, it is very likely that people will resent this change in environment, increasing the risk of the failure of such a move.

From this perspective, claims about the impact of facilities on culture should be viewed with a critical eye. Even so, it remains highly interesting to look at the role workplace design can play in cultural change processes. This is because buildings are very visible, concrete and touchable entities and thus much more tangible than other means of organisational communication. For example, putting a CEO in an open-plan work area is probably one of the most effective ways to demonstrate that the company wants to create a culture of openness and equality (leaving out the question whether it is really nice for people to sit next to the CEO or not). In that sense, the physical work environment can help to send out strong signals about desired norms and values. In short: workplace design cannot change or shape culture by itself, but it can be an effective means of communication as part of a wider change programme.

Dimensions of culture

There are many factors that influence the culture of an organisation – for instance, its history and its founders, national culture and occupational culture. Furthermore, organisational cultures tend to consist of all sorts of subcultures on business unit level or team level. This makes it difficult to analyse culture in a structured way, especially in relation to CREM and FM. At the same time, probably every real estate manager or facility manager is likely to affirm the everyday impact in their work of invisible yet strong obstacles and drivers, that can probably best be characterised as 'cultural issues'. Just think of discussions with managers about their private offices, or nicely designed social areas that remain empty, because employees are afraid of being perceived as not working.

In order to be able to investigate the impact of various factors on organisational culture, various authors have made a distinction between different dimensions of culture. Many studies rely on the work of the Dutch researcher Geert Hofstede who is considered one of the leading thinkers on this topic. He distinguishes between five dimensions of national culture (Hofstede et al., 2010); see Table 7.1.

Hofstede was a social scientist who was mainly interested in national cultures. His work was never intended to be used for research in relation to the physical environment. The only place where Hofstede makes an explicit reference to buildings is the dimension of power distance, where he mentions office design as a potential status symbol.

The same can be seen in other studies on culture. For example, Cameron and Quinn (2006) focused on organisational culture and made a distinction between

Table 7.1 Key dimensions of national culture according to Hofstede et al. (2010)

Dimension	Content
Small versus large power distance (PDI)	The extent to which the less powerful members of institutions and organisations within a country expect and accept that power is distributed unequally.
Collectivism versus individualism (INV)	Individualism pertains to societies in which the ties between individuals are loose; everyone is expected to look after himself or herself and his or her immediate family. Collectivism is the opposite, pertaining to societies in which people from birth onward are integrated into strong, cohesive in-groups, which continue to protect people throughout their lifetimes in exchange for unquestioning loyalty.
Femininity versus masculinity (MAS)	A society is called masculine when emotional gender roles are clearly distinct; men are supposed to be assertive, tough, and focused on material success, whereas women are supposed to be more modest, tender and concerned with the quality of life. A society is called feminine when emotional gender roles overlap.
Weak versus strong uncertainty avoidance (UAI)	The extent to which the members of a culture feel threatened by ambiguous or unknown situations. This feeling is, among other things, expressed through a need for predictability and clear rules.
Long-term versus short-term orientation (LTO)	Long-term orientation stands for the fostering of virtues oriented towards future rewards, whereas short-term orientation stands for the fostering of virtues related to the past and present.

four types of organisational culture: clan culture, adhocracy culture, hierarchy culture and market culture. Igo and Skitmore (2006) linked organisational culture to dominant organisational characteristics such as the degree of teamwork and sense of belonging, the level of creativity and dynamism, focus on goals and competition, reliance upon systems, leadership style, management of employees, bonding mechanisms that hold the organisation together, organisational strategy drivers, and criteria for success. These are all thorough studies on the topic of culture, but they are mainly people oriented and ignore the relation between culture and physical facilities.

This may be the reason why studies that have used these dimensions in relation to workplace design are struggling to find clear correlations between these dimensions and the shape and use of the physical environment. We will come back to this issue in the section on empirical research findings.

Dimensions of culture and the physical environment

Because of the lack of empirical research on the relation between culture and facilities, we have taken the liberty of using the dimensions of Hofstede and other researchers in a rather loose way and have come up with a new set of dimensions that are most likely to impact the work of real estate and facility managers, especially when developing and implementing new work environments. These are: the degree of hierarchy, formality, individuality, trust, openness to change, and uniformity. This set of dimensions is partly based on prior research and has been extended on the basis of experiences from practice. Below, each dimension is briefly discussed with examples of how it relates to the design and management of the physical work environment.

Hierarchy

Everybody is familiar with the almost clichéd image of self-important managers seated in spacious corner offices on the building's top floor. It is a very vivid image of how office design can express certain cultural ideas about power and status. In recent years, however, such overt markers of hierarchy in the office have become rarer. The overall trend is towards more flexible, egalitarian work spaces in which management is – in terms of space allocation – treated like everyone else. However, there are still plenty of exceptions to this trend. In China, for example, hierarchy is still firmly rooted in many corporate cultures, which is reflected in spatial features such as executive offices located on floors separate from those of other workers, and separate elevator systems for executives (Gensler, 2015). In more egalitarian cultures, such obvious status symbols are frowned upon. However, it would be an illusion to say that hierarchy is absent in even the most egalitarian organisations. Almost any facility manager is probably familiar with the sensitivities around the private offices from managers. Managers tend to claim that they need these for the many confidential meetings they hold, which is perhaps true, but it is very likely there is also a cultural component of status anxiety to it.

Formality

Contemporary organisational cultures seem to move away from formality towards more informal and casual ways of working and interacting. Leading in this trend are technology companies like Google and Facebook, where not only the employees but also top management go round in t-shirts and trainers and seem to have dropped all notions of business formality. This casual culture is also reflected in the facilities of these companies. Google's offices, with slides and game rooms, are well-known extremes that signify a wider trend. The usual greys and beiges of office design are being replaced by bold colours and pop-art style prints; kitchenettes and vending machines seem to give way to espresso bars and micro-kitchens; and bland conference rooms are transformed into casual meeting spots with bean bags and lounge chairs. Obviously the extreme examples in the Google offices are not for everyone. In particular, more traditional companies like banks, insurance companies and law firms may not want to come across as too hip or progressive. Such companies are expected to be professional, smoothly-running bureaucracies, and they want to come across as embodying cultures of trustworthiness, efficiency and professionalism. In these more formal cultures workplace design is likely to be more 'neat' and 'clean', with high levels of standardisation and unobtrusive design.

Individuality

The way people relate to each other is at the core of the concept of organisational culture. A distinction can be made between very individualistic, task-oriented cultures, and more group-oriented cultures with a strong focus on establishing and maintaining relations. This cultural difference can have a large impact on how work environments are used and perceived in daily practice. For example, the difference is very relevant in relation to the continuous discussion about open-plan offices. In a group-oriented culture, open plans may be bustling, vivid places – as intended – with people actively using the openness to collaborate and maintain their social relations. In a more task-oriented culture, however, open planning may be awfully quiet as people focus on their work, spend less time on maintaining social relations, and try to disturb their co-workers as little as possible. The same difference is also likely to determine whether communal spaces, such as kitchenettes or lounge areas, will be truly used or remain 'dead spaces'. In individual cultures there may be little use for such spaces as people prefer to focus on their work and may even see chatting as a waste of time. In contrast, group cultures are more likely to make good use of such social places, using them as gathering points where knowledge, ideas and ordinary gossip are exchanged.

Trust

A distinction can be made between 'low trust' and 'high trust' cultures (Fukuyama, 1995). In low trust cultures, employees and their managers see themselves as different castes with different, conflicting interests. Managers believe that their staff should be closely supervised, and the staff eye the intentions of management with a certain degree of suspicion. In contrast, high trust cultures emphasise common interests and the autonomy of employees. This cultural dimension may have an impact on the way facilities are managed and used. For instance, the success of new ways of working in settings with shared use of activity-based workplaces depends on managers being able to manage on output and define clear targets regarding employees' performance, and accept that employees to a large extent decide for themselves where, when and how they want to work. Clearly, this only works if there is a culture of trust: managers have to trust their employees' ability and willingness to perform, while employees in their turn must trust their managers to evaluate them based on their output rather than their presence at the office (Baane et al., 2010; Bach, 2015).

Openness to change

Change processes, including workplace change, are never easy, but in some organisations change seems more difficult than in others. To a large extent, this may relate to the organisation's culture. Some organisations have a culture that is very open to change and experimentation. They seem to thrive on a continuous stream of new ideas and business practices, following the popular mantra of "change is the only constant". Such organisations may have a greater appetite for all sorts of new, flexible and innovative workplace concepts than organisations with a culture that is focused on stability and structure. Organisations with such cultures may actually suffer from the implementation of radical new workplace concepts as they disrupt the existing order. In such organisations, gradual changes may be more successful than grand transformation programmes. Furthermore, they are likely to prefer workplace solutions that have been tried and tested elsewhere, with a higher certainty of success, rather than more experimental concepts.

Uniform versus pluriform cultures

Organisations are never monolithic. Especially in large organisations, there may be considerable variance in the culture of business units, departments and teams. Think of differences between the legal department and the sales department, or the marketing group and the engineering group. The question is whether such variance should be promoted or toned down. A strong, united culture decreases the chances for internal conflicts and fragmentation within an organisation (Deal and Kennedy, 1982). Diversity in subcultures, however, creates alternative lines of thinking in an organisation, which helps to spur new ideas and even innovation (Becker, 2007). Corporate facilities can play a role in this, in the sense that they can express organisational uniformity or diversity. Organisations that strive for a strong uniform culture may have a preference for standardised work environments, using the same colours, design and furniture to express an image of homogeneity and unity. Companies that promote, or allow, cultural diversity may opt for a model in which different organisational units get more freedom to customise their work environment to their specific needs.

Empirical research findings

We assume that the discussed cultural dimensions play a large role in the real-life practice of developing and managing physical work environments. However, it remains unclear how this works and to what extent cultural change can be attributed to physical changes alone. As we indicated earlier, only a limited amount of empirical research has been conducted into the reciprocal relationship between facilities and organisational culture.

De Jonge and Rutte (1999) discussed the transformation of the Interpolis insurance company in the Netherlands from a traditional cellular office into a much more open setting with modern furniture and shared workspaces in the 1990s. According to their study, this resulted in a more collaborative culture and a perceived increase of autonomy among the employees. However, it is important to point out that the physical change was part of a much larger corporate restructuring. Other evidence comes from research by Van Meel (2000) that compared the floorplans of offices in the UK, Germany, Sweden, Italy and the Netherlands and concluded that national culture is one of the factors that influences office design. For instance, the egalitarian culture of the Netherlands results in a high level of end-user involvement in the design and management of office buildings. Because most people like to sit near a window due to daylight and outside views, deep buildings with an open-plan office concept are not popular in this country.

Van der Voordt et al. (2003) conducted various case studies that showed how modern web-designers try to express a culture of being young, creative and innovative by 'hip' and 'cool' design, colourful materials, luxurious facilities such as gyms or lounge areas and gimmicks such as jukeboxes and pool tables. Gall (2009) presented the results of an exploratory study on understanding the influence of national culture on the design, use and attribution of space. The main theory used in this book is the theory of Hofstede. The five dimensions of national culture are used to compare five countries around the globe on the same five themes.

Rothe et al. (2011) found that people from Finland and the Netherlands prioritise different aspects as most important to be able to work in a pleasant and productive way. Both groups of respondents shared the same five most frequently chosen attributes: functionality and comfort of the workplace, opportunities to concentrate, accessibility of the building, indoor climate, and opportunities to communicate. However, with 55% of the Finnish sample marking opportunities to concentrate as one of the three most important aspects versus 37% of the Dutch sample, the Finnish seem to be more concerned about this issue. Privacy is a little higher on the list compiled by Finnish users: 21% of all Finnish respondents versus 12% of the Dutch respondents marked this topic in their top three most important issues. Adjacency and locality of spaces and openness and transparency of the work environment were more important for the Dutch office users.

Riratanaphong and Van der Voordt (2012) compared these data with the responses from Thai office users. Whereas high masculinity and individualism are reflected in the national culture and the organisational culture of the Thai case was perceived as a hierarchical culture, a clan culture was preferred. The cultural context seemed to have an impact on the appraisal and prioritising of aspects of the work environment. The hierarchical culture contributed to the move to a single tenant building in order to control the expanding responsibilities. The typical workplace layout in Thailand – providing less variety in the spaces for socialising – and limited opportunities for sharing ideas about the working environment may reflect the hierarchical culture as well. However, other factors such as the former and current work environment, the work processes, the external context, and the implementation process seemed to be more influential.

Khanna et al. (2013) showed how seven multinationals use real estate as a means to brand values such as trust, being people-oriented, transparency and sustainability; see Chapter 6. Plijter et al. (2014) found that the different national cultures of Germany, the Netherlands and the UK were partly reflected in different workplace concepts and the interior design of offices of two companies with locations in these three countries. However, whereas all interviewed companies had their real estate portfolio to some extent aligned to the local national culture, none had a strict central policy about this issue. Differences in workplace characteristics were mainly caused by the involvement of local people in workplace design. Using Hofstede's cultural categories the case studies showed relationships between the masculinity of a culture and the expression of status, and between uncertainty avoidance and openness to innovation, whereas no relationships were found in relation to differences in power distance and short-/long-term orientation. The organisational culture showed to be dominant in both corporate real estate management of multinationals and local workplace design.

Based on a literature review and interviews Bach (2015) concluded that ICT-enabled time and place independence from work affects the organisational culture by inducing less hierarchy and more informal behaviour, a shift from presence-oriented to output-oriented style of control, increasing employee autonomy, and decreasing organisational attachment due to employees being less present in the organisation.

All these studies are mainly focused on how real estate and other facilities express cultural values and are affected by them (type A in Figure 7.1) and much less on how facilities affect organisational culture (type B). Besides, most studies do not incorporate their effects on organisational performance.

Benefits and costs

Table 7.2 shows some typical interventions to support or change the organisational culture by CREM or FM. For interventions to express cultural values and corporate identity by facilities we refer to Chapter 6 on Image.

Table 7.2 Typical interventions with management, benefits and costs

Interventions	Management	Benefits	Costs
Creating more openness	Replacing cellular offices by open settings	Stimulation of social interaction and exchange of values	Cost of change Distraction
Sharing of spaces and other facilities	Replacing personal desks and private spaces by shared desks and shared spaces	Stimulation of social interaction and exchange of values	Cost of change Loss of privacy, territoriality and personal control
Same places for similar activities, not related to status	Re-arranging the spatial layout, with more similar workplaces and spaces	Flexibility Sense of less hierarchy and power distance	Resistance of (top) managers
Allowing to work time and place independent	Facilitating place and time independent working by modern IT facilities Management on output	More autonomy and freedom for individual staff	Loss of managerial control Loss of social cohesion if contact is not managed or natural
Moving to a more sustainable building	Change management Facilitating the move	Raising and supporting awareness of the need for sustainability	Cost of change.
Providing different types of food	Differentiation of food and places to serve food	Alignment to multicultural composition of staff	Extra cost
Freedom to dress	Change of behavioural rules	Increase of informal behaviour	Loss of authority

How to measure and manage

In the previous sections we have touched on how several cultural dimensions affect the design and management of corporate facilities. The review of the literature and practical experiences show that organisational culture is a highly elusive concept. Its relation with FM and CREM is even more difficult to pin down. Whereas most other chapters in Part II discuss clear and managerial 'key performance indicators' such as the use of square metres per person, occupancy costs per workplace or CO_2 emission per m², the subject of culture does not really lend itself to being captured in quantitative ratios, figures or indicators. Even so, there is a need to know whether workplace interventions have the desired effect on the organisation's culture. Especially when organisations are investing heavily in new types of work environments, spending a lot of money on furniture, technologies and an intensive implementation process, they may want to know whether it contributes to a more productive company culture.

From this perspective, Key Performance Indicators should indicate the extent to which people believe that their building and other facilities and services support the desired culture. For example, if the goal is to create a more collaborative culture, the KPI should indicate the percentage of people that believe that the new environment supports collaboration. Likewise, if the goal is to create a more egalitarian culture, the KPI should indicate the percentage of people that think that the work environments are less hierarchical than before. To get a grip on the relation between culture and FM/CREM, we suggest the use of quantitative tools such as questionnaires. Such an approach has also been used by cultural researchers such as Hofstede (2008) who developed the Value Survey Module (VSM), and Cameron and Quinn (2006), who developed the Organisational Culture Assessment Instrument (OCAI). Survey questions can be formulated to get an idea of:

- A company's culture, e.g. "How do you rate your organisation in terms of hierarchy?"
- The match between a company's culture and existing or desired facilities, e.g. "Do you think that the building and spaces match the culture of your organisation?" or "Do you think that transforming this cellular office into a more open setting with shared activity-based workplaces would support a culture of trust?"
- To what extent employees agree with certain statements, e.g. "To what extent do you agree with the statement that managers should have their own office?"

A tricky issue can be the gap between the actual culture and the desired or perceived culture of an organisation. It may very well be that a work environment has been designed according to egalitarian principles – entirely free of status symbols and hierarchical differences – but that the culture of the organisation has not changed accordingly, with managers behaving just as hierarchically as before. This reflects a mismatch between the intended and the actual culture. Another vital KPI therefore concerns the degree to which the work environment matches with the *actual* organisation culture.

The outcomes of such surveys should be seen as proxies, giving an indication of an organisation's culture rather than in-depth understanding. It can be argued that true understanding of cultural issues can only come from experiencing and observing how organisations operate in daily life and talking to a lot of people, e.g. by conducting interviews, organising workshops and observations in situ. That may all sound awfully 'touchy feely' to practical facility managers, but culture is a factor to be reckoned with in any FM or CRE project. It seems quite obvious that an organisation's culture can throw a spanner into the implementation of new facility concepts, no matter how well designed or prepared they are.

Perspectives

As work gets increasingly more mobile and digital, the nature and function of the work environment is likely to change. It is difficult to predict exactly which changes will take place, but the general expectation is that more and more work will take place outside the office and the office itself will become more of a meeting space than a workplace. It could be argued that this will make the cultural relevance of the physical work environment more important. More than before, the office should become a place where employees feel a part of the company and a place where 'acculturation' processes can take place. In part, such social processes are driven by the interaction with colleagues and managers, for which places should be provided such as meeting areas, coffee points and work areas. In addition, the building may be used to express a certain message by its proportions, materialisations, inventory and services. Google is probably the best known – almost a clichéd – example of a company that puts enormous effort in using workplace design to promote a particular type of culture. All over the world, Google offices follow a fit-out formula of bold colours, lots of graphics and lavish amenities. Google sees their offices as a hallmark of their corporate culture and a logical outcome of the company's overarching philosophy "to create the happiest, most productive workplace in the world" (Stewart, 2013; Van Meel, 2015).

Further research is needed to understand if and to what extent investing in creating such a particular atmosphere, culture or identity in the office environment adds value in the long run and supports the organisational performance. Another research topic is to find out whether this approach might also be effective for other companies and in other sectors. Twenty years ago the Google concept was mostly seen as a strategy for advertising agencies and hip consultancy firms, but nowadays all major companies seem to use the physical work environments as a means for establishing a productive corporate culture. An obvious question is how, and to what extent, that can be done. So far this question cannot be answered in an exact way, because correlations between culture and workplace design will be intermediated by many other factors. Indicating that culture type A needs to be accommodated in workplace type B, would be an underestimation of the complexity of the topic, and an overestimation of the impact that buildings can have on people. However, this doesn't alter the fact of an untapped potential in the development process of workplace solutions as a vehicle to discuss and develop cultural issues in a company, in an attempt to create a shared vision of what a good work environment entails. Though the cultural dimensions that were presented above have a tentative status, they can be used to further explore the impact of workplace design on organisational culture and vice versa. At the same time, these dimensions should be considered as hypothetical, requiring confirmation from further research.

References

Baane, R., Houtkamp, P. and Knotter, M. (2010) *Het nieuwe werken ontrafeld. Over bricks, bytes and behaviour*, Assen: Koninklijke Van Gorcum BV.

Bach, R. (2015) *Influence of ICT-enabled time and place independent working styles on the organizational structure/culture and the academic office workplace*, MSc Thesis, Wageningen University.

Becker, F. (2007) 'Organizational ecology and knowledge networks', *California Management Review*, 49, pp. 42–61.

Cameron, K.S. and Quinn, R.E. (2006) *Diagnosing and changing organizational culture*, San Francisco, CA: Jossey-Bass.

Deal, T.E. and Kennedy, A.A. (1982) *Corporate culture: The rites and rituals of corporate life*, Reading, MA: Addison-Wesley.

De Jonge, J. and Rutte, C. (1999) 'Een quasi-experimenteel veldonderzoek naar de psychologische effecten van een flexibel kantoor concept', *Gedrag and Organisatie* 12 (6), pp. 427–444.

Fukuyama, F. (1995) *Trust: The social virtues and creation of prosperity*, London: Hamish Hamilton.

Gagliardi, P. (Ed.) (1990) *Symbols and artefacts: Views of the corporate landscape*, New York: Aldine de Gruyter.

Gall, C. (2009) *Office Code: Building connections between cultures and workplace design*, Leinfelden-Echterdingen, Gesellschaft für Knowhow-transfer in Architektur und Bauwesen. Steelcase.

Gensler (2015) *Modernity vs. hierarchy: China's evolving commercial office building market*, available online at http://www.gensleron.com/work/2014/1/13/modernity-vs-hierarchy-chinas-evolving-commercial-office-bui.html

Hassell (2015) *Workplace delivers cultural change and lower costs*, available online at http://www.hassellstudio.com/en/cms-news/new-look-for-qantas-headquarters

Hofstede, G. (2008) *Value Survey Module 2008*, available online at http://www.geerthofstede.nl/research--vsm/vsm-08.aspx

Hofstede, G., Hofstede, G.J. and Minkov, M. (2010) *Cultures and organizations: Software of the mind*, New York: McGraw-Hill.

Igo, T. and Skitmore, M. (2006) 'Diagnosing the organisational culture of an Australian engineering consultancy using the competing values framework', *Construction Innovation*, 6 (2), pp. 121–139.

Khanna, C., Van der Voordt, D.J.M. and Koppels, P. (2013) 'Real Estate mirrors Brands. A conceptual framework and practical applications', *Journal of Corporate Real Estate*, 15 (3/4), pp. 213–230.

Miller, R., Casey, M. and Konchar, M. (2014) *Change your space, change your culture: How engaging workspaces lead to transformation and growth*, Chichester: John Wiley and Sons.

Plijter, E., Van der Voordt, D.J.M. and Rocco, R. (2014) 'Managing the workplace in a globalized world. The role of national culture in workplace management', *Facilities*, 32 (13/14), pp. 744–760.

Riratanaphong, C. and Van der Voordt, D.J.M. (2012) 'Performance measurement of workplace change: A comparative analysis of data from Thailand, the Netherlands and Finland', in Jensen, P.A., Van der Voordt, T. and Coenen, C. (Eds.) *The added value of facilities management: Concepts, findings and perspectives*, Lyngby, Denmark: Polyteknisk Forlag.

Rothe, P.M., Beijer, M. and Van der Voordt, T.J.M. (2011) 'Most important aspects of the work environment. A comparison between two countries', conference paper, *European Facility Management Conference EFMC 2011*, Vienna, 23–25 May.

Schein, E. (2004) *Organizational culture and leadership*, San Francisco, CA: Jossey-Bass.

Stewart, J.B. (2013) 'A place to play for Google staff', *The New York Times*, 16 March, B1.

Van Meel, J. (2000) *The European office: Office design and national context*, Rotterdam: 010 Publishers.

Van Meel, J. (2015) *Workplaces today*, Rotterdam/Copenhagen: ICOP/Centre for Facilities Management.

Van der Voordt, T., Van Meel, J., Smulders, F. and Teurlings, S. (2003) 'Corporate culture and design. Theoretical reflections on case-studies in the web design industry', *Environments by Design*, 4 (2), pp. 23–43.

Interview 3: Ole Emil Malmstrøm, Real-FM, Denmark

Ole Emil Malmstrøm has twenty-five years of experience in FM. He was educated in civil engineering. A major part of his experience was obtained as director of Real Estate Operation, Maintenance and Development in a Danish building and real estate company. He now runs his own FM consultancy. Ole Emil has been chairman and board member of the Danish Facilities Management Association. He was one of the initiators and the first chairman of NordicFM, and he chaired the NordicFM working group on "Highlighting the added values for core business provided by FM". Ole Emil is also the former treasurer and a board member of EuroFM ,and he is now an honourable member of EuroFM.

Added value in general

Q. What does the term "Added Value" mean to you?

A. *The term is derived from Value, which has both an economic meaning and meanings related to feelings and other subjective and qualitative aspects. The Added Value of FM is related to making the Core Business (CB) more effective. It concerns the innovative part of FM with impacts on the primary activities. I often compare FM with IT. The development in IT creates added value for users and companies by making things easier. It relates to the laziness of people. Engineers focus on rationalising to get away with doing more with less effort.*

Q. How do you see the relation between "Added Value" and cost reduction?

A. *Added value and cost reduction do not necessarily have so much to do with each other. The cost has to do with the provider's agreement. For instance, if a provider can save cost by use of the output-based cleaning standard INSTA 800, he might be able to offer a more competitive tender price and/or make a higher profit, but it does not have any effects on the CB processes. Cost reduction is a natural part of professional FM.*

Q. Is "Added Value" mostly treated at a strategic, tactical or operational level?

A. *Added value is mostly treated at a strategic level, but is of relevance at all levels and for everybody in the FM organisation. It should be part of the organisational culture. It is more innovative at the strategic level.*

Q. In which context or dialogue is "Added Value" mostly considered in your experience?

A. *In real estate administration the main dialogue related to added value was with the clients, who were the real estate owners, and with the tenants, who were the clients' customers. Added value was particularly important in relationships, where the communication was good. The dialogue with the clients was the easiest, because the owner's interest was mostly on economic issues and the dialogue was strategic – for instance, focusing on retaining tenants. The tenants were very uneven with different CB, which made their interests varied and the dialogue more difficult. A typical way of creating added value for tenants was to create higher use value, for instance by changes in office layout and by making a safe and comfortable environment for the staff.*

Benefits and limitations of "Added Value"

Q. What are the benefits of considering and talking about "Added Value"?

A. *The main advantage is that the dialogue is moved away from the contractual agreement and the SLAs. It makes the customer feel that you are interested in his business and not just in submitting the next bill. It makes it possible to raise the level of the whole FM provision.*

Q. What are the limitations of considering and talking about "Added Value"?

A. *The main problem is that added value is difficult to document. Added value concerns things that cannot be measured immediately, for instance in economic terms. It involves feelings and subjective perceptions. A major pitfall is that one needs to keep one's feet on the ground. The overall objectives should be kept in focus. For instance, in a real estate administration, if you have a dialogue with a tenant, you should be careful not to create expectations that the owner is not willing to accept or fulfil. The focus must stay on the needs of the core business.*

Management of "Added Value"

Q. What are your top five main values to be included in the management of accommodations, facilities and services? Could you mention a few examples of concrete FM interventions to attain these added values, and KPIs to measure them?

A. *See table below.*

Prioritised values in FM	Concrete FM interventions
1. CB objectives	Ongoing focus on space utilisation
2. Innovation	Changing the size of work desks to create more workplaces, more intense communication and less paper mess
3. Coherent strategy between CB and FM	Example of close collaboration between tenant and real estate administration to adapt offices to changes in tenants' business activities.
4. Productivity of CB	Knowledge about how to create wellbeing, safety and comfort for staff
5. Communication	Guidelines for staff, when they move in to a new tenancy Use of "cartoons" with drawings of rebuilding projects

Q. Do you measure whether you attain the targetted added values?

A: *No.*

Q. What methods have you used to document "Added Value"?

A. *The real estate administration produced reports (annually or quarterly) to owners, which documented economy, rental degree and building conditions, but also gave strategic advice on developing the portfolio, the individual properties and selected rentals – for instance, physical upgrading and advertisement campaigns. Satisfaction surveys were made regularly among housing tenants (every three to four years in a specific housing estate). Among office tenants, annual customer meetings were held at strategic level. After Ole Emil left the real estate administration internet-based communication with tenants was implemented with a tenants' portal and a graphical user interface with access to drawings etc.*

Last comment

Q. Are there other topics that you see as important in connection to the concept of "Added Value"?

A. *The interface between core business and FM can be debatable and can cause difficulties in identifying and agreeing on the added value that can be seen as a result of FM. In the NordicFM group on added value there were examples of sorting metal waste at Volvo and changing cooling temperature for production machines at LEGO, which were on the borderline.*

8 Health and safety

Per Anker Jensen and Theo van der Voordt

Introduction

This chapter concerns the relation between buildings, facilities and services and the health and safety (H&S) of people. The relationship with health regards the prevention and reduction of work fatigue, occupational stress, burnout, and health problems such as headaches, migraine, irritation of eyes, nose or throat, increased levels of blood pressure, getting a cold, or worse diseases. Input factors include a well-considered HRM policy and providing a healthy indoor environment that supports thermal comfort, healthy indoor air quality (IAQ) without chemical and biological agents, sound lighting and acoustics, ergonomic furniture to support physical comfort and to prevent CANS (complaints of the arm, neck and/or shoulder), and avoiding hazardous materials, harmful substances and radiation. Safety regards the prevention or reduction of accidents that may hurt, damage, or even kill people, and also the prevention or reduction of theft, burglary, violent behaviour, and fraud. H&S are relevant values by themselves, but are also related to other values such as productivity, employee satisfaction, Corporate Social Responsibility, sustainability, profitability and risk.

H&S is an area that is strongly regulated by authorities, e.g. in Health and Safety Acts and by national and international standards. In the European standard on FM taxonomy (CEN, 2011) H&S is a sub-product of Health, Safety, Security and Environment (HSSE). H&S is described as "providing health and welfare of people in their workplace". It is further sub-divided in Workplace Safety and People Occupational Health. Workplace Safety is specified as "providing safety in workplaces, especially in the production, mining, transport and construction industries", while People Occupational Health is specified as "providing health and welfare of people such as: healthcare such as a company doctor, fiscal or manual therapists, safe working practices, policies on health and welfare facilities, and special food and beverages".

The chapter will discuss how FM/CREM can contribute to the H&S of end users. Overall the research on H&S in connection with FM and CREM is rather limited, and is diverse in both scope and methods. Here we will mainly focus on office environments and the effects of indoor climate and different workplace solutions. After a description of state of the art research we will discuss possible benefits and costs of particular interventions. Finally the chapter discusses how to measure H&S by the use of various Key Performance Indicators, how to manage H&S, and future perspectives. Elaboration of technical issues regarding fire safety, construction safety and protection against harmful materials and substances goes beyond the scope of this chapter. For these aspects we refer to the many handbooks, guidelines and standards in this field.

State of the art

H&S in FM and CREM textbooks

The treatment of H&S in FM textbooks varies considerably. Some authors, like Atkin and Brooks (2009) and Jensen (2008), combine H&S with Environment (understood as external environment/environmental protection), similar to the FM taxonomy mentioned above but without the inclusion of Security. The textbook edited by Alexander (1995) has a chapter on Environmental Management which includes H&S as a main aspect, while Park (1998) has a separate chapter on H&S but no chapter concerning Environment. The very comprehensive Facilities Manager's Desk Reference by Wiggins (2010) surprisingly does not include separate chapters on either H&S or Environment, but instead presents specific chapters on First Aid at Work, Asbestos, Water Supplies and Water Safety, Fire Safety and Legislation, Electrical Supplies and Electrical Safety, and Energy Management. The recent textbook by Barker (2013) only has a brief treatment of H&S in connection to outsourcing, and Barrett and Finch (2014) does not even include Health and Safety in the index.

CREM related textbooks on ergonomics (Wijk and Luten, 2001) and architecture in use (Van der Voordt and Van Wegen, 2005) discuss both physical wellbeing and safety. Physical wellbeing is related to lighting, acoustics and indoor climate. Safety is split into ergonomic safety, public safety (defined as low risk of crime such as theft, burglary and violence, and no fear of crime), fire safety, construction safety (strength, rigidity, stability), chemical safety (particularly relevant in industrial plants due to risk of explosion, hazardous substances, polluted water) and traffic safety.

H&S in recent FM and CREM research papers

A search in recent (post-2010) volumes of the scientific FM journals *Facilities*, *Journal of Facilities Management* and *International Journal of Facility Management* using the words Health and Safety (in combination and for each word) in article titles and keywords only revealed a few results. Hon et al. (2011 and 2014) investigated safety in relation to repair, maintenance, minor alteration and addition (RMAA) works. Leung et al. (2013) presented a survey-based study on the relationships between FM, risks and the health of elderly residents in care and attention homes. Hebert et al. (2013) conducted a case study on safety lighting for exterior illumination. Rogers et al. (2013) presented a study based on archival research on reducing the fall fatality rate by managing the risk associated with working at heights.

A search in recent (post-2010) volumes of the *Journal of Corporate Real Estate* using the words Health and Safety in article titles and key words revealed only two papers, one by Smith and Pitt (2011a) and one by Feige et al. (2013). Smith and Pitt (2011a) conducted a comprehensive literature review of the history of sustainable development in the built environment and its rationale. The authors argue that sustainable construction contributes not only to environmental sustainability but also to improved health, satisfaction and wellbeing among building users. Feige et al (2013) conducted an empirical study of eighteen office buildings using physical measurements, interviews and survey data from almost 1,500 employees. It was concluded that in addition to other influencing parameters such as job design and social work environment, building characteristics have a clear impact on the comfort level of the end users. In particular the positive impacts of operable windows and the absence of air conditioning were clearly identified. Comfort and work engagement showed to be correlated, and as such a high user comfort may reduce the turnover rate of employees. Productivity did not show to be directly correlated with comfort levels.

Some other JCRE papers from this period are related to H&S issues as well. Too and Harvey (2012) discuss the harmful impact of what they call "toxic workplaces". The sources of toxicity can vary from the physical dimensions of the building, the barriers to free flow of employees, obstacles to face-to-face communications and electronic contact, and the lack of personal privacy. Each of these dimensions can have an impact, but combined they can form a debilitating set of forces affecting the wellbeing of employees. In particular bullying and destructive leadership are mentioned as negative forces. Although this behaviour is mainly related to the social and organisational environment, environmental conditions that hinder personal wellbeing and development – e.g. lack of space, lack of privacy, poor ventilation, excessive noise, frequent technical breakdowns, lack of personal control – have an impact as well.

In an earlier JCRE paper, Smith and Pitt (2009) tried to identify and demonstrate the benefits of plants in offices in contributing to employee health and wellbeing by conducting a literature review and applying the insights to study a working office. The comprehensive literature study showed the importance of indoor plants in office environments, first through physically improving the air quality and removing pollutants and second in improving employee wellbeing through psychological benefits. The perception survey showed a general preference for plants in offices. The occupants of planted offices felt more comfortable, more productive, healthier, more creative and felt less pressure than the occupants of non-planted offices. In a related paper in *Facilities*, Smith and Pitt (2011b) compared two offices in the same building, one with plants and one without plants. Daily tests were undertaken for relative humidity, carbon dioxide, carbon monoxide and volatile organic compounds (VOCs). The results showed that relative humidity increased following the introduction of plants and more significantly following additional hydroculture plants being installed. Carbon dioxide was slightly higher in the planted office for the majority of the trial. Carbon monoxide levels reduced with the introduction of plants and again with additional plants. VOC levels were lower in the non-planted office.

Indoor climate

Indoor climate is a compound field, which is highly dependent on both the design and operation of buildings and the activities within them. The indoor climate is also important for both physical and psychological H&S. The influences from the indoor climate are experienced differently by different persons. As such, it is a strongly multidisciplinary field and related to architecture, engineering and medical expertise.

Indoor climate can be divided into six sub-areas (Valbjørn et al., 2000):

- Thermal conditions (including air temperatures, radiation temperatures, air velocity and air humidity)
- Air quality (including the content of pollutants such as dust, air humidity, gases, vapours and uncomfortable smells)
- Static electricity (including the charging of persons)
- Light conditions (including luminosity, luminous colour, contrasts and reflections)
- Sound conditions (including volume and spread of frequency)
- Ionizing radiation (including the radon concentration).

Some typical health problems that may be related to a poor indoor climate are:

- Irritation of eyes, nose, throat (mucosa irritation)
- Reddening and dryness of the skin and unspecified allergic reactions
- Headache or feeling of heaviness in the head
- Unnatural tiredness or problems with concentration
- Nausea and dizziness.

Such symptoms do not in themselves point in a certain direction, and they may be caused by other things than the indoor climate. However, when a large group of people in the same building show identical symptoms, factors of the indoor climate may be an important reason. The probability is much increased if the symptoms disappear for many of the people when they are outside the building for shorter or longer times (Valbjørn et al., 2000).

In 1983 the World Health Organisation defined a pathological picture termed *Sick Building Syndrome* (SBS). This is defined by an increased frequency of symptoms like the above among the people in a certain building. There are many inconclusive questions in relation to its origin, and about which influences can result in which symptoms. Based on cross-sectional data from 4,052 participants Marmot et al. (2006) concluded that the physical elements of office buildings were less important than features of the psychosocial work environment in explaining differences in the prevalence of ten health symptoms such as headache, cough, itchy dry eyes, blocked, runny nose and so on. Even in buildings where many people show symptoms, the reason is rarely that the measured influences are above specific thresholds and standards. Therefore the symptoms must be caused by a series of influences which in themselves are acceptable, but which in combination cause the observed symptoms. Hence it must be recommended to reduce each unwanted influence as much as possible (Skov et al., 1989). An interesting finding is that employees near windows experienced lower levels of SBS symptoms than those located further away from windows, even though windows were not operable and the effect cannot be due to increased ventilation (Fisk and Rosenfeld, 2005).

Rasila and Jylhä (2015) tried to assess the influence of noise from a holistic perspective. Based on a literature review and interviews with forty-eight workers in a telecommunications company in Finland, they list seven dimensions of office noise: physiological symptoms, psychological symptoms, social wellbeing, knowledge transfer, socialisation, activation and noise masking. The authors found that the perception of the noise environment is closely related to the specific job type and personal traits of individual workers. For example, because employees in contact centres are constantly and simultaneously on the phone, the noises of individual conversations become blurred and masks the noises with harmful information content. This would not happen in an office environment with people doing lots of silent work and few telephone discussions. The interviewees noted that it is sometimes not easy to know what feature of an environment causes what consequences. For example, tiredness could be influenced by noise but also by a bad IAQ. The information content and quality of the noise matters a lot. Noise with no information content is less disturbing than noise with information. The interviewees were most bothered by the intelligible noises that were irrelevant to them.

Rasila and Jylhä (2015) also discuss various studies on how furniture, partitions and surface materials affect acoustics and user satisfaction with the noise environment. Many studies conclude that office workers prefer silent and less noisy environments to more noisy environments. Office noises may cause stress, tiredness and lack of motivation, but do not increase the number of sick leave days. Various studies tested different kinds of sound masking. Speech may also be used in sound masking if there are the right numbers of individuals speaking at the same time.

An example of a cost-benefit assessment of improved ventilation can be found in a paper by Franchimon et al. (2008). They performed an economic assessment to determine the Incremental Cost-Effectiveness Ratio (ICER = the amount of money needed to produce one healthy life year) for a full-scale ventilation upgrade of building stock in the Netherlands, to increase the healthy lifespan of citizens by preventing and diminishing COLD (Chronic Obstructive Lung Disease), lung cancer and asthma. The upgrade included a capacity increase of ventilation systems in dwellings and schools, as well as demand-driven ventilation control. Current and upgraded ventilation systems were compared for operating costs, healthcare costs and DALYs (Disability Adjusted Life Years). The calculations resulted in an Incremental Cost-Effectiveness Ratio for one extra healthy year (DALY) of €18,000, which is perceived as an acceptable amount for a healthy life year in the Netherlands.

A disease specifically related to buildings and facilities is Legionnaire's disease. This is a lung disease caused by infection of the bacteria Legionella. Usually the infection is spread from building installations like hot water and air-conditioning systems to people via the indoor climate. As such, one can consider Legionnaire's disease to be an indoor climate related disease. In spite of its name, which may suggest associations to the old days and exotic places, it is a new disease which was first discovered in 1976 during the American Veteran Organisation's congress (American Legion – hence the name) where twenty-nine people died. The bacteria grew in a cooling tower close to the air intake for the building's air-conditioning system, and it was spread through the ventilation air. In Denmark there are on average approximately one hundred cases of Legionnaire's disease recorded annually, of which 10% to 20% are fatal (Valbjørn et al., 2000).

A number of other diseases may be related to or aggravated by the indoor climate, such as bronchial infections, asthma, bronchitis and sinusitis (Valbjørn et al., 2000). Some indoor climate problems are caused by lack of maintenance. For instance, leakage from roofs can lead to very unpleasant indoor climate problems derived from fungal attacks etc. In Denmark, many municipalities have been forced to close down and renovate badly maintained schools due to attacks of mould (Jensen, 2008).

Which requirements must be specified to obtain a healthy indoor climate depends to a great extent on the activities which take place in a building or part of a building. A good indoor climate should not only be defined negatively, minimising influences which are experienced to be unpleasant or which can be pathogenic, but also in such a way that the indoor climate contributes to positive sense impressions, for instance in terms of light and acoustics. For an extensive overview of possible interventions see, for instance, Wargocki and Seppänen (2000), Franchimon (2009), Bluyssen (2010, 2015) and Bluyssen et al. (2013).

Workplace layout

Based on a review of the literature De Croon et al. (2010) conclude that not much research is available about the impact of the office location (e.g. telework office versus a conventional office), the office layout (e.g. open plan office versus cellular office), and the office use (e.g. fixed versus shared workplaces). Out of 1,091 hits, forty-nine relevant studies were identified and searched for evidence about the impact of these three office dimensions on the office worker's job demands (cognitive workload, working hours), job resources (communication, work autonomy, privacy, interpersonal relations at work), short-term reactions (physiological responses like endocrine and autonomous reactions, and psychological responses such as job satisfaction and stress due to crowding), and long-term reactions on health and performance. The results provide strong evidence that working in open workplaces reduces privacy and job satisfaction. Limited evidence is available that working in open workplaces intensifies cognitive workload and worsens interpersonal relations. A more important factor is the distance between workstations: a close distance intensifies cognitive workload and reduces privacy. Due to a lack of studies no evidence was obtained for an effect of the three office dimensions on long-term reactions. The results suggest that ergonomists involved in office innovation could play a meaningful role in safeguarding the worker's job demands, job resources and wellbeing. Particular attention should be paid to the effects of workplace openness by providing sufficient acoustic and visual protection.

In a Dutch study by Meijer et al. (2009) the researchers followed a company's move from cell-offices to flex-offices over a period of fifteen months. It was found that employees reported better general health and fewer complaints concerning their upper extremities after the move, probably due to the extra effort put into the ergonomics of the workstations in the new flex-offices. In a longitudinal study by Pejtersen et al. (2011) employees in cell-offices reported lower rates of sick leave than those working in open offices with more than six people.

Bodin Danielsson et al. (2014) analysed the effect of office type on sickness absence among 1,852 office employees working in cell-offices, shared-room offices, small, medium-sized and large open-plan offices, flex-offices and combi-offices. They used the responses from the same employees to the Swedish Longitudinal Occupational Survey of Health (SLOSH) in 2010 and 2012 and studied the association between the exposure to office type in 2010 and sickness absence during the twelve months before measurement in 2012. Sick leaves were self-reported as the number of short and long (medically certified) sick leave spells and the total number of sick leave days. Multivariate logistic regression analysis with adjustment for background factors showed a significant excess risk for short sick leave spells in the open-plan offices. In the gender separate analyses this remained for women, whereas men had a significantly increased risk in flex-offices. For long sick leave spells, a significantly higher risk was found among women in large open-plan offices and for total number of sick days among men in flex-offices. The authors refer to Bodin Danielsson and Bodin (2008) who found the best health among employees in flex-offices and cell-offices, and the worst in medium-sized open-plan offices. The equally good health in the former two, very different office types might be explained by the higher level of personal control, albeit through different means.

Based on their review of the literature, Bodin Danielsson et al. (2014) concluded that apart from a few excellent studies, much office research shows substantial shortcomings. In most cases the definitions are vague regarding the office environments studied, and the studies don't recognise that different types of open-plan offices exist which vary substantially in their spatial and functional arrangements. In addition, most studies are Post-Occupancy Evaluations after relocation, or else are cross-sectional. A limitation in the former is that the actual shift of environment in itself may have an impact on the outcomes which "disturbs" the influence of the office environment per se. In cross-sectional studies no causal relationships can be established.

Sitting for extended durations is found to increase the risk of health problems (Bankoski et al., 2011). Karokolis and Callaghan (2014) reviewed fourteen papers on the effectiveness of sit-stand workstations in reducing worker discomfort without causing a decrease in productivity. Seven studies reported either local, whole body or both local and whole body subjective discomfort scores. Six studies indicated that implementing sit-stand workstations in an office environment led to lower levels of reported subjective discomfort. This review concluded that sit-stand workstations are probably effective in reducing perceived discomfort. Eight studies also reported a productivity outcome. Three studies reported an increase in productivity during sit-stand work, four reported no effect on productivity, and one reported mixed productivity results.

Benefits and costs

Table 8.1 shows a number of possible interventions to support H&S and related benefits and sacrifices.

Table 8.1 Benefits and costs of various FM/CREM interventions

Interventions	Management	Benefits	Sacrifices
Changing the physical environment			
Thermal comfort Improving the users' ability to control temperature and air according to individual or group preferences For instance, improvements may include more intense zoning of the work space	Implement changes in building automation systems, possibly in air-conditioning system, and training users	Improved wellbeing and reduced risk of health problems like colds and flu	Higher investment cost and possibly increased energy consumption and CO_2 emissions
Lighting Removing unpleasant exposure to light, for instance direct sunlight, glare and strong contrasts, and/or increasing individual control of lighting Improvements may include changes in workplace layout, installation of indirect lighting, solar screens, blinds and individual desk lamps	Arrange analysis of lighting conditions and implement improvements	Improved wellbeing and reduction of the risk of health problems like headaches	Higher investment cost

Table 8.1 continued

Interventions	Management	Benefits	Sacrifices
Noise Reduce noise annoyance, for instance by reduction of noise, reduction of reverberation time (e.g. by avoiding parallel walls and smooth, 'hard', sound-reflecting surfaces, and by the use of sound-absorbing material), or by creating quiet zones and spatial separation from busy places	Rearranging the spatial layout and removing disturbing sound sources	Prevention of health problems such as fatigue and headaches Improvement of speech intelligibility	Costs of change Costs of sound absorbing materials
Ergonomics Replace workplace furniture with new furniture with higher ergonomic quality, for instance height adjustable work desks which can be adapted to individual preferences and allow changes in body positions over the work day	Implement procurement and installation of new furniture and instruction of the users	Prevention of health problems like aching back and other muscular tensions	Higher investment cost
Fitness facilities Providing fitness facilities offered for use by the staff, e.g. a gym with fitness equipment, dressing rooms and baths	Establish fitness facilities and develop policy for use	Improve health of staff and prevent health problems	Higher investment cost, unproductive use of space, and risk of injuries from inadequate training and wrong use of fitness equipment

Changing the facilities services

Interventions	Management	Benefits	Sacrifices
Healthy food Changing the catering service to include more healthy food, for instance food with less fat and sugar	Arrange changes in catering service provision	Better general health of the staff	Possibly higher catering cost
Free fruit Offering free fruit to the staff during working hours	Arrange delivery and distribution of fruit	Better general health of staff	Cost of fruit provision
Healthcare provision Offering healthcare services for the staff, for instance company doctors, therapists, massaging and vaccinations	Arrange provision of healthcare service	Healthcare provision might improve the general health of staff and prevent illnesses and accidents	Cost of healthcare provision

How to measure and manage

Key performance indicators

To be able to define the extent to which complaints about health problems may be (partially) caused by building characteristics or technical services, the Dutch Government Buildings Agency developed the so-called 'healthy building quality' method (Bergs, 1993). This tool builds on the Healthy Building Quality tool from Canada (Visscher, 1989). These checklists regard the input side of the Value Adding Management model.

KPIs that define the objective performance of buildings and facilities are laid down in standards and guidelines, both national and international. Commonly used in the Netherlands are the Dutch NEN standards and the Dutch Working Conditions Act (Arbeidsomstandighedenwet, abbreviated as Arbo). A new Arbo Handbook is published annually (Zwaard et al., 2015). For example, in order to provide sufficient daylight in rooms where people spend a significant amount of time, the total surface of windows should be at least one twentieth of the floor area. Regarding IAQ, the concentration of CO_2 should not exceed 0.1% by volume. The Dutch Arbo law requires at least 30m³ of fresh air per person per hour in case of light work, and at least 50m³ in case of heavy work. A rule of thumb is a minimal refreshment of two to three times per hour. Relative humidity should preferably be in the range of 30% to 70%, or even better between 40% and 60%, to discourage the growth of micro-organisms.

Possible output indicators regarding H&S are:

- Actual health, e.g. measured by an annual professional medical check
- Absenteeism and sick leave (percentage of staff, number of days)
- Number of accidents, per week, per month, or annually
- Absence of employees due to accidents (percentage of staff, number of days)
- Number of complaints about health and safety that are submitted monthly to a complaints box or mentioned in an end-user survey
- Self-measurement of health and health supportive behaviour, e.g. by using wearables and apps to measure the number of steps per day, heart rate, calories, sleep etc.
- Self-reported health and safety in employee surveys, summarised in KPIs such as the percentage of (dis)satisfied employees with H&S issues in work environment surveys like the comfortmeter, an online comfort survey tool (http://www.comfortmeter.eu/).

How to manage

To identify H&S problems, avoid escalation, prevent problems and amend the working environment, it is important that a company has a well-functioning system to capture and react to the possible complaints from staff. An essential part of such a system is *workplace assessments*, which all companies in Europe with employees are obliged to implement based on an EU directive from 1989. The specific content of workplace assessments depends on the company's character; there is in principle freedom of method. As a minimum a workplace assessment must include the following four steps: Mapping, Assessment, Prioritised plan of action, and Follow-up (Jensen, 2008). In Denmark there is a broad consensus that workplace assessment has initiated more activities concerning H&S in relation to the working environment than any other tool. It has contributed to making H&S more visible, because it has provided something concrete to work with. Several companies have by the implementation of workplace assessment solved a number of problems which had been recognised in the companies for a long time. However, many companies still have difficulties in tackling more significant problems, including problems that require large investments in technology (Hasle and Thoft, 2000).

Lowe (2004, 2010) links H&S management to change management. He suggests the application of a well-balanced and integrated top-down and bottom-up approach that should include four steps: 1) create enabling conditions; 2) design a dynamic change process based on participation and learning; 3) determine the scope and depth of the change interventions; and 4) measure the results for the employees, the organisation and the company.

Perspectives

Although many handbooks on ergonomics and indoor environment are available to design healthy, safe and supportive offices, furniture and IT devices, and many standards and guidelines have been developed to create healthy and safe environments, more research is needed on input-output/outcome relationships. This research should address questions such as: what is the impact of new ways of working on sick leave? Does providing fitness facilities, healthy food, good acoustics and sufficient thermal comfort contribute to healthier employees and less burnout? If yes, to what extent? What are the investment costs and running costs? Do the benefits from a human resource point of view and the positive impact on productivity and profit counterbalance the costs and sacrifices from a business point of view? How can we measure the costs and benefits in a valid and reliable way? Advanced research that applies different research methods, both qualitatively and quantitatively, is needed to be able to answer these complex questions. IT devices such as apps and sensor techniques may help in both data collection and to support responsive interventions.

References

Alexander, K. (Ed.) (1995) *Facilities Management – Theory and Practice*. London: E&FN Spon.
Atkin, A. and Brooks, A. (2009) *Total Facilities Management*, Chichester: Wiley-Blackwell.

Bankoski, A., Harris, T., McClain, J., Brychta, R., Caserotti, P., Chen, K., Berrigan, D., Troiano, R. and Koster, A. (2011) 'Sedentary activity associated with metabolic syndrome independent of physical activity', *Diabetes Care*, 34, pp. 497–503.

Barker, I.C. (2013) *A Practical Guide to Facilities Management*, Caithness: Whittles Publishing.

Barrett, P. and Finch, E. (2014) *Facilities Management: The Dynamics of Excellence*, Chichester: Wiley-Blackwell.

Bergs, J.A. (1993) *Evaluatie onderzoek kantoorgebouwen. Handleiding voor onderzoek met de GBK-methode.* [Guide to Assess Healthy Buildings]. Amersfoort: DHV Bouw.

Bluyssen, P. (2015) *All you need to know about indoor air: A simple guide for educating yourself to improve your indoor environment*, Delft: Delft Academic Press.

Bluyssen, P. (2009) *The Indoor Environment Handbook. How to make buildings healthy and comfortable*, New York: Earthscan.

Bluyssen, P.M., Oostra, M. and Meertins, D. (2013) 'Understanding the indoor environment: How to assess and improve indoor environmental quality of people?' *Proceedings of CLIMA 2013, 11th REHVA World Congress and 8th International Conference on IAQVEC., "Energy efficient, smart and healthy buildings"*, Praag: Guarant.

Bodin Danielsson, C.B., Chungkham, H.S., Wulff, C. and Westerlund, H. (2014) 'Office design's impact on sick leave rates', *Ergonomics*, 57 (2), pp. 139–147.

Bodin Danielsson, C., and L. Bodin. (2008) 'Office-Type in Relation to Health, Well-being and Job Satisfaction Among Employees', *Environment and Behavior*, 40 (5), pp. 636–668.

CEN (2011) *Facility Management – Part 4: Taxonomy, Classification and Structures in Facility Management*. European Standard EN 15221-2. European Committee for Standardisation.

De Croon , E., Sluiter, J., Kuijer, P.P. and Frings-Dresen, M. (2005) 'The effect of office concepts on worker health and performance: A systematic review of the literature', *Ergonomics*, 48 (2), pp. 119–134.

Feige, A., Wallbaum, H., Janser, M. and Windlinger, L. (2013) 'Impact of sustainable office buildings on occupant's comfort and productivity', *Journal of Corporate Real Estate*, 15 (1), pp. 7–34.

Fisk, W.J. and Rosenfeld, A.H. (1997) 'Estimates of improved productivity and health from better indoor environments'. *Indoor Air*, 7, pp. 158–172.

Franchimon, F., Ament, A.H.J.A., Pernot, C.E.E., Knies, J. and Van Bronswijk, J.E.M.H. (2008) 'Preventing chronic lung disease in an aging society by improved building ventilation: An economic assessment', *Gerontechnology*, 7 (4), pp. 374–387.

Franchimon, F. (2009) *Healthy building services for the 21st century*, PhD Thesis, Eindhoven: Eindhoven University of Technology.

Hon, C.K.H., Chan, A.P.C. and Chan, D.W.M. (2011) 'Strategies for improving safety performance of repair, maintenance, minor alteration and addition (RMAA) works', *Facilities*, 29 (13/14), pp. 591–610.

Hon, C.K.H., Hinze, J., and Chan, A.P.C. (2014) 'Safety climate and injury occurrence of repair, maintenance, minor alteration and addition works: A comparison of workers, supervisors and managers', *Facilities* 32 (5/6), pp. 188–207.

Hasle, P. and Thoft, E. (2000) *Arbejdspladsvurdering i fremtiden – Motivation og kvalitet (Workplace assessment in the future – Motivation and quality)*. LOKE no. 2.

Hebert, P., Kang, M. and Kramp, J. (2013) 'Safety lighting: exterior illumination at an existing US government facility', *Journal of Facilities Management*, 11 (3), pp. 210–225.

Jensen, P.A. (2008) *Facilities Management for Practitioners and Students*, Centre for Facilities Management – Realdania Research, DTU Management Engineering.

Karakolis, T. and Callaghan, J.P. (2014) 'The impact of sit-stand office workstations on worker discomfort and productivity: A review', *Applied Ergonomics*, 45 (3), pp. 799–806

Leung, M.-y., Chan, I.Y.S and Olomolaiye, P. (2013) 'Relationships between Facility Management, Risks and Health of Elderly in Care and Attention Homes', *Facilities*, 31 (13/14), pp. 659–680.

Lowe, G.S. (2004) *Healthy workplace strategies: Creating change and achieving results*, Kelowna: Graham Lowe Group.

Lowe, G.S. (2010) *Creating Healthy Organizations: How Vibrant Workplaces Inspire Employees to Achieve Sustainable Success*, Rotman-UTP Publishing, University of Toronto Press.

Marmot, A.F., Eley, J., Stafford, M., Stansfeld, S.A., Warwick, E. and Marmot, M.G. (2006) 'Building health: An epidemiological study of "sick building syndrome" in the Whitehall II study', *Occupational and Environmental Medicine*, 63, pp. 283–289.

Meijer, E. M., Frings-Dresen, M. H., and Sluiter, J. (2009) 'Effects of Office Innovation on Office Workers' Health and Performance', *Ergonomics*, 52 (9), pp. 1027–1038.

Park, A. (1998) *Facilities Management – An Explanation*. Basingstoke: Palgrave.

Pejtersen, J. H., Feveile, H., Christensen, K.B. and Burr, H. (2011) 'Sickness Absence Associated with Shared and Open-Plan Offices – A National Cross Sectional Questionnaire Survey', *Scandinavian Journal of Work, Environment and Health*, 37 (5), pp. 376–582.

Rasila, H. and Jylhä ,T. (2015) 'The many faces of office noise – case contact center', *Facilities*, 33, 7/8, pp. 454–464.

Robertson, M.M., Ciriello, V.N. and Gabaret, A.M. (2013) 'Office ergonomics training and a sit-stand workstation: Effects on musculoskeletal and visual symptoms and performance of office workers', *Applied Ergonomics*, 44, pp. 73–85.

Rogers, J.W., Schneider, J. and Radio, F. (2013) 'Reducing the Fall Fatality Rate by Managing the Risk Associated With Working at Heights', *International Journal of Facility Management*, 4, 1.

Skov, P., Valbjørn, O., Gyntelberg, F. and DISG (1989) *Rådhusundersøgelsen – Indeklima i kontorer (The town hall investigation – Indoor climate in offices)*, The Health and Safety Fund.

Smith, A. and Pitt, M. (2011a) 'Sustainable workplaces and building user comfort and satisfaction', *Journal of Corporate Real Estate*, 13 (3), pp. 144–156.

Smith, A. and Pitt, M. (2011b) 'Healthy workplaces: Plantscaping for indoor environmental quality', *Facilities* 29 (3/4), pp. 169–187.

Smith, A. and Pitt, M. (2009) 'Sustainable workplaces: Improving staff health and well-being using plants', *Journal of Corporate Real Estate*, 11 (1), pp. 52–63.

Too, L. and Harvey, M. (2012) 'TOXIC workplaces: The negative interface between the physical and social environments', *Journal of Corporate Real Estate*, 14 (3), pp. 171–181.

Valbjørn, O., Laustsen, S., Høwisch, J., Nielsen, O. and Nielsen, P.A. (Eds.) (2000) *Indeklimahåndbogen. (The Indoor Climate Handbook)*. SBI-direction 196, The Danish Building Research Institute.

Van der Voordt, D.J.M. and Van Wegen, H.B.R. (2005) *Architecture in use: An introduction to the programming, design and evaluation of buildings*, Oxford: Elsevier, Architectural Press.

Vischer, J.C. (1989) *Environmental Quality in Offices*, New York: Van Nostrand Reinhold.

Wargocki, P. and Seppänen, O. (2000) *Indoor Climate and Productivity in Offices. How to integrate productivity in life-cycle cost analysis*, Guidebook 6, REHVA, Federation of European Heating and Air-Conditioning Associations.

Wiggins, J.M. (2010) *Facilities Manager's Desk Reference*, Chichester: Wiley-Blackwell.

Wijk, M. and Luten, I. (2001) *Tussen mens en plek: Over de ergonomie van de fysieke omgeving*, Delft: IOS Press.

Zwaard, W., Van der Steeg, M. and Ronner, S. (Eds.) (2015) *Arbo Jaarboek. Preventiegids veilig en gezond werken.* Deventer: Vakmedianet.

Interview 4: Wim Ledder, Skenn BV, the Netherlands

Wim Ledder has over twenty-five years of experience in FM. He worked at the Governmental Building Agency, the Embassy and various banks on real estate driven FM and accommodation management. Currently, together with his daughter Krista he owns a consultancy firm from which he is planning to retire in 2016. Wim Ledder is a former chair of Facility Management Netherlands (FMN) and current chair of the Association of Large-scale Consumers of Postal Services (Vereniging Grootgebruikers Postdiensten, VGP). His educational background includes the study of building sciences at a University of Applied Sciences, FM at Nijenrode, advanced FM education in Groningen (AOG) and various FM certificates.

Added value in general

Q. What does the term "Added Value" mean to you?

A. *Added Value (AV) of FM refers to its contribution to organisational goals and objectives. I use the AV concept in 80% of all discussions regarding the impact of FM interventions on strategic goals, image, return, profit, costs etc.*

Q. Are there other related terms that you prefer to use rather than "Added Value"?

A. *No.*

Q. How do you see the relation between "Added Value" and cost reduction?

A. *Cost reduction is an added value in itself. All interventions should be assessed on benefits and costs i.e. investment costs, running costs and depreciation costs in a business case, both financially and non-monetary.*

Q. Is "Added Value" mostly treated at a strategic, tactical or operational level?

A. *Mainly at a strategic level, e.g. in discussions with heads of departments, not at an operational level. Actually FM is not a strategic issue in most organisations, CEOs are not really interested. FM is mainly a tactical issue derived from the organisational strategy. FM can be strategic in itself. FM depends on the context. For instance, in a period of economic growth the focus will be on avoiding dissatisfaction and commotion. In times of economic decline the focus will be on cost reduction. FM also depends on the size of the company as well. In small firms FM is mainly operational. In service providers FM is core business.*

Q. In which context or dialogue is "Added Value" mostly considered in your experience?

A. *Mainly between inhouse FM organisations and their clients, between clients and consultants, and between clients and deliverers, e.g. of IT systems and equipment.*

Q. How do you see the relation between innovation and "Added Value"?

A. *We need sustainable innovation, both as a value in itself and to attain other values. Be clear about what innovation actually means and its effects on the short and long run.*

Benefits and limitations of "Added Value"

Q. What are the benefits of considering and talking about "Added Value"?

A. *The added value concept can help to focus on the strategic aspects of FM which is necessary to add value by FM. The concept itself is not that important. Most important is the discussion of costs and benefits. Listen to top managers: what do they want and what do they find important. Then discuss possible added values of FM in connection to their interests. FM should be able to both speak the language of the 'general' and the 'soldiers'.*

Q. What are the limitations of considering and talking about "Added Value"?

A. *Added value is perceived differently by different people and can only be discussed with people with an academic background and not at an operational level. The most important problem is the difficulty of making added value concrete and operational.*

Management of added value

Q. What are your top five main values to be included in management of accommodations, facilities and services? Could you mention a few examples of concrete FM interventions to attain these added values, and KPIs to measure them?

A. *See table below.*

Prioritised values	Concrete interventions	KPIs
1. Cost reduction	Reduce m²	Cost/m² Cost/fte (depends on product, e.g. accommodation, IT, catering)
2. Improve core business/ productivity	E.g. digitalisation of all documents → improved efficiency New ways of working → better work climate → increased productivity Ditto indoor air quality and comfort	Activities: what, by whom? Workplace occupancy Staff absence
3. Health	Provision of healthy food	Sick leave Health complaints

Business continuity is a top priority. Issues such as fire safety and data safety are often just incorporated in response to accidents. Adaptability is important as well in order to be able to cope with future changes. Change occurs faster than ever. Always take into account the costs of being flexible. Client image of FM is also important, e.g. regarding appropriate responses from the help desk, corporate or product familiarity, and sustainability. Here, too, one has to assess both the financial costs and benefits. Topics such as innovation and sustainability are only important when supported by the top management. A sustainable image is usually not key, unless the products have a high environmental impact. The impact of the surroundings can be relevant as well. For instance, unemployment rate could be included in site selection. It can be cost effective due to lower wages. Mobility is another issue to include in FM policy, e.g. try to reduce travel time between work and home. Integration could be supported by providing a religious room. Usually CEOs are not interested in FM: never has any CEO joined a members' meeting of Facility Management Netherlands. Probably they sometimes attend sustainability conferences. Young people are interested in gadgets such as apps to allow personal choices.

Q. Do you measure whether you attain the targeted added values? If so, what Key Performance Indicators do you apply for each of the added values you mentioned before?

A. *See table above. Most issues are not really measured, e.g. the impact of new ways of working on private life. Large companies make use of customer panels and digital surveys. Measuring also depends on the context. Many organisations hate surveys (too time consuming, less claimable hours, no added value) and focus more on external customers than on internal customers. Business cases show a huge variety: many of them just focus on financial costs and benefits. Some clients don't calculate business cases at all, others calculate on a detailed level.*

Q. Do you benchmark your data with data from other organisations?

A. *Only in a limited way. Large companies use the Netherlands Facility Cost index (NFC) and monitor customer satisfaction, internally and externally.*

Q. How is "Added Value" included in your communication with your stakeholder?

A. *In discussions.*

Last comment

Q. Are there other topics that you find important in connection with the concept of "Added Value"?

A. *Unfortunately, usually value adding management is not explicitly included in strategy reports. More attention is needed to disentangle the abstract concept of Added Value in different values, concrete interventions, and how to measure the added value of FM per product (accommodation, services etc.). We also need best practices of added value by FM interventions like digitisation of document management. A topic for further research is how to improve the impact of FM at a macro level. For instance, the privatised delivery of post seems to be very inefficient. Another example is how to cope with growing vacancy. What is or could be the role of FM when interventions result in lower space demand?*

9 Productivity

Iris de Been, Theo van der Voordt and Barry Haynes

Introduction

According to the Oxford English Dictionary, (economic) productivity means the effectiveness of productive effort, especially in industry, as measured in terms of the rate of output (e.g. goods, products, services) per unit of input (labour, materials, equipment etc.). In scientific literature, productivity is comparably defined as the relationship between output and input, between results and sacrifices (Aronoff and Kaplan, 1995). Output can relate to the number and quality of products or the operating result, for example expressed as the net profit or market share. Input involves resources, i.e. production factors such as labour, capital, technology, information and facilities.

There are three ways to increase the ratio between output (the 'numerator') and input (the 'denominator') (Keizer and Eijnatten, 2000; Van der Voordt, 2003):

1   Producing more output with the same input (higher numerator, same denominator)
2   Producing the same output with less input (same numerator, smaller denominator)
3   Increasing the output more strongly than the input (proportionally, the numerator increases more than the denominator).

Productivity is also linked to effectiveness and efficiency. A work process is effective if the right things are being done: all activities contribute to achieving the established goals and purpose and the achieved result is as similar as possible to the intended result. Efficiency means doing things properly: the intended result is achieved with as few resources as possible. With respect to productivity, effectiveness is mainly linked to the output (the best possible results) whereas efficiency is linked to the input (as few resources as possible). The costs of buildings and facilities are typically considerably lower (10%) than the costs of staff (80%) (Hanssen, 2000). This implies that if productivity of knowledge workers can be encouraged by improving facilities, it is potentially very cost effective to do so.

The ratio between the total output and total input is called the total factor productivity (Frankema, 2003). If the ratio regards only a particular part of the input, it is called partial productivity. For example, labour productivity is the output produced per unit of labour (Christopher and Thor, 1993). In this chapter we will first discuss the state of the art of current knowledge and empirical evidence regarding the impact of the physical environment on labour productivity. We will focus on four physical characteristics of the work environment: physical conditions, space, ergonomics, and aesthetics. Then we present various possible interventions to increase labour productivity with possible benefits and costs. The next section discusses how to measure the impact of the work environment on labour productivity and suggests a shortlist of Key Performance Indicators. The chapter ends with future perspectives.

State of the art

According to Bakker (2014), to ensure a knowledge worker is optimally productive and happy, it is important that he or she can attain personal objectives and that facilities and services fit with personal needs. An appropriate physical environment should optimally facilitate different job activities, ranging from communication to concentration, informal and formal meetings, and different moods, from being calm and relaxed to being stressed or excited. Due to the possible impact of many different variables, it is rather difficult to define the relative importance of the external conditions compared to other factors. A recent study showed that satisfaction with the organisation plays a more substantial role in the perceived productivity support of the work environment than the office concept (layout and use) itself (De Been and Beijer, 2014). However, the office concept did have a significant effect on satisfaction with the work environment and perceived productivity support. Many studies came up with empirical data that confirm the assumed relationship between external conditions and productivity. For example, Batenburg and Van der Voordt (2008) revealed a significant correlation between satisfaction with facilities and the perceived support of productivity: the more satisfied with the facilities, the higher the rating of productivity support by the working environment. Further analyses revealed that both functional aspects and psychological aspects of the working environment – such as adequate privacy and an inspiring office design – positively affect the perceived support of labour productivity.

The significance of the workplace is backed up by the research of Brill and Weidemann (2001), who after analysing a dataset of 13.000 respondents concluded that the physical workplace contributes 5% to individual performance and 11% to team performance. Another study came up with even higher impact figures for interventions that were undertaken at different organisations, including a productivity rise of 10–38% as a result of improved ergonomic furniture, an effect of 6–11% by improved lighting and an increase of 39% after implementing private offices, more comfortable chairs and advanced computer hardware (Kleeman et al., 1991). Based on thirty case studies, Kaczmarczyk et al. (2001) found productivity improvements of 2–58% after the introduction of teleworking, 15% due to high quality design, 22% due to ergonomic furniture, 9–13% due to high quality lighting and 7–10% due to noise reduction. These percentages are remarkably high. It is not completely clear if these impacts have been measured in a valid and reliable way and if the findings are interrelated.

Research undertaken by Barrett et al. (2013) evaluated the impact of a physical classroom on the academic progress of 3,766 pupils from 153 classrooms in 27 schools. Using multilevel regression modelling the researchers identified seven key design parameters that impacted on the students' learning progress. The seven factors were light, temperature, air quality, ownership, flexibility, complexity and colour.

Cause-effect relationships: the CIBS model

Whereas most people are able to indicate whether a specific environment has a positive or negative effect on their production, they find it difficult to define the exact *relationship* between their production (output) and particular resources used by the organisation (input). Defining cause-effect relationships is also difficult for researchers, especially when conducting field studies. In the period before and after an intervention such as a renovation, the introduction of New Ways of Working, or a new IT system, often other variables have changed as well, such as the composition of the staff, management style, or contextual factors such as the labour market or the economy. For this reason, some environmental psychology research is conducted in artificial settings. This makes it possible to isolate the impact of a single factor. However, findings cannot always be generalised to real life settings.

Many authors have tried to visualise the assumed impact of various variables and constructs on labour productivity. Most models confirm that many variables including buildings and facilities, work processes, organisational characteristics, personal characteristics and the external context may have an impact on labour productivity (Clements-Croome, 2000; Van der Voordt, 2003; Batenburg and Van der Voordt, 2008; Mawson, 2002; Haynes, 2007). In the CIBS model it is assumed that in addition to many other variables, physical conditions, space, ergonomics and aesthetics have an effect on satisfaction with the environment, on motivation, on job satisfaction and consequently on performance and productivity; see Figure 9.1 (Mawson, 2002). We will use these four external condition factors to discuss empirical evidence regarding the relationship between the physical work environment and labour productivity.

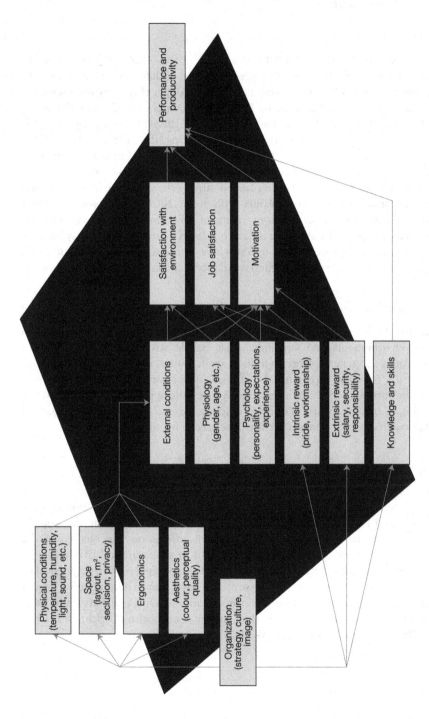

Figure 9.1 Impact of various variables on performance and productivity (Mawson, 2002)

Physical conditions

Indoor climate

An uncomfortable warm or cold temperature in the office can have a negative effect on the productivity of employees (Lan et al., 2009; Niemelä et al., 2002). This seems especially true for long tasks lasting longer than sixty minutes often performed by knowledge workers. Dorgan and Dorgan (2005) state that if the Indoor Air Quality (IAQ) is not at the right level, this will have an impact on the occupants' health and productivity. Other factors affect the valuation of the indoor climate as well, such as the local climate, social and cultural habits (e.g. regarding clothing) (Kurvers and Leijten, 2013) and organisational factors like managing expectations, explaining and visualising the use of the installation and responding adequately to complaints (Pols et al., 2009). Thompson and Jonas (2008) found that a general improvement of office environments can increase productivity up to 15%. An increase of fresh air to dilute pollutants may increase productivity of employees by up to 3%, whereas productivity may drop up to 6% when users are relocated from natural ventilation to air-conditioned environments.

Light (daylight, windows, lighting)

Galasiu and Veitch (2006) reviewed over sixty research studies on daylight in office environments and concluded that people strongly prefer daylight in workplaces. In general, larger windows are favoured. Fully automated systems receive low occupant acceptance. Individual control over lighting systems showed to be important to building users, and is especially appreciated if these systems are simple and easy to use. Preferred illuminance levels and discomfort glare in offices with daylight were differently experienced by different persons.

Greenery

Various studies showed that the presence of greenery in the work environment can increase employee wellbeing, psychological comfort and productivity (Knight and Haslam, 2010; Smith and Pitt, 2009; Bakker and Van der Voordt, 2010). Plants have a number of measurable effects on the quality of the indoorenvironment by their influence on light, temperature, relative humidity, air quality, noise and static electricity. The reactions of people can be physical or physiological (e.g. lower blood pressure and fewer headaches), affective (e.g. a positive mood), or cognitive (better concentration). Plants may have an indirect positive effect on productivity by their influence on health (Ulrich, 1984; Van den Berg, 2005) and behaviour (Wolf, 2002).

Sound

Among the most distracting sounds in the work environment are overheard conversations of others (Sundstrom et al., 1994; see also Chapter 8 on health and safety). This may not only be noise from conversations; also the presence of (uncontrollable) background music seems to worsen performance compared to working in a silent environment (Furnham and Strbac, 2002). The layout is an important factor when it comes to distraction and influence on productivity (see next sub-section).

Personal control

Various researchers revealed that the ability to personally control environmental factors, like temperature, air quality, light and noise levels, has an impact on (self-reported) productivity (Leaman, 1995; Pols et al., 2009; Boerstra et al., 2014). Based on objective productivity measurements, Wyon (1996) concluded that the productivity effect of ideal personal control over the thermal environment is +2.7% for logical thinking, +7% for typing, +3.4% for skilled office work and +8.6% for repetitive office work. Boerstra et al. (2014) found a 6–10% higher perceived productivity when full personal control is experienced, compared to no control at all. Leaman and Bordass (1999) claimed that in seven out of eleven buildings a significant association was found between self-assessed productivity and perceived control. The lack of environmental control showed to be the most important concern for office occupiers. This finding is supported by Whitley et al. (1996) who found that people like to have an internal *"locus of control"* and report that they are more productive when they perceive that they have control over their physical environment.

Space

Many studies have shown an impact of the office concept on (perceived) productivity (Hedge, 1982; Brill and Weidemann, 2001; Haynes, 2008; De Been and Beijer, 2014). Based on their survey of 13.000 office employees working in different settings, Brill and Weidemann (2001) concluded that the two factors with the largest impact on performance and satisfaction are: 1) the ability to work distraction-free, and 2) the possibilities for interaction with co-workers (especially spontaneous interaction). This finding was confirmed by Haynes (2008) who states that interaction and distraction have the largest impact upon perceived productivity. In a Dutch survey study with over 7,000 respondents, similar results came to the fore: satisfaction with the ability to concentrate at work was the most important predictor of perceived individual productivity support, whereas satisfaction with communication possibilities was the most important predictor for the perceived support of team productivity (Maarleveld and De Been, 2011). The accomplishment of concentrated tasks as well as high quality teamwork are key elements in organisational performance (Hua et al., 2010). It is a challenge to create an environment which supports both activities.

Spatial arrangements favouring spontaneous interaction (Brill and Weidemann, 2001) and collaboration (Strubler and York, 2007) are shown to be important to enable productivity. The large amount of visual accessibility in open work environments can facilitate effective communication among colleagues (Becker and Sims, 2000) and seems to lead to a higher frequency of interaction with co-workers (Bouttelier et al., 2008; Becker and Sims, 2000). Much interaction between colleagues occurs spontaneously (e.g. in a corridor, canteen or shared service area) (Backhouse and Drew, 1992; Hua et al., 2010). Appel-Meulenbroek (2014) showed that proximity, visibility and flow positively affect the number of interactions between knowledge workers and as such are supposed to support innovation; see also Chapter 11 on innovation and creativity. Moments of communication seem to be more frequent but also shorter in open office environments compared to enclosed cell environments (Bouttelier et al., 2008; Becker and Sims, 2000).

The presence of dedicated meeting spaces does stimulate the amount of communication (Peponis et al., 2007; Oseland et al., 2011) and also the perceived support of collaboration, provided that these spaces are located close to the workstations (Hua et al., 2010). People prefer using meeting spaces which are conveniently located and which offer a certain level of privacy (Oseland et al., 2011).

There are indications that working in activity-based office environments with various types of unassigned workplaces and possibilities for remote working, even when including many places for communication, can also have a negative effect on communication, e.g. due to difficulties with finding each other and less social bonding (De Been et al., 2015).

Working in an open setting often leads to distraction and disruption, resulting in lower perceived support of productivity compared to working in a cell-office (Hedge, 1982; Brill and Weidemann, 2001; Haynes, 2008; De Been and Beijer, 2014; Seddigh et al., 2014). The openness can lead to distraction, particularly when conducting work which requires concentration (Hua et al., 2010) but also when doing creative work, since much creative thought takes place alone (Oseland et al., 2011). Open spaces, meeting spaces and communication areas need to have good acoustic enclosure and should be strategically positioned in order to avoid distracting employees working at nearby workstations (Hua et al., 2010; De Been et al., 2015).

Task complexity and personal characteristics

The activities performed by the building users and the accompanying task complexity seem to play an important role in the relationship between the environment and productivity. Whereas people working on complex tasks were found to be more satisfied and productive in a private office, those performing simple tasks appeared to perform better in a non-private setting (Block and Stokes, 1989; Haynes, 2008). Introverts seem to have even more difficulties with working on complex tasks while being distracted compared to extraverts (Furnham and Strbac, 2002).

Ergonomics

The physical design of workstations and other office furniture has great importance for ergonomics, which can influence both health and productivity. For instance, a literature review by Karakolis and Callaghan (2014) showed that implementing sit-stand workstations in an office environment will likely result in lower levels of body discomfort and possibly also have a positive effect on performance. Research by Barber (2001) showed that, besides variables related to control, concentration and indoor climate, other variables are also considered important for productivity by office employees, such as ergonomic chairs, advanced technology (supporting IT facilities) and adequate (electronic) filing space. Ergonomics, enough space for items and access to technology are also mentioned as important influencing factors in the extensive research of Brill and Weidemann (2001).

Where office users use the same desk every day for their work activities, specific ergonomic considerations can be given to their desk and chair design. Therefore, an individual ergonomic solution can be designed for a particular office user (Sauter et al., 1991). However, the trend in today's modern office environment is more towards group interactions with collaborations being undertaken at multi-user workstations. Given that these multi-user workstations might be used by a number of different people throughout a working day, consideration needs to be given to an ergonomic design that provides an optimal fit for a range of users (Mahoney et al., 2015).

Aesthetics

Aesthetics are influenced by the architectural design of the exterior and the interior. Colour is one of the factors that affect wellbeing and mood (Mahnke, 1996; Kaya and Epps, 2004; Bakker, 2014) and people's behaviour (Elliot and Maier, 2007) and as such may also affect people's productivity (Bakker, 2014). Research has shown that the colour blue can enhance performance on creative tasks whereas red improves performance on detail-oriented tasks (Mehta and Zhu, 2009). However, a real-life experiment conducted by Bakker et al. (2013) showed no significant effect of a red, blue or neutrally coloured meeting room on the perceived outcomes of the meeting, the social cohesion of the group and wellbeing. Quite a large number of participants responded that the colour of the meeting space had no effect on productivity (65 per cent of participants), collaboration (58 per cent) or wellbeing (33 per cent). By contrast, Barrett et al. (2013) identified colour as impacting on students' performance.

Benefits and costs

Table 9.1 shows a number of typical interventions that FM/CREM can implement to increase labour productivity. It also shows that it is not easy to create the best possible outcome because the benefits of various interventions and design choices are counterbalanced by negative impacts and risks.

Table 9.1 Interventions, management, benefits and sacrifices

Interventions	Management	Benefits	Sacrifices
Create a more open work environment	Make the office more transparent, e.g. by using transparent materials, breaking down (some) current walls etc.	Support of social interaction and collaboration	Cost of project Risk: less privacy, more distraction
Create places to enable concentration	Create enclosed concentration cells or silent concentration zones and/or carry out acoustic measures, especially when the work environment is (relatively) open	Increase of concentration opportunities	Cost of planning and re-designing the work environment Risk: too much enclosure which can possibly lead to less knowledge sharing
Create an optimal indoor climate	Take care of a climate installation that results in a comfortable temperature, sufficient ventilation Pay attention to lighting and comfortable acoustics	Increase of employee satisfaction Improved health and wellbeing	Cost of extra facilities
Give people more personal control over their environment	Give employees (more) freedom in where, when and how to work Create opportunities to personalise the environment Create opportunities to personally control the indoor climate	Increased employee satisfaction and commitment Improved health and wellbeing	Managers should manage their employees in a different way (results-oriented) Personalisation may lead to appropriation of personal desks Individual control of indoor climate may lead to conflicts between employees
Provide ergonomic furniture	Replace non-ergonomic furniture by ergonomic, adjustable chairs and tables	Improved health and wellbeing, less risks of fatigue or illness	Cost of ergonomic furniture
Provide supporting IT facilities and an (electronic) filing system	Provide IT facilities and filing systems that suit work processes and (expected) flexibility of employees	Support of work processes in general Essential prerequisite of flexible working	Cost of facilities

How to measure

Nowadays the economy of developed countries is strongly based on the productivity of knowledge workers. In a knowledge society it is important not only to focus on the quantity but also to measure the quality of the outcomes (Blok et al., 2011). Measuring the productivity of knowledge workers – at organisational, team and individual levels – is quite difficult (Davenport and Prusak, 2000). For instance: how to measure the contribution of employees to innovative ideas that resulted in larger profits, or how to measure the quality of a research report? Various organisations use inventive methods for measuring the output of their organisation. For example, some universities measure teaching productivity by the ratio between the number of students that have successfully completed a particular course component and the number of hours spent by lecturers (Van der Voordt, 2003). Research productivity is often measured by means of the number of publications per fte, with a weighting factor to take into account the scientific status and the impact factor of the publication. In spite of the rational ideas behind these systems, they often evoke criticism due to their limited validity and reliability.

Because it is rather difficult to measure the productivity of knowledge workers, most studies on the impact of buildings, other facilities and services on labour productivity measure the *perceived* (impact of facilities on) productivity. Besides, more objective measurement methods are being used, if possible at all. According to the literature on environmental psychology, corporate and public real estate management, facility management and business administration, the impact of the work environment can be measured in various ways (Van der Voordt, 2003; Sullivan et al., 2013):

- *Actual output versus actual input.* For example:
  - The number of translated words per team or per employee per unit of time (translation agency), the number of phone calls per day (call centre), or the number of manufactured cars per fte (automobile industry)
  - The impact of facilities on the outcomes of cognitive performance tests (e.g. working memory, processing speed, concentration)
- *Actual input*, for example monitoring the computer activity (keystrokes, mouse clicks) that is used to produce the output
- *Amount of time spent or saved*, for instance the amount of time gained by implementing a new computer system which makes logging on less time consuming, or the opposite: the amount of time lost by having to log on frequently to a time consuming computer system
- *Absenteeism due to illness* or for other reasons and as such being non-productive, or the opposite: *presence*. Connected topics include the reported frequency of health issues. See also Chapter 8 on health and safety
- *Satisfaction*, based on the assumption that a happy worker is a productive worker (Halkos and Bousinakis, 2010). Connected topics include job

satisfaction, job engagement, satisfaction with facilities, or the intention to stay or to quit. See also Chapter 5 on satisfaction
- *Perceived productivity support,* i.e. the perceived support of productivity by the current work environment, measured on a Likert scale (Vos and Dewulf, 1999; Maarleveld et al., 2009), the estimated percentage of time being productive (Batenburg and Van der Voordt, 2008), the perceived productivity gain when all facilities would be excellent (Von Felten et al., 2015) or the perceived increase or loss of productivity after a change (Leaman and Bordass, 1997)
- *Indirect indicators,* for instance the extent to which people are able to concentrate properly, the frequency of being distracted, the ease with which employees can solve a problem, or the lack of knowledge through insufficient interaction with colleagues.

While some studies have found significant correlations between subjective and objective measures of performance (i.e. Oseland, 1999), these measures appear to be weakly correlated in general. Therefore, it is likely that self-reported and objectively measured productivity measure different aspects of performance. After an extensive review, Sullivan, Baird and Donn (2013) suggest that occupant surveys are the best method to measure the influence of the work environment on productivity. An added benefit is that conducting surveys is a relatively time and cost efficient method. It seems the best option to ask occupants *directly* about the effect of the work environment on productivity. Objective methods can complement occupant surveys, especially if they measure important organisational outcomes such as absenteeism (Sullivan et al., 2013).

The complexity of measuring office occupiers' productivity is not only caused by the large number of possibly influential variables, but is also the result of the lack of a clear definition as to what actually constitutes an output in productivity terms. Office occupiers undertake a range of different activities and each activity may have its own specific output. Therefore, the start of understanding any productivity measure is to define the different work processes that are undertaken in the office environment (Greene and Myerson, 2011; Haynes, 2008). Work processes that are largely routine and repetitive, such as process working, lead to a more clearly defined output. This form of office output can lead to a more mechanistic measurement of office productivity such as output/input (Greene and Myerson, 2011). Defining and measuring where and how knowledge is created and transferred in the office environment is a more complex issue (Oseland et al., 2011, Appel-Meulenbroek, 2014).

Key Performance Indicators

Whereas many environmental aspects may have an impact on (perceived) productivity, it seems most important to users of an office building that the facility supports their current activities. A short list of activities is created based on what people consider most significant for their productivity: 1) support of

Table 9.2 Proposed shortlist of KPIs

The extent to which the work environment and facilities support the following activities (scores on a five-point Likert scale ranging from 1 – very unsupportive to 5 – very supportive):

- Focused concentrated work
- Knowledge sharing
- Social interaction
- Your individual productivity
- Your team productivity.

This list could be extended with questions regarding the perceived productivity support by important environmental factors such as the indoor climate, personal control, ergonomics, IT facilities and interior design.

If possible, the subjective measurement of productivity should be completed with objective measures, e.g.:

- The actual output per employee (related to the sector and organisation, e.g. the number of students getting a diploma within the regular study period, or the number of transactions in a call centre)
- The percentage of sick leave
- Quality of the output (e.g. client satisfaction about the delivered output).

concentration, and 2) support of communication. The activities identified in Table 9.2 could be subdivided into a number of other sub-level activities. By also measuring the extent to which the work environment is perceived as supportive for the individual and team productivity, connections can be made with the support of the different activities. As such, the impact of the environment on the overall productivity can be evaluated.

Perspectives

Although much research has been conducted into the impact of buildings and facilities on labour productivity, much work still has to be done, in particular regarding:

- The differentiation in understanding individual needs and preferences of different groups, classified, e.g. by gender, age, psychological and profile
- Productivity outcomes per work type, e.g. classified by job title, type of work and flexibility of the work
- Defining a typology of work environments. Clear definitions need to be made with regards to the different spaces so that research findings can be made more meaningful when extrapolated to a wider context
- Operationalisation of input and output factors, preferably in a quantitative way
- Interpretation and explanation of cause-effect relationships and the impact of intermediary variables, for instance by in-depth interviews, focus groups and expert meetings

- Statistical analysis of quantitative data in search of correlations and weights of the relative contributions of different input variables on (perceived) productivity
- A combination of cross-sectional and longitudinal studies. The cross-sectional studies allow for comparisons across a number of buildings at a certain instant in time, whereas the longitudinal studies allow for pre-evaluation, an intervention and post-evaluation
- Cross-case analyses of different settings (offices, educational facilities, retail and leisure, healthcare) using standardised research methods.

References

Aronoff, S. and Kaplan, A. (1995) *Total Workplace Performance: Rethinking the Office Environment*, Ottawa: WDL Publications.

Appel-Meulenbroek, R. (2014) *How to measure added value of corporate real estate and building design. Knowledge sharing in research building*, PhD thesis, TU Eindhoven.

Backhouse, A. and Drew, A. (1992) 'The design implications of social-interaction in a workplace setting', *Environment and Planning B: Planning and design*, 19 (5), pp. 573–584.

Bakker, I. (2014) *Uncovering the secrets of a productive work environment. A journey through the impact of plants and colour*, PhD Thesis, Faculty of Industrial design, TU Delft.

Bakker, I., Van der Voordt, Th., Vink, P. and De Boon, J. (2013) 'Red or blue meeting rooms: Does it matter? The impact of colour on perceived productivity, social cohesion and wellbeing', *Facilities*, 31 (1/2), pp. 68–83.

Bakker, I., and Van der Voordt, D.J.M. (2010) 'The influence of plants on productivity. A critical assessment of research findings and test methods', *Facilities*, 28 (9/10), pp. 416–439.

Barber, C. (2001) 'The 21st-Century Workplace', in Kaczmarczyk, S. et al. (Eds.), *People and the Workplace*, Washington DC: GSA Office of Governmentwide Policy.

Barret, P., Zhang, Y., Moffat, J. and Kobbacy, K. (2013) 'A holistic, multi-level analysis identifying the impact of classroom design on pupils' learning', *Building and Environment*, 59 (0), pp. 678–689.

Batenburg, R. and Van der Voordt, D.J.M. (2008) 'Do Facilities Matter? Effects of Facility Satisfaction on Perceived Productivity', *Proceedings of the European Facility Management Conference*, Manchester, 10–11 June, pp. 139–150.

Becker, F. and Sims, W. (2000) *Offices That Work: Balancing Cost, Flexibility, and Communication*. New York: Cornell University International Workplace Studies Program (IWSP).

Block, L.K. and Stokes, G.S. (1989) 'Performance and satisfaction in private versus non-private work settings', *Environment and Behavior*, 21, pp. 277–297.

Blok, M., Groenesteijn, L., van den Berg, C. and Vink, P. (2011) 'New ways of working: A proposed framework and literature review', in Robertson, M.M. (Ed.), *Ergonomics and health aspects*, Heidelberg: Springer-Verlag, pp. 3–12.

Boerstra, A.C., Loomans, M.G.L.C. and Hensen, J.L.M. (2014) 'Personal control over indoor climate and productivity', *Proceedings of Indoor Air 2014*, 7–12 July, Hong Kong, China, pp. 1–8.

Bouttelier, R., Ullman, F., Schreiber, J. and Nael, R. (2008) 'Impact of office layout on communication in a science-driven business', *R&D Management,* 38 (4), pp. 372–391.

Brill, M. and Weidemann, S. (2001*) Disproving widespread myths about workplace design.* Jasper, IN: Kimball International.

Christopher, W.F. and Thor, C.G. (Eds.) (1993) *Handbook for Productivity Measurement and Implementation.* Portland: Productivity Press.

Clements-Croome, D. (2000) *Creating the Productive Workplace.* New York: Taylor and Francis.

Davenport, T.H. and Prusak, L. (2000) *Working knowledge,* Boston, MA: Harvard Business School Press.

De Been, I. and Beijer, M. (2014) 'The influence of office type on satisfaction and perceived productivity support', *Journal of Facilities Management,* 12 (2), pp. 142–157.

De Been, I., Beijer, M. and Den Hollander, D. (2015) 'How to cope with dilemmas in activity based work environments: Results from user-centred research', *European Facility Management Conference EFMC 2015,* Glasgow, 2–3 June.

Dorgan, C.E. and Dorgan, C.B. (2005) 'Assessment of link between productivity and indoor air quality', in Clements-Croome, D. (Ed.) *Creating The Productive Workplace,* London: E and FN Spon, pp. 113–135.

Elliot, A. and Maier, M.A. (2007) 'Colour and psychological functioning', *Current Directions in Psychological Science,* 16 (5), pp. 250–254.

Frankema, E.H.P. (2003) *Kantoorinnovatie in economisch perspectief.* Delft: Centre for People and Buildings.

Furnham, A. and Strbac, L. (2002) 'Music is as distracting as noise: The differential distraction of background music and noise on the cognitive test performance of introverts and extraverts', *Ergonomics,* 45(3), 203–217.

Galasiu, A.D. and Veitch, J.A. (2006) 'Occupant preferences and satisfaction with the luminous environment and control systems in daylit offices: A literature review', *Energy and Buildings,* 38, pp. 728–742.

Greene, C. and Myerson, J. (2011) 'Space for thought: Designing for knowledge workers', *Facilities,* 29 (1), pp. 19–30.

Halkos, G. and Bousinakis (2010) 'The effect of stress and satisfaction on productivity', *International Journal of Productivity and Performance Management,* 59 (5), pp. 415–431.

Hanssen, S.O. (2000) *Economics vs. Indoor Climate – A Basis for Deciding Appropriate Measures and Rational Decision Models in Building Administration,* Oslo: Research Council of Norway.

Haynes, B.P. (2007) 'An evaluation of office productivity measurement', *Journal of Corporate Real Estate,* 9 (3), pp. 144–155.

Haynes, B.P. (2008) 'Impact of workplace connectivity on office productivity', *Journal of Corporate Real Estate,* 10 (4), pp. 286–302.

Haynes, B.P. (2012) 'Conversational Networks in Knowledge Offices', in: Alexander, K. and Price, I. (Eds), *Managing Organizational Ecologies: Space, Management and Organization,* New York: Routledge, pp. 201–212.

Hedge, A. (1982) 'The open-plan office. A systematic investigation of employee reaction to their work environment', *Environment and Behavior,* 14, pp. 519–542.

Hua, Y., Loftness, V., Kraut., R. and Powell, K. M. (2010) 'Workplace collaborative space layout typology and occupant perception of collaboration environment', *Environment and Planning B: Planning and Design,* 37, pp. 429–448.

Huang, L., Zhu, Y., Ouang, Q. and Cao, B (2012) 'A study on the effects of thermal, luminous, and acoustic environments on indoor environmental comfort in offices', *Building and Environment*, 49, pp. 304–309.

Karakolis, T. and Callaghan, J.P. (2014) 'The impact of sit-stand office workstations on worker discomfort and productivity: A review', *Applied Ergonomics*, 45, pp. 799–806.

Kaya, N. and Epps, H.H. (2004) 'Relationship between colour and emotion, a study of college students', *College Student Journal*, 38, pp. 396–405.

Kaczmarczyk. S., Barber, C., Vega, G., Simpson, G.T., Thormalen, A., Cohn, D., Michael, W. and Shore, J. (2001) *People and the Workplace*. Washington DC: GSA Office of Governmentwide Policy.

Keizer, J. and Van Eijnatten, F. (2000) *Improving productivity via workspace design. An exploratory study*. TU Eindhoven: Department of Technology Management.

Kleeman, W.B., Duffy, F., Williams, K.P. and Williams, M.K. (1991) *Interior design of the electronic office: The comfort and productivity payoff*. New York: Van Nostrand Reinhold.

Knight, C. and Haslam, S.A. (2010) 'The relative merits of lean, enriched, and empowered offices: An experimental examination of the impact of workspace management strategies on well-being and productivity', *Journal of Experimental Psychology: Applied*, 16 (2), pp. 158–172.

Kurvers, S.R. and Leijten, J.L. (2013) 'Thermisch comfort: huidige en toekomstige normen', *TVVL Magazine*, 06/2013, pp. 46–51.

Lan, L., Lian, Z., Pan, L. and Ye., Q. (2009) 'Neurobehavioral approach for evaluation of office workers' productivity: The effects of room temperature', *Building and Environment*, 44, pp. 1578–1588.

Leaman, A. (1995) 'Dissatisfaction and office productivity', *Facilities*, 13 (2), 13–19.

Leaman, A. and Bordass, B. (1997) 'Productivity in Buildings: the 'killer' variables'. Paper presented at the *Workplace Forum*, London, October 1997.

Leaman, A. and Bordass, B. (1999) 'Productivity in buildings: The 'killer' variables', *Building Research and Information*, 27 (1), pp. 4–19.

Maarleveld, M. and De Been, I. (2011) 'The influence of the workplace on perceived productivity', *European Facility Management Conference*, Conference Paper, May 2011.

Maarleveld, M., Volker, L. and Van der Voordt, T.J.M. (2009) 'Measuring employee satisfaction in new offices – the WODI toolkit', *Journal of Facilities Management*, 7 (3), pp. 181–197.

Mahnke, F. (1996) *Color, Environment, Human Response*. New York: Van Nostrand Reinhold.

Mahoney, J.M., Kurczewski, N.A. and Froede, E.W. (2015) 'Design method for multi-user workstations utilizing anthropometry and preference data', *Applied Ergonomics*, 46, Part A (0), pp. 60–66.

Mawson, A. (2002) *The Workplace and its Impact on Productivity*, London: Advanced Workplace Associates Ltd.

Mehta, R. and Zhu, R.J. (2009) 'Blue or Red? Exploring the Effect of Colour on Cognitive Task Performance', *Science Express*, 3223 (5918), pp. 1226–1229.

Niemelä, R., Hannula, M., Rautio, S., Reijula, K. and Railio, J. (2002) 'The effect of air temperature on labour productivity in call centres – a case study', *Energy and Buildings*, 34, pp. 759–764.

Oseland, N. (1999) *Environmental Factors Affecting Office Worker Performance: A Review of Evidence*. London: Chartered Institution of Building Services Engineers: DETR.

Oseland, N., Marmot, A., Swaffer, F. and Ceneda, S. (2011) 'Environments for successful interaction', *Facilities*, 29 (1/2), pp. 50–62.

Peponis, J., Bafna, S., Bajaj, R., Bromberg, J., Congdon, C. and Rashid, M. (2007) 'Designing space to support knowledge work', *Environment and Behavior*, 39 (6), pp. 815–840.

Pols, J.P., Karels, M. and Ten Bolscher, G.H. (2009) 'Praktijkonderzoek in tien energiezuinige gebouwen. Gezond binnenklimaat niet vanzelfsprekend', *VV+*, November, pp. 642–649.

Sauter, S.L., Schleifer, L.M. and Knutson, S.J. (1991) 'Work Posture, Workstation Design, and Musculoskeletal Discomfort in a VDT Data Entry Task', *Human Factors: The Journal of the Human Factors and Ergonomics Society*, 33 (2), pp. 151–167.

Seddigh, A., Berntson, E., Bodin Danielson, C. and Westerlund, H. (2014) 'Concentration requirements modify the effect of office type on indicators of health and performance', *Journal of Environmental Psychology*, 38, pp. 167–174.

Smith, A. and Pitt, M. (2009) 'Sustainable workplaces: Improving staff health and wellbeing using plants', *Journal of Corporate Real Estate*, 11 (1), pp. 52–63.

Strubler, D.C. and York, K.M. (2007) 'An exploratory study of the Team Characteristics Model using organizational teams', *Small Group Research*, 38 (6), pp. 670–695.

Sullivan, J., Baird, G. and Donn, M. (2013) *Measuring productivity in the office workplace. Final report*. Victoria University of Wellington, New Zealand: Centre for Building Performance Research.

Sundstrom, E., Town, J.P., Rice, R.W., Osborn, D.P. and Brill, M. (1994) 'Office noise, satisfaction, and performance', *Environment and Behavior*, 26, pp. 195–222.

Thompson, B. and Jonas, D. (2008) *Workplace Design and Productivity: Are They Inextricably Linked?* London: Royal Institute of Chartered Surveyors.

Ulrich, R.S. (1984) 'View through a window may influence recovery from surgery', *Science*, 224, pp. 420–1.

Van den Berg, A.E. (2005) *Health impacts of healing environments: A review of the benefits of nature, daylight, fresh air and quiet in healthcare settings*. Groningen: Foundation 200 Years University Hospital Groningen.

Van der Voordt, D.J.M. (2003) *Costs and benefits of innovative workplace design*. Delft: Centre for People and Buildings.

Von Felten, D., Böhm, M. and Coenen, C. (2015) 'Multiplier Effects through FM services: A survey-based analysis of added value in FM', *European Facility Management Conference EFMC 2015*, Glasgow, 2–3 June.

Vos, P.G.J.C. and Dewulf, G.P.R.M. (1999) *Searching for data: a method to evaluate the effects of working in an innovative office*. Delft: Delft University Press.

Whitley, T.D.R., Makin, P.J. and Dickson, D.J. (1996) 'Job satisfaction and locus of control: Impact on Sick Building Syndrome and self-reported productivity', *7th International Conference on Indoor Air Quality and Climate*, Nagoya, Japan.

Wolf, K.L. (2002) 'The impact of nature in and around shop areas: Creation of an environment specifically suited to a consumer', paper presented at the *People/Plant Symposium*, Amsterdam.

Wyon, D.P. (1996) 'Individual microclimate control: required range, probable benefits and current feasibility', *Proceedings of the 7th International Conference on Indoor Air Quality and Climate*, 1, pp. 1067–1072.

Interview 5: Peter Prischl, Reality Consult, Austria

Peter Prischl has nineteen years of experience in FM consulting, and previously worked for three years with building technology and six years in corporate real estate. His educational background is in Business Administration, and he has a MBA from Vienna University of Economics and Business. He is managing director of Reality Consult, which has fifteen employees and offices in Vienna and Frankfurt. The company is a partner in an international real estate consultancy group.

Added value in general

Q. What does the term "Added Value" mean to you?

A. *I use "Added Value" or similar terms like "Benefits" and the German terms "Wertschöpfung" and "Nutzen" in my daily work. I base it on a micro-economic definition of economic value and value chain.*

Q. How do you see the relation between "Added Value" and cost reduction?

A *Cost reduction is one side of increasing value and revenue. It concerns minimising input and the other side is maximising output. You cannot do both at the same time.*

Q. Is "Added Value" mostly treated at a strategic, tactical or operational level?

A. *In most businesses added value is pushed down from top management. It is a top down process with a breakdown into profit centres. Most top managers aim at many profit centres and only a few cost centres. This was particularly a strong trend earlier. ABB started a process in the 1980s resulting in four thousand internal profit centres. Now the trend is to some degree going the other way. The reason is that it creates internal competition and it can create incentives to focus more on selling to other internal profit centres rather than to external customers. There are examples of CREM/FM units that have been changed from profit centres and back to cost centres. Reality Consult is divided into two profit centres – one for each office with reference to Peter as managing director.*

Q. In which context or dialogue is "Added Value" mostly considered in your experience?

A. *For us as consultants it is mostly used in our dialogue with clients.*

Q. How do you see the relation between innovation and "Added Value" in FM?

A. *The relation is not good. Too high pressure on added value can limit the development of an experimental and creative environment, which is important for innovation. It limits innovation to mostly focus on cost reductions derived from lack of resources. According to the book "The Innovator's Dilemma" about disruptive innovation by Harvard professor Clayton Christensen, there are three types of innovation with a focus on:*

- *Market creation*
- *Extension (maintenance of market share within existing markets)*
- *Cost reduction (efficiency)*

Benefits and limitations of "Added Value"

Q. What are the benefits of considering and talking about "Added Value"?

A. *There are many benefits, but the most important is to create tangible advantages for the client.*

Q. What are the limitations of considering and talking about "Added Value"?

A. *Added value is difficult to calculate. According to the economist and Nobel Prize winner Joseph Stiglitz, the more information you have about a product, the more you are willing to pay. This is a problem for FM. For example, in a bank which Reality Consult worked for, the top management was very concerned about the internal rent charged by the FM department being higher per m^2 than the rent they could see advertised for commercial property. It was not made clear to them that the internal rent was based on net area and not gross area and included the cost of a lot of FM services not included in the advertised rents.*

Management of Added Value

Q. What are your top five main values to be included in management of accommodations, facilities and services? Could you mention a few examples of concrete FM interventions to attain these added values and KPIs to measure them?

A. *See table below.*

Prioritised values in FM	Concrete FM interventions	KPIs
1. Person to person services	Bringing lost item to my work station Bring food into a meeting	Customer satisfaction Reduction of complaints
2. Removal of noticeable barriers	Reprogramming/ optimising of elevators	Productivity measures Output (e.g. call centre)
3. Noise reduction		Sickness
4. Chairs	Ensure optimal chairs (office, crane, machine, truck)	Long-term health
5. Climate comfort	Optimal heat, humidity and air change rate	Sickness

Q. Do you measure whether you attain the targeted added values?

A: *We use KPIs as shown above and make documentation of observations. We produce weekly status reports on projects.*

Q. Do you benchmark your data with data from other organisations?
A. *If clients ask for it, we will do benchmarking, but I think benchmarking is overrated.*

Q. Which aspects do you find most important to include in your value adding management approach?
A. *The most important is output orientation and to ensure recognition.*

Q. Do you experience any dilemmas you have to cope with in relation to added value?
A. *People are lazy and cowards. That goes both for top and middle management. They want improvements but are not willing to make the necessary changes.*

Q. Have you noticed a change in mind-shift regarding "Added Value"?
A. *There is much sarcastic realism with a focus on cost reduction. It is not just a result of the financial crisis. It has been so since the 1980s. It has to do with the challenges to the welfare state.*

Q. Any final comments?
A. *Added value is one of the most important tools and strategies for FM to develop.*

10 Adaptability

*Rob Geraedts, Nils O.E. Olsson and
Geir Karsten Hansen*

Introduction

Real estate is a product with a high economic value, a long technical life cycle and a large spatial-physical impact. That is why it is of great societal importance to use real estate as efficiently as possible. To enable a high-quality use and a high occupancy rate, a building must be able to keep up with qualitative and quantitative changes in demands. In recent decades the interest in flexible building, also called adaptive building, has grown substantially. In many countries this interest is mainly caused by the structural vacancy of real estate, in particular office buildings, the economic crisis, the congestion of the housing market and the increased awareness of and interest in sustainability (Arge and Blakstad, 2010; Cairns, 2010; Horgen, 2010; Hansen and Olsson, 2011; Van Meel, 2015). A direct connection can be made between adaptive building and sustainability (Wilkinson et al., 2009; Wilkinson and Remøy, 2011). Market developments show increased demands for flexibility and sustainability by users and owners as well as a growing understanding of the importance of a circular economy (Eichholtz et al., 2008).

Different actors may have different interests and needs regarding adaptability:

- Users: accommodation that is adaptable to a changing primary process
- Owners: a building with the highest possible profitability during ownership and adaptability to organisational changes such as growth or shrinkage and market change; sometimes the user is the owner as well
- Society: real estate that contributes to an attractive and sustainable living and working environment.

There are three basic ways to act when a building no longer meets the users' needs: adapt the location, building and/or unit (transformation/conversion), design and construct a new building, or move to another and more suitable existing building; see Figure 10.1.

This chapter first discusses the state of the art regarding the concepts of adaptability and flexibility and the need for adaptable buildings. Then it presents possible benefits and costs of typical interventions. The next section discusses

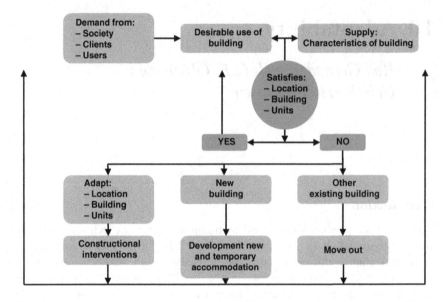

Figure 10.1 The accommodation cycle of real estate and the demand for change (Geraedts et al., 2014)

ways to measure the adaptability capacity of buildings and includes both a long list and a short list of KPIs. The assessment criteria for adaptability are based on previous research by Geraedts and Van der Voordt (2007), Remøy and Van der Voordt (2007), Wilkinson et al. (2009), DGBC (2013) and Geraedts et al. (2014). The chapter ends with some suggestions for further research.

State of the art

Definitions and relations

According to Hermans et al. (2014) the adaptive capacity of a building includes all characteristics that enable it to keep its functionality during its technical life cycle in a sustainable and economic profitable way, withstanding changing requirements and circumstances. The adaptive capacity is considered a crucial component when scrutinising the sustainability of the real estate stock.

Flexibility in a project can be associated with the decision process or the final product (Gill et al., 2005; Olsson, 2006). Flexibility in the decision process is based on an approach where commitments in projects are made sequentially. This philosophy is developed further by combining the concept of layered building designs (Brand, 1994) and adaptability to a layered decision process (Blakstad, 2001; Arge, 2005). The adaptive capacity can be split into three different aspects (see Figure 10.2):

- *Organisational flexibility*: the adaptive capacity of an organisation or user to respond adequately to the changing demands of the built environment, including strategic, structural and operational flexibility (Volberda, 1997). Strategic flexibility of an organisation is one of the most crucial assets to respond to changing market circumstances like increasing unemployment in the sector, technological innovations, economic competition, new legislation and changed relationships with customers. Structural flexibility incorporates the ability to transform the aims of the organisation to adapt to changing circumstances on a more or less permanent basis – for instance, by producing multifunctional teams or collaboration with suppliers in the design. Operational flexibility mainly incorporates changes in the activities of an organisation. Examples are the creating of stock, the use of temporary human resources or the reservation of capacity at suppliers. Financial and contractual flexibility are both also strongly related to organisational flexibility. Financial flexibility is related to the financial situation and the arrangements of owners and users of the building and in the real estate market in general. Contractual flexibility depends on the types and length of contracts and the ruling practices in the market, own/lease strategies, as well as on available alternatives (both for owners and occupants) (Gibson, 2000; Blakstad, 2001).
- *Process flexibility*: the adaptive capacity of a process that enables a building to react to changing circumstances, wishes or demands during the initial, design and construction phases, including planned flexibility and response flexibility (Olsson, 2006; Genus, 1997; Verganti, 1999).
- *Product flexibility*: the adaptive capacity of a building (the product) that enables it to respond to changing circumstances, wishes or demands during the use phase of the building without complex and costly technical interventions.

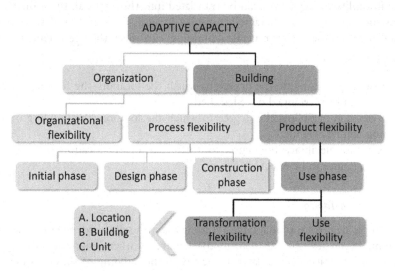

Figure 10.2 Different aspects of adaptive capacity and the focus within this chapter (marked dark) (Geraedts et al, 2014)

Adaptability is considered to be a value. Flexibility is considered to be a set of possible measures that could be taken to create the value adaptability.

Demand for adaptability: use dynamics and transformation dynamics

Changing market demand is a driver for flexibility measures, and as such it precedes the input part of the Value Adding Management model that was presented in Chapter 1. The focus of the adaptable capacity determination method that is presented later in this chapter is exclusively product flexibility during the use phase of buildings. The target here is the translation of the demand into transformation and use dynamics on three different levels: location, building and unit – see Figure 10.3. The split in goals (demand) and means (supply) is also described by Van der Voordt and Van Wegen (2000; 2005).

A building must be able to cope with changing user demands. This may lead, for instance, to a demand in a brief or programme of requirements that the building must be parcelled into smaller units (e.g. compartmentalisation into smaller units), or that specific facilities must be added to the units or building. This is called use dynamics (end-user perspective).

Transformation dynamics concern the demands that a building must be able to accommodate totally different user groups or different functions in the near future. This may lead to specific demands for the ability to rearrange the building for different user groups. This is called transformation dynamics (client perspective).

The supply: rearrange, extension and rejection flexibility

The flexibility of the supply can be translated into three spatial/functional and construction/technical characteristics (also including the technical services as installations). They determine if a building can meet the requirements; see Figure 10.3:

- Rearrange flexibility: the degree to which the location, the building or the unit can be rearranged or redesigned
- Extension flexibility: the degree to which the location, the building or the unit can be extended
- Rejection flexibility: the degree to which (part of) the location, the building or the unit can be rejected.

Adaptive building

Designing flexible and adaptable buildings is an established practice; see for example Brand (1994). The main approaches are to either design the building to allow cost-effective changes later on, or design the building to allow flexible use of it without having to make adaptations. As discussed by De Neufville et al.

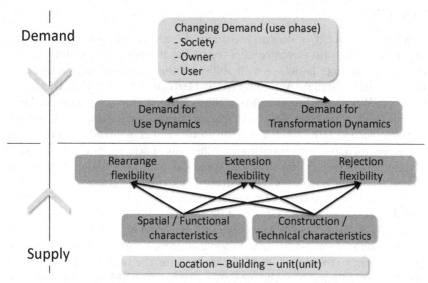

Figure 10.3 Framework of the adaptive capacity method for the demand (use and transformation dynamics) and supply (rearrangement, extension and rejection flexibility) on three different levels (location, building and unit) (Geraedts et al., 2014)

(2008) and Miller and Swensson (2002), flexibility is a key design criterion for buildings. With a life span of forty years or more, it is likely that changes will be required before the building is abandoned. Such flexibility can, according to Pati et al. (2008) and Bjørberg and Verweij (2009), be achieved through different approaches.

The adaptive capacity of a building has been addressed in the literature to a very limited extent. More attention is paid to aspects like flexibility, extendibility, multifunctionality, reusability or removability. These aspects often have strongly overlapping meanings (Geraedts, 2013). Schuetze (2009) defines the adaptive capacity of a building as being easily adaptable to different functions or changing requirements, constructed with components and products which allow re-use and recycling with a minimum of effort and loss of quality. It mainly concerns function-neutral buildings, which have a user-related transformation potential. Use has been made of reusable construction components, based on the different life cycles of the components. Richard (2010) states that nobody can predict the demands, wishes or different tastes of the current and the future users of a building during its lifespan. Adaptable systems are necessary to meet this challenge, to develop user-friendly buildings open to change with freedom of choice for the first generation of users and possibilities to adapt for the next generations of users.

Flexibility

Flexibility is generally related to managing the effects of uncertainty. Bahrami and Evans (2005) list eleven concepts related to flexibility: adaptability, agility, elasticity, hedging, liquidity, malleability, mobility, modularity, robustness, resilience and versatility. According to the Merriam-Webster Online Dictionary (Merriam Webster Online Dictionary 2006), being flexible is 'characterised by a ready capability to adapt to new, different, or changing requirements'. According to Schneider and Hill (2007) flexibility enables real estate to adapt to changing needs and patterns, both socially and technically. These needs can be personal (e.g. an expanding family situation), practical (e.g. as a result of aging) or technical (e.g. the renewal of old services). The changing patterns can be caused by demographic, economic or environmental changes. Flexible housing means that developments are adaptable to different users during their whole lifespan. According to Groák (1992) flexibility is accomplished by changing the physical composition to combine or expand rooms or units. Adaptability deals with the use of rooms and units, while flexibility concerns the technical aspects. Prins (1992) defines flexibility as the physical property of a building to be able to be adaptable or changeable for technical, spatial and social purposes.

Habraken (1961) proposed a subdivision of a building into different layers or levels: the support level with construction components with a long lifespan, and the infill level with construction components with a short life span. This was based on a distinction between collective components for the community to decide (support level) and individual components for the individual user to decide (infill level). Flexibility in the decision process is based on an approach where commitments in projects are made sequentially. This philosophy is further developed by combining the concept of layered building designs (Brand, 1994) and adaptability to a layered decision process (Blakstad, 2001; Arge, 2005). Another perspective on process flexibility is found in real options theory. In real options, the ability to wait to commit to an investment until more information is available can serve as one option and have a corresponding value (Brach, 2003). An option is the right, but not the obligation, to take an action in the future (Amram and Kulatlaka, 1999). Uncertainty can increase the value of a project as long as flexibility is preserved and resources are not irreversibly committed. Van Reedt Dortland (2013) has shown how real options can be used in strategic real estate management decision making related to healthcare facilities.

Organisational flexibility

Changing circumstances and demands in an organisation don't necessarily need to lead to changes in a building. A new good match between organisation and building can also be found by adapting the organisation instead of the building. This could be another way of working, a different distribution of activities within the available spaces, or by outsourcing certain activities. In this case we speak about organisational flexibility (Geraedts, 2013). According to Volberda (2007) organisational flexibility concerns the extent to which an organisation is capable of responding adequately to changing demands. The concept of flexibility in hospital buildings has recently been developed further, and now includes a variety of principles and solutions. According to Rechel et al. (2009), flexibility should encompass a wide range of building characteristics, supporting infrastructure such as transport links and relationships to other parts of the healthcare system, and it should also occur in relation to financing.

Process flexibility

Process flexibility relates to the whole lifespan of buildings: development, construction, operation and use, from the first initiative and briefing phase until the final use. Decision-making processes need to be structured in such a way that the freedom for changes is maximized (Geraedts, 1998). Lean construction emphasises the concept of "last responsible moment". According to Ballard (2000), the last responsible moment is the 'point at which failing to make the decision eliminates an alternative'. This means that decisions must be made within the lead-time for realising alternatives. The term last responsible moment implies that it would be irresponsible to delay a decision further. On the other hand, a decision should not be made until it has to be made, in order to allow for flexibility (Ballard and Howell, 2003). A common tool for achieving flexibility in projects is the use of option-based contracts, which enable a continuous locking of the projects. Mahmoud-Jouini and Midler (2004) point out that a key factor in creating win-win situations between the stakeholders in Engineering, Procurement and Construction (EPC) contracts lies in the flexibility created between the contracts. Flexibility also has to do with politics. In general, stakeholders are less likely to favour a continued flexible decision process when an initial decision has been taken in their favour. Consequently, a continued flexible decision process is valued by those who do not prefer an initial decision. Flexibility options might be used as a tool for stakeholders who want decisions to be remade. Flexible decision processes can be used to justify that decisions do not need to be taken or can be revised, thus becoming a tool for irresoluteness.

Product flexibility

Product or object flexibility concerns the changeability of the product. If the product is a building, then the product flexibility concerns the changeability of that building, not only during the development phase, but also in particular during the exploitation or use phase. Product flexibility concerns the spatial and technical properties of the product, the building. Product flexibility enables use flexibility and transformation flexibility (Geraedts, 1998).

Pati et al. (2008) made a distinction in three types of product flexibility: adaptability, convertibility, and expandability. A similar classification has been used by Arge (2005) and Bjørberg and Verweij (2009). Adaptability can be defined as a building's ability to meet shifting demands without physical changes.

Convertibility can be defined as the possibility for construction and technical changes with minimum cost and disturbance. The third type of flexibility, expandability, can be defined as the ability to add (or reduce) the size of a building. In this chapter, all three characteristics are seen as subsets of flexibility. Flexibility can be related to modularity. Modularity refers to the possibility of dividing a project into more or less independent sub-units. Design modularity is a common approach to achieve flexibility by decoupling physical and functional components (Thomke, 1997: Hellström and Wikström, 2005). The two dimensions of flexibility of the design process and in the product can be applied independently, but also in combination. This has been analysed by researchers previously (Brand, 1994; Gill et al., 2005). Flexibility in the decision process and the product may interact for any given project.

Benefits and costs

Added value comes with the possibility to adapt to changing market or user demands, to reduce the risk of future vacancy of the building, lower adaptation costs, higher rental income, happier users, less pollution of the environment (or a more sustainable building), and possible re-use of building construction components. The largest unknown factor is estimating whether provisions made for future adaptability will actually be used in a given period. If not, most of the accompanying extra investment costs will not be profitable. Table 10.1 shows an overview of seventeen flexibility interventions in connection with the shortlist of most important Key Performance Indicators that is presented in the next section, and which could lead to different forms of future added value (benefits) against certain costs (sacrifices).

How to measure

Flex 2.0: indicators of transformation dynamics and use dynamics

Flex 2.0 is a method to formulate the demand for flexibility in the briefing and design phase, and to assess the supply of flexibility of buildings in the use phase. Due to the lack of a widely accepted method with criteria or indicators for the potential for adaptation during the life cycle of a building (to be used in the briefing and design phase as well as the use phase), an extensive international literature survey has been conducted, resulting in 147 indicators that are more or less connected to flexibility of buildings (Geraedts, 2015). Seven important transformation dynamics indicators from the perspective of the owner of a building were formulated. They can be used to express the owner's wishes and demands for the adaptive capability of the building and the user units; see Figure 10.4. Also, seven use dynamics indicators from the perspective of the users of abuilding were developed to express their wishes and demands of the adaptive capability of the units and the building; see Figure 10.5.

Reallocate / Redesign Concerning (1) the change in size or division of user units within a building (combine, split up or rearrange); wishes/demands concerning the possibilities of (2) changing the layout on building level and/or the possibilities of (3) changing the functions on building level.	
Grain size The number of user units in a building (increasing or decreasing).	
Facilities Change of the facilities in the building (1) and/or outside the building at location level (2).	
Quality Change of the layout and finishing per user unit (1) or per building (upgrading) (2).	
Expansion To which extent the use surface of a building can be increased in the future (horizontal and/or vertical).	
Rejection To which extent the use surface of a building can be decreased in the future (horizontal and/or vertical).	
Transfer Whether or not the building can be transferred to another location .	

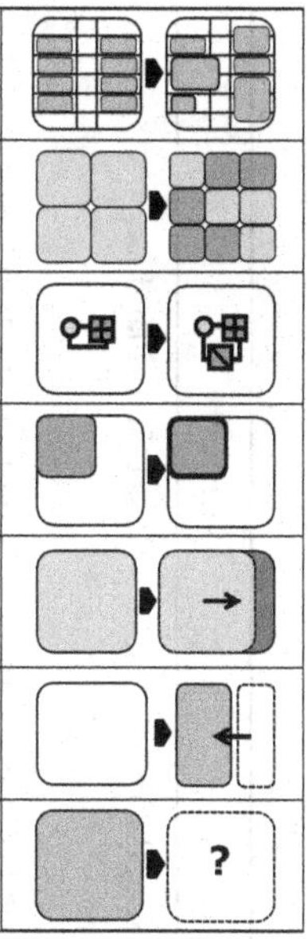

Figure 10.4 Seven important transformation dynamics indicators from the perspective of the owner of a building (Geraedts et al., 2014)

Redesign Changing the layout of the user units in a building (1) and/or changing the functions of the user units in the building (2).	
Reallocate Internal Changing the location of the user units in a building.	
Relation Internal Changing the internal relation with other users/stakeholders in the building.	
Quality Changing the layout and finishing (look and feel) of the user unit in a building.	
Facilities Changing the facilities in the user units (1), in the building (2) and/or at location level (3).	
Expansion To which extent the use surface of a user unit in a building should be extendable in the future (horizontal and/or vertical).	
Rejection To which extent the use surface of a user unit should be contractible in the future (horizontal and/or vertical).	

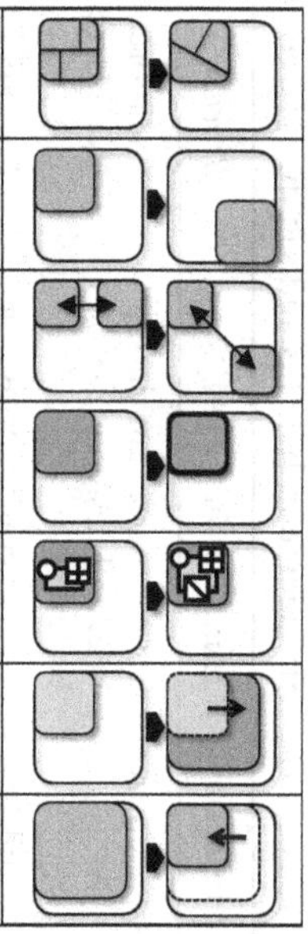

Figure 10.5 Seven important use dynamics indicators from the perspective of the users of a building (Geraedts et al., 2014)

Table 10.1 Interventions, management, benefits and sacrifices (Geraedts, 2015)

Flexibility Interventions	Management	Benefits	Sacrifices
1 Surplus of site space	Take into account during brief and design phase of building	The more surplus space on site, the more expandable the building.	Less efficient landuse
2 Surplus of building space/floor space	Take into account during brief and design phase of building	The building can be rearranged or transformed to other functions more easily; less vacancy, less adaptation costs, less environmental pollution.	Constructing more space than initially needed; higher investment costs
3 Surplus free of floor height	Take into account during design and construction phase of building	The building can be rearranged or transformed to other functions more easily; less vacancy, less adaptation costs, less environmental pollution, sustainable	More floor height than initially needed; higher investment costs
4 Decentralised access to building: location of stairs, elevators, core building	Take into account during design and construction phase of building	The building entrance system can be used independently by different user groups; less vacancy, less adaptation costs, less environmental pollution.	Higher investment/construction costs
5 Surplus of load-bearing capacity of floors	Take into account during design and construction phase of building	The larger the load-bearing capacity of the floors, the more easily a building can be rearranged or transformed to other functions; less vacancy, less adaptation costs.	Higher investment/construction costs (use of more steel and concrete)
6 Extendible building/unit horizontal	Take into account during design and construction phase of building. Dismountable (part of) the facade	The more easily a building or unit can be expanded for new or larger existing functions, the more easily a building can be rearranged or transformed to other functions or expanded; less vacancy/adaptation costs/environmental pollution, sustainable	Higher investment/construction costs (dismountable facade components)

7	Extendible building/unit vertical	Take into account during design and construction phase of building	The more easily a building or unit can be vertically expanded with new floors or basement, the more easily it can be rearranged or transformed to other functions; less vacancy/adaptation cost/environmental pollution, sustainable	Higher investment/construction costs (extra foundation parts; fontanel constructions)
8	Dismountable facade	Take into account during design and construction phase of building. Dismountable (part of) the facade	The more facade components are easily dismountable, the more easily a building can be rearranged or transformed to other functions; less vacancy, less adaptation costs, less environmental pollution.	Higher investment/construction costs (dismountable facade components)
9	Customisability and controllability of facilities	Take into account during design and construction phase of building. Small grain size of facilities	The more easily facilities are customisable and controllable to respond to changing functional requirements, the easier a building can be rearranged or transformed to other functions; less vacancy/adaptation costs/pollution.	Higher investment/construction costs (more facilities than initially needed)
10	Surplus facilities, shafts and ducts	Take into account during design and construction phase of building	The more surplus space the facilities shafts and ducts have, the easier a building can be rearranged or transformed to other functions; less vacancy, less adaptation costs, less environmental pollution.	Higher investment/construction costs (more facilities parts than initially needed)
11	Surplus capacity of facilities	Take into account during design and construction phase of building	The more surplus capacity the facilities have, the easier a building can be rearranged or transformed to other functions; less vacancy, less adaptation costs, less environmental pollution.	Higher investment/construction costs (more capacity than initially needed)
12	Disconnection of facilities components	Take into account during design and construction phase of building	The more facility parts can be disconnected or demounted, the easier a building can be rearranged or transformed to other functions; less vacancy, less adaptation costs, less environmental pollution.	Higher investment/construction costs (dismountable instead of fixed facilities parts)

Table 10.1 continued

	Flexibility Interventions	Management	Benefits	Sacrifices
13	Distinction between support - infill (fit-out)	Take into account during design and construction phase of building	The more construction components belong to the infill domain, the easier a building can be rearranged or transformed to other functions; less vacancy, less adaptation costs, less environmental pollution.	No higher investment or construction costs required
14	Access to building: horizontal routing, corridors, gallery	Take into account during design and construction phase of building	The more the horizontal disclosure of the units is limited by a central core, the more easily units in a building can be rearranged or transformed to other functions; less vacancy, less adaptation costs, less pollution.	No higher investment or construction costs required
15	"Removable, relocatable units in building	Take into account during design and construction phase of building	The more building units are constructed with demountable and reusable components, the easier it is to relocated user units in the building or elsewhere; less vacancy, less adaptation costs, less environmental pollution.	Higher investment/construction costs (dismountable instead of fixed building parts)
16	Removable, relocatable interior walls in building	Take into account during design and construction phase of building	The more inner walls are easily removable/reusable, the better a building can be rearranged or transformed to other functions; less vacancy, less adaptation costs, less environmental pollution.	Higher investment/construction costs (removable/relocatable interior walls)
17	Disconnecting/detailed connection interior walls; hor/vert.	Take into account during design and construction phase of building	The easier it is to dismount interior walls, the easier a building can be rearranged or transformed to other functions; less vacancy, less adaptation costs, less environmental pollution.	Higher investment/construction costs (disconectable/demountable building parts)

Layers and reduction of indicators

Blyth and Worthington (2010) and Hansen and Olsson (2011) discuss layered design and a layered design process. In order to identify the independent responsibilities of parties involved in the transformation of a built environment Habraken (1998) introduced levels of control/decision making, while Duffy (1998) and Brand (1994) defined functional levels within a building in order to identify functions with different rates of change (Durmisevic, 2006). To structure, cluster and reduce the large number of possible indicators, use has been made of the distinctions between different layers of the building and its environment according to Brand (1994). Brand's model consists of site, structure, skin, services, space plan, and stuff:

- Site: the urban location; the legally defined lot whose context lives longer than buildings. According to Brand and Duffy, the site is eternal
- Structure: the foundation and load-bearing elements, which last between thirty and three hundred years. However, few buildings last longer than fifty years
- Skin: the exterior finishing, including roofs and façades. These are upgraded or changed approximately every twenty years
- Services: the HVAC (heating, ventilating, and air conditioning), communication, and electrical wiring. These wear out after seven to fifteen years
- Space plan: the interior layout including vertical partitions, doors, ceiling, and floors. According to Brand, commercial space can change every three years
- Stuff: the furniture that is moved daily, weekly or monthly. In Italian furniture is called *mobilia*, for good reason.

By analogy with the model of Brand (1994), different levels were created in Flex 2.0 to structure, locate and cluster all possible flexibility indicators that had been found previously. Figure 10.6 gives an overview of the five layers and the eighty-three flexibility performance indicators that influence the adaptive capacity of buildings. The term services has been replaced by facilities, and the space plan and stuff have been combined and rephrased into space plan/finishing.

Flexibility scores

Next to a column with the layers, sub-layers and performance indicators, Figure 10.6 includes a column to mark the score or level of the specific indicator concerned and a column to allocate a weight to this score. The multiplication of the assessment value and the weighting factor for that specific indicator leads to the flexibility score. Figure 10.7 shows an example of how this works.

The addition of all scores of the individual indicators results in a total score of the adaptive capacity of a building. With this method it is possible to develop a score range, from a theoretical minimum flexibility/adaptability score up to a

LAYER	Sub-layer	Nr.	FLEX 2.0 Flexibility Performance Indicator	Weighting 1	2	3	Score / Level 1	2	3	4
GENERAL REQUIREMENTS		a	Development plan: deviation/modification							
		b	Development plan: extensions							
		c	National building regulations							
		d	General technical condition of building							
		e	Latest renovation of building							
		f	Age of building							
		g	Last user type of building							
1. SITE		1	Multifunctional site							
		2	Surplus of site space							
		3	Expandable site							
		4	Rejectable (part of) site							
2. STRUCTURE	2.1 Measurements	5	Surplus of building space / floor space							
		6	Available floor space of building							
		7	Size of building floors							
		8	Number of floors							
		9	Ground surface of building							
		10	Vertical exchangeability of building floors							
		11	Surplus free floor height							
		12	Measurement system; modular coordination							
		13	Measurement system facade							
		14	Horizontal canvas measurement							
		15	Horizontal zone division / layout							
		16	Shape of floor-plan							
	2.2 Access	17	Access to building: location of stairs, elevators, core building							
		18	Presence of stairs and/or elevators							
		19	Vertical extension: access							
		20	Extension / reuse of stairs and elevators							
	2.3 Construction	21	Surplus load bearing capacity of floors							
		22	Load bearing floors							
		23	Self supporting facade							

			a	b	c	d	e	f	g
	24	Shape of columns							
	25	Positioning obstacles / columns in load bearing structure draagstructuur							
	26	Presence of fontanel constructions							
	27	Positioning of facilities zones and shafts							
	28	Fire resistance of main load bearing construction							
	29	Extendible building / unit horizontal							
	30	Extendible building / unit vertical							
	31	Rejectable part of building / unit horizontal							
	32	Rejectable part of building vertical							
	33	Vertical extension: construction / foundation							
	34	Horizontal extension: construction							
	35	Interruption of load bearing structure							
	36	Detailed connection between foundation and facilities							
	37	Construction technique for main load bearing construction							
	38	Insulation between floors and units							
3. SKIN									
3.1 Entrance	39	Visibility of main entrance of building							
	40	Social safety of main entrance of building							
3.2 Facade	41	Possibility of balconies at facades							
	42	Dismountable facade							
	43	Reuse facade windows							
	44	Facade windows to be opened							
	45	Placement bottom side of facade windows							
	46	Location and shape of daylight facilities							
	47	Day light facilities							
	48	Insulation of facade							
	49	Detailed connection between facade / end wall components							
3.3 Roof	50	Outdoor space on roof							
	51	Own identity on roof / facade of building							

Figure 10.6 Overview of the general requirements (a–g) and the eighty-three performance indicators that influence the adaptive capacity of buildings (Geraedts, 2015)

4. FACILITIES	**4.1 Measure & Control**	52	Measure and control techniques						
		53	Customisability and controllability of facilities						
		54	Control of sun screens						
		55	Adaptability of elevators						
	4.2 Dimensions	56	Overdesign/surplus of facilities shafts and ducts						
		57	Overdesign/surplus of capacity of facilities						
		58	Overdesign capacity public facilities						
		59	Number of connecting points facilities (electricity and ICT)						
		60	Modularity of facilities						
		61	General-purpose of facility components						
		62	Independence of user units						
	4.3 Distribution	63	Distribution of facilities (heating, electricity)						
		64	Location sources of facilities (heating, cooling)						
		65	Disconnection of facilities components						
		66	Reachableness facilities components						
		67	Independence user unit						
5. SPACE PLAN / FINISHING	**5.1 Functional**	68	Multifunctional building						
		69	Multifunctional units						
		70	Distinction between support - infill (fit-out)						
		71	Exchangeability of (infill) construction components						
		72	Size of units						
	5.2 Access	73	Horizontal routing, corridors, access						
		74	Access of units						
		75	Personal access of user units						
		76	Relocation of building / unit access						
	5.3 Technical	77	Disconnectability / portability / movability of units						
		78	Disconnectability / portability / movability of interior walls						
		79	Disconnectability / detailed connection between interior walls; ho./vert.						
		80	Possibility of suspended ceilings						
		81	Possibility of raised floors						
		82	Individual infill / finishing						
		83	Barrier-free access of building / units						

Figure 10.6 continued

11 **Surplus of free floor height**	**Assessment values of the free floor height**	**Remark**	**Weighting**	**Score**
How much is the net free floor height?	< 2.60 m (Bad) 2.60–3.00 m (Normal) 3.00–3.40 m (Better) > 3.40 m (Best)	The higher the free floor height, the better a building can be rearranged or transformed to other functions, the better a building can meet to changing demands of facilities and the quality of the building or units.	1 = less important 2 = important 3 = very important	Score = assessment × weighting

Figure 10.7 An example of how the assessment method works: indicator no. 11 (surplus of free floor height) has four assessment values from bad to good

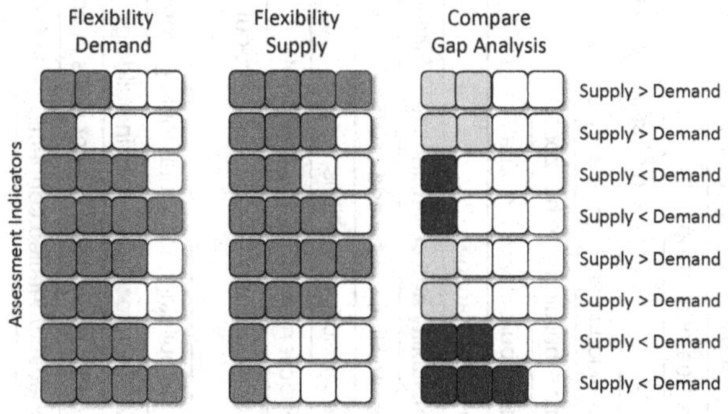

Figure 10.8 Example of a demand and supply profile with a comparative gap analysis to evaluate if there could be a match between the demands for future adaptability and the supplied flexibility by the building (Geraedts et al., 2014)

Flex 2.0 Light: 17 most important indicators

As a default setting with the possibility to switch to other priorities and another selection if needed by the user, the seventeen most crucial indicators are shown in Figure 10.9.

Figures 10.10a and 10.10b show a detailed overview of the assessment values and related remarks regarding these seventeen most important performance indicators.

Layer	Sub-layer	Nr.	Flexibility Performance Indicator	Weighting 1	2	3	Score/Level 1	2	3	4
1. Site		01(2)	Surplus of site space							
2. Structure	2.1 Measurements	02(5)	Surplus of building space/floor space							
		03(11)	Surplus free of floor height							
	2.2 Access	04(17)	Access to building: location of stairs, elevators, core building							
	2.3 Construction	05(21)	Surplus of load bearing capacity of floors							
		06(29)	Extendible building/unit horizontal							
		07(30)	Extendible building/unit vertical							
3. Skin	3.2 Facade	08(42)	Dismountable façade							
4. Facilities	4.1 Measure & control	09(53)	Customisability and controllability of facilities							
	4.2 Dimensions	10(56)	Surplus facilities shafts and ducts							
		11(57)	Surplus capacity of facilities							
		12(65)	Disconnection of facilities components							
5. Space plan/finishing	5.1 Functional	13(70)	Distinction between support – infill (fit-out)							
	5.2 Access	14(73)	Access to building: horizontal routing, corridors, gallery							
	5.3 Technical	15(77)	Removable, relocatable units in building							
		16(78)	Removable, relocatable interior walls in building							
		17(79)	Disconnecting/detailed connection interior walls; hor/vert.							

Figure 10.9 Flex 2.0 Light, a default setting with a limited number of the most important indicators to assess the adaptability of a building; the numbers in brackets refer to the number of the indicator in the long list (see Figure 10.7) (Geraedts, 2015)

Indicator / Question	Assessment values	Remark
01. Surplus of site space Does the site have a surplus of space and is the building located at the centre?	**Assessment values surplus of site space** 1. No, the site has no surplus of space at all 2. 10-30% surplus 3. 30-50% surplus 4. The site has a surplus space of more than 50	**Remark** The more surplus space on site, the better the building is expandable.
02. Surplus of building space / floor space Does the building or the user units have a surplus of the needed usable floor space?	**Assessment values oversized building space in % of oversize** 1. No, the building or user units have no surplus of floor space at all 2. 10-30% surplus 3. 30-50% surplus 4. The building has a surplus of floor space of more than 50%	**Remark** The more surplus space a building or user units have (for instance by the use of a zoning system with margin space), the more easily a building can be rearranged or transformed to other functions, the easier the grain size can be changed, the better a building can meet to changing demands, the easier parts of the building or units can be rejected and the easier the building or units are expandable.
03. Surplus of free floor height How much is the net free floor height?	**Assessment values of the free floor height** 1. < 2.60 m 2. 2.60 - 3.00 m 3. 3.00 - 3.40 m 4. > 3.40 m	**Remark** The higher the free floor height, the better a building can be rearranged or transformed to other functions, the better a building can meet to changing demands of facilities and the quality of the building or units.
04. Access to building: location of stairs, elevators, core building To what extent is a centralized and/or decentralized building entrance (location entrances, cores, stairs, elevators) has been implemented?	**Assessment values access to building** 1. Decentralized and separated building entrance and core. 2. Decentralized and combined building entrance and core. 3. Building divided in different wings, each with a centralized/combined entrance/core. 4. Building with one centralized entrance, divided in different wings, each with a centralized/combined entrance and core.	**Remark** The more a building entrance system can be used for a more independent use by different user groups the easier a building can be rearranged or transformed to other functions, the better a building is horizontal expandable, the better parts of a building can be rejected.
05. Surplus of load bearing capacity of floors How large is the load bearing capacity of the floors in the building?	**Assessment values of load bearing capacity of floors** 1. < 3 kN/m2 2. 3 - 3,5 kN/m2 3. 3,5 - 4 kN/m2 4. > 4 kN/m2 and several areas > 8 kN/m2.	**Remark** The larger the load bearing capacity of floors, the easier a building can be rearranged or transformed to other functions, the better a building can meet to changing user demands, the better possibilities for vertical expansion of the building, the more possibilities to change the location of user units within the building.

Figure 10.10a Overview of the assessment values of indicators 1 to 8 (Geraedts, 2015)

		Remark
06. Extendible building / unit horizontal extension Is it possible to expand the building horizontally for new extension to the building or user units?	**Assessment values horizontal extension building / unit** 1. Horizontal extension is not possible at all. 2. Horizontal extension of building/unit is very limited possible (f.i. only at one side). 3. Horizontal extension of building/unit is limited possible (f.i. only at more sides). 4. Horizontal extension of building/unit is easily possible at all sides.	**Remark** The more a building or unit can be expanded for new or larger existing functions, the easier a building can be rearranged or transformed to other functions or expanded, the better a building can meet the changing user quality demands, the more possibilities to expand the space of the units in the building.
07. Extendible building / unit vertical Is it possible to expand the building vertically, for adding new floors (topping) or a basement?	**Assessment values vertical extension building** 1. Vertical extension is not possible at all. 2. Vertical extension of building is limited possible, only possible for a few units in the building. 3. Vertical extension of building with added floor and basement is possible at more units after total rearrangement of building. 4. Vertical extension of building with new floors and basement and individual vertical extension of the user units is rather easy without disturbing other user units (implementation of zoning-margin system and fontanel constructions in supporting floors.	**Remark** The more a building or unit can be vertically expanded with new floors or basement, the easier a building can be rearranged or transformed to other functions or expanded, the better a building can meet the changing individual user quality demands, the more possibilities to expand the space of the units in the building, including new extra internal stairs/elevators.
08. Dismountable facade To what extent can facade components be dismantled in case of transformation of the building?	**Assessment values dismountable facade in % of dismountable.** 1. Facade components are not or hardly dismountable and have to be fully demolished and removed (<20%). 2. A small part of the facade components is dismountable (between 20 en 50%). 3. A large part of the facade components is dismountable (between 50 en 90%). 4. All facade components are easily dismountable (> 90%).	**Remark** The more facade components are easily dismountable, the easier a building can be rearranged or transformed to other functions, the better a building can meet the changing individual user quality demands, the more a building can be horizontally extended.

Figure 10.10a continued

09. Customisability and controllability of facilities To what extend can facilities (heating, cooling, electricity, ICT) respond to changing functional requirements?	**Assessment values customisability and controllability of facilities** 1. Not customisable and individual controllable (mono functional of fixed use). 2. Limited customisable and individual controllable. 3. Partly customisable and individual controllable. 4. Easy customisable and individual controllable.	**Remark** The more facilities are customisable and controllable to respond to changing functional requirements, the easier a building can be rearranged or transformed to other functions, the easier the grain size can be changed, the better a building can meet to changing demands of facilities and the quality of the building or units, the easier parts of the building or units can be rejected and the easier (parts of) the building or units are expandable.
10. Surplus of facilities shafts and ducts Do the facilities shafts and ducts have a surplus of space (heating, cooling, electricity, ICT)?	**Assessment values surplus of facilities shafts and ducts** 1. Shafts and ducts have no surplus at all 2. 10-30% surplus 3. 30-50% surplus 4. Surplus of space of more than 50%	**Remark** The more surplus space facilities shafts and ducts have, the easier a building can be rearranged or transformed to other functions, the easier the grain size can be changed, the better a building can meet to changing user demands, and the easier the building or units are expandable.
11. Surplus capacity of facilities Does the capacity of (the sources of) the facilities have a surplus (heating, cooling, electricity, ICT)?	**Assessment values surplus of capacity of facilities** 1. Capacities of facilities have no surplus at all 2. 10-30% surplus 3. 30-50% surplus 4. The surplus capacities of facilities > 50%	**Remark** The more surplus capacity the facilities have, the easier a building can be rearranged or transformed to other functions, the easier the grain size can be changed, the better a building can meet to changing user demands, and the easier the building or units are expandable.
12. Disconnection of facilities components Can the components of the facilities be disconnected?	**Assessment values of disconnection of facilities** 1. Facility (parts) can not be disconnected or unmounted; 'wet' connections 2. Hardly be disconnected, unmounted 3. Partly be disconnected, unmounted 4. Facility (parts) can be disconnected very easily (completely demountable, pluggable).	**Remark** The more facility parts can be disconnected or demounted, the easier a building can be rearranged or transformed to other functions, the easier the grain size can be changed, the better a building can meet to changing user demands of facilities and the quality of the building or units, and the easier the building or units are expandable.
13. Distinction between support - infill What is the % of the application of project independent produced infill construction components (and therefore well demountable and exchangeable), with respect to the % of support construction parts is used?	**Assessment values in % of application of project independent produced infill construction components** 1. < 10% 2. 10 - 50% 3. 50 - 90% 4. > 90%	**Remark** The more construction components belong to the infill domain, the easier a building can be rearranged or transformed to other functions, the easier the grain size can be changed, the better a building can meet to changing user demands of facilities and the quality of the building or units, and the easier the building or units are expandable.

Figure 10.10b Overview of the assessment values of indicators 9 to 17 (Geraedts, 2015)

	Assessment values horizontal routing	Remark
14. Horizontal routing, corridors units In what way is the horizontal disclosure of the units in the building accomplished (for instance by internal corridors, single double, gallery)?	1. Disclosure of the building by only a single internal corridor. 2. Disclosure by a double internal corridor. 3. All disclosures directly by a central core in the building with a surrounding corridor. 4. All disclosures directly by a central core in the building, or an external gallery	The more the horizontal disclosure of the units is limited by a central core the more easily units in a building can be rearranged or transformed to other functions.
15. Removable, relocatable units in building To what extend are the user units in a building removable, relocatable?	**Assessment values removable, relocatable units** 1. Units are not removable/reusable 2. Units are removable/reusable as a whole with drastic constructional and costs consequences 3. Units are pretty removable/reusable; constructed with demountable 3D modules or components 4. Units are good removable reusable. **Remark**	The more building units are constructed with demountable and reusable, the user units can be easier relocated in the building or elsewhere.
16. Removable, relocatable interior walls in building To what extent are the interior walls easily removable, relocatable?	**Assessment values removable, relocatable inner walls** 1. Interior walls are not removable, reusable without drastic, expensive constructional interventions. 2. Interior walls are not replaceable, but good destructible. 3. Interior walls are removable/reusable by dismantle them and rebuild them at another location. 4. Interior walls are easily removable, reusable without radical, expensive constructional interventions (f.i. system walls). **Remark**	The more inner walls are easily removable/reusable, the easier a building can be rearranged or transformed to other functions, the better a building can meet the changing user demands of facilities and the quality of the building or units, and the easier parts of the building or units can be rejected and the easier the building or units are expandable.
17. Disconnecting/detailed connection interior walls; hor./vert. Which detailed construction is applied between the connection of interior walls and support structure, columns, façade, floor and ceiling?	**Assessment values detailed connection interior walls** 1. Penetrating connections 2. Wet connections (like mortar, sealant, glue). 3. Specific project bound produced connection elements 4. Project unbound dismountable connections **Remark**	The easier the connection of interior walls can be dismounted, the easier a building can be rearranged or transformed to other functions, the easier the grain size can be changed, the better a building can meet to changing user demands of facilities and the quality of the building or units, and the easier the building or units are expandable.

Perspectives

Just a few methods exist to map the costs and benefits of flexibility measures for future adaptability. One of them is called Flexcos (Geraedts, 2001). Using several scenario/strategy combinations, this method allows organisations to consider the effects of different construction strategies at an early planning stage. The relative costs of investing in future flexibility and possible future savings can be compared with each other. Further research is needed to collect empirical data about the cost-benefit effects of various adaptability interventions. The largest unknown factor in cost-quality research is estimating whether provisions made for future flexibility will actually be used in any given period, and the trade-off between the costs of flexibility measures and financial benefits. Case studies of projects in which extra measures were provided for future flexibility could offer added insights. Which measures have been used successfully and which not? What can we say about the initial costs versus the benefits? The Flex adaptive capacity method is a first important step in the development of instruments to formulate adaptive demands and to assess adaptive supplies of buildings. In the next steps this method has to be discussed and evaluated with building owners, users and construction companies. Further developments will also look into the implementation of this method for small and simple projects and large complex projects as well. The method could be specified for different sectors within construction (hospitals, schools, office buildings and residential housing). Finally, further research is necessary on the drivers behind change and the need for adaptability.

References

Amram, M. and Kulatlaka, N. (1999) *Real Options: Managing Strategic Investment in an Uncertain World*, Boston: Harvard Business School Press.

Arge, K. (2005) 'Adaptable office buildings: theory and practice', *Facilities*, 23 (3/4), pp. 119–127.

Arge, K. and Blakstad, S.H. (2010) 'Briefing for adaptability', in Blyth, A. and Worthington, J. (Eds.), *Managing the Brief for Better Design*, London and New York: Routledge.

Bahrami, H. and Evans, S. (2005) *Super-Flexibility for Knowledge Enterprises*. Berlin: Springer.

Ballard, G. (2000) 'Positive vs. negative iteration in design', *8th Annual Conference of the International Group for Lean Construction (IGLC-8)*, Brighton: University of Sussex.

Ballard, G. and Howell, G.A. (2003) 'Lean project management', *Building Research and Information*, 31 (2), pp. 119–133.

Bjørberg, S. and Verweij, M. (2009) 'Life-cycle economics: Cost, functionality and adaptability', in: Rechel, B., Wright, S., Edwards, N., Dowdeswell, B. and McKee, M. (Eds.) *Investing in hospitals of the future*, Copenhagen, WHO Regional Office for Europe.

Blakstad, S.H. (2001) *A Strategic Approach to Adaptability in Office Buildings*. PhD thesis, NTNU, Trondheim,

Blyth, A. and Worthington, J. (2010) *Managing the Brief for Better Design*, London and New York: Routledge.

Brach, M.A. (2003) *Real options in practice*, Chichester: John Wiley & Sons.

Brand, S. (1994) *How buildings learn; what happens after they're built*, New York: Viking.

Cairns, G. (2010) 'Briefing for the future', in Blyth, A. and Worthington, J. (Eds.) *Managing the Brief for Better Design*, London and New York: Routledge.

De Neufville, R., Lee, Y.S. and Scholtes, S. (2008) 'Flexibility in hospital infrastructure design', *IEEE Conference on Infrastructure Systems*, Rotterdam.

DGBC (2013) *Concept flexibility assessment module*, Rotterdam: Dutch Green Building Council.

Duffy, F. (1998) *Design for change, The Architecture of DEGW*, Basel: Birkhauser

Durmisevic, E. (2006) *Transformable Building Structures*, Delft: Delft University of Technology.

Eichholtz, P., Kok, N., Quigley, J.M. and Berkeley, C.A. (2008) *Doing Well by Doing Good? Green Office Buildings*, Berkeley Program on Housing and Urban Policy: W08.

Genus, A. (1997) 'Managing large-scale technology and inter-organisational relations: The case of the Channel Tunnel', *Research Policy*, 26 (2), pp. 169–189.

Geraedts, R. (1998) *Flexis; communicatie over en beoordeling van flexibiliteit tussen gebouwen en installaties - Communication about and assessment of flexibility between buildings and facilities*, Rotterdam: Stichting Bouwresearch.

Geraedts, R. (2001) 'Design for Change; Flexcos compares costs and benefits of flexibility', *CIB World Building Congress 2001*, Wellington: New Zealand.

Geraedts, R. (2013) *Adaptief Vermogen; brononderzoek – literatuurinventarisatie*. Delft: Centre for Process Innovation in Building & Construction.

Geraedts, R. (2015) *Flex 2.0 Light; definition and asessment of spatial/functional and construction/technical quality*. Internal document, Delft University of Technology, Faculty of Architecture, Department of Real Estate and Housing.

Geraedts, R., Remøy, H., Hermans, M. and Van Rijn, E. (2014) 'Adaptive Capacity of Buildings: a determination method to promote flexible and sustainable construction', *Proceedings of UIA2014 Durban Architecture Otherwhere*, pp 1034–1068.

Geraedts, R.P. and Van der Voordt, D.J.M. (2007) *A Tool to measure opportunities and risks of converting empty offices into dwellings*. ENHR; Rotterdam: Sustainable Urban Areas.

Gibson, V. (2000) 'Property Portfolio Dynamics: The Flexible Management of Inflexible Assets', in Nutt, B. and McLennan, P. (Eds.) *Facility Management: Risks and Opportunities*. Oxford: Blackwell Science.

Gill, N., Tommelein, I.D., Stout, A. and Garrett, T. (2005) 'Embodying product and process flexibility to cope with challenging project deliveries', *Journal of Construction Engineering and Management*, 131 (4), pp.439–448.

Groák, S. (1992) *The idea of building: thought and action in the design and production of buildings*, London: E&F Spon.

Habraken, N. (1961) *De dragers en de mensen, het einde van de massawoningbouw*, Eindhoven: Stichting Architecten research.

Habraken, N. (1998) *The Structure of the Ordinary* Boston: First MIT Press.

Hansen, G.K. and Olsson, N.O.E. (2011) 'Layered project – layered process. Lean thinking and flexible solutions', *Architectural Engineering and Design Management*, 7 (2), pp. 70–85.

Hellström, M. and Wikström, K. (2005) 'Project business concepts based on modularity – Improved maneuverability through unstable structures', *International Journal of Project Management*, 23 (5), pp. 392–397.

Hermans, M., Geraedts, R., Van Rijn, E. and Remoy, H. (2014) *Bepalingsmethode Adaptief Vermogen van gebouwen ter bevordering van flexibel bouwen*. Leidschendam: Brink Groep.

Horgen, T. (2010) 'Briefing for the changing workplace', in Blyth, A. and Worthington, J. (Eds.) *Managing the Brief for Better Design*, London and New York: Routledge.

Mahmoud-Jouini, S. and Midler, C (2004) 'Time-to-market vs. time-to-delivery. Managing speed in Engineering, Procurement and Construction projects', *International Journal of Project Management*, 22 (5), pp. 359–367.

Merriam Webster Online Dictionary (2006) http://www.merriam-webster.com/

Miller, R.L. and Swensson, E.S. (2002) *Hospital and healthcare facility design*, New York: W. Norton.

De Neufville, R., Lee, Y.S. and Scholtes, S. (2008) 'Flexibility in hospital infrastructure design', *IEEE Conference on Infrastructure Systems*, Rotterdam.

Olsson, N.O.E. (2006) 'Management of flexibility in projects', *International Journal of Project Management*, 24 (1), pp. 66–74.

Olsson, N.O.E. (2011) 'Flexibility: implications on projects and facilities management', *CFM Nordic Conference*, Copenhagen.

Pati, D., Harvey, T. and Cason, C. (2008) 'Inpatient unit flexibility design characteristics of a successful flexible unit', *Environment and Behavior*, 40 (2), pp. 205–232.

Prins, M. (1992) *Flexibiliteit en kosten in het ontwerpproces; een besluitvormingondersteunend model*. PhD thesis, Eindhoven University of Technology.

Rechel, B., Wright, S., Edwards, N., Dowdeswell, B. and McKee, M. (Eds.) (2009) *Investing in hospitals of the future*, Copenhagen: WHO Regional Office for Europe.

Remøy, H. and Van der Voordt, T. (2007) 'A new life – conversion of vacant office buildings into housing', *Facilities*, 25 (3/4), pp. 88–103.

Richard, R.B. (2010) *Four Strategies to Generate Individualised Buildings with Mass Customisation New Perspective in Industrialisation in Construction; A state of the art report*. Zürich: IBB Institut für Bauplanung und Baubetrieb: 10.

Schneider, T. and Hill, J. (2007) *Flexible Housing*, Oxford: Elsevier Architectural Press.

Schuetze, T. (2009) 'Designing Extended Lifecycles', *3rd CIB International Conference on Smart and Sustainable Built Environment*. Delft: Delft University of Technology.

Thomke, S.H. (1997). 'The role of flexibility in the development of new products: An empirical study', *Research Policy*, 26 (1), pp. 105–119.

Van der Voordt, D.J.M. and Van Wegen, H.B.R. (2000) *Architectuur en gebruikswaarde. Programmeren, ontwerpen, evalueren*, Bussum: Uitgeverij Thoth.

Van der Voordt, D.J.M. and Van Wegen, H.B.R. (2005) *Architecture in use. An introduction to the programming, design and evaluation of buildings*, Oxford: Elsevier, Architectural Press.

Van Meel, J. (2015) *Workplaces Today*, Centre for Facilities Management. Lyngby: Technical University of Denmark.

Van Reedt Dortland, M. (2013) *Cure for the future: The real options approach in corporate real estate management. An exploratory study in Dutch health care*. PhD thesis, Enschede: Twente University of Technology.

Verganti, R. (1999) 'Planned flexibility: Linking anticipation and reaction in product development projects', *Journal of Product Innovation Management*, 16 (4), pp. 363–376.

Volberda, H.W. (1997) 'Building flexible organizations for fast-moving markets', *Long Range Planning*, 30 (2), pp. 169–183.

Volberda, H.W. (2007) *De Flexibele Onderneming*, Deventer: Kuwer Bedrijfs Informatie (KBI).

Wilkinson, S.J., James, K. and Reed, R. (2009) 'Using building adaptation to deliver sustainability in Australia', *Structural Survey*, 27 (1), pp. 46–61.

Wilkinson, S. J. and Remøy, H. (2011) *Sustainability and within use office building adaptations: A comparison of Dutch and Australian practices*. Gold Coast: Pacific Rim Real Estate Society, Bond University.

Interview 6: Bart Voortman, Achmea, the Netherlands

Bart Voortman has over 16 years of experience in FM and CREM. He has an educational background in music. 37 years ago he started to work at Centraal Beheer, an insurance company. He studied business administration and followed various courses. Since 1998 he is director of FM and Document Logistics at Achmea, a care insurance company. He is responsible for all facilities and buildings in use by the company itself, but not for investing in commercial real estate. Bart Voortman has a special interest in user experiences.

Added value in general

Q. What does the term "Added Value" mean to you?

A. *I use the term both regarding added value of our FM/CREM division on strategic level and regarding added value of FM/CREM interventions. The added value of FM/CREM is its contribution to cope with customers' interests i.e. the people that pay care premium. The idea is to create good buildings that support effective and efficient working. This benefits the customers and helps to deliver acceptable products for lowest costs. Acceptable is related to quality. The desired level of quality - e.g. of cleaning - fluctuates with the economic context and the market. Internal clients are called business partners. Achmea wants to be the most reliable and trusted care insurance company. This is expressed by openness and transparency. We try to use facilities such as IT to support efficiency and speed of responses to customers.*

Q. Are there other related terms that you prefer to use rather than "Added value"?

A. *Not really*

Q. How do you see the relation between "Added value" and cost reduction?

A. *Both are related: AV is about improvements in connection to costs.*

Q. Is "Added value" mostly treated on strategic, tactical or operational level?

A. *It is a concept on strategic level, i.e. steering on optimal performance for lowest costs, with clear procedures, appointments and arrangements, measuring of outcomes, and a focus on clear decisions instead of too lengthy discussions. On operational level talking about added value will lead to nothing. On this level FM/CREM is a dissatisfier. You have to deliver good services in an efficient way, but don't expect high added value of basic services on customer satisfaction.*

Q. In which context or dialogue is "Added value" mostly being considered from your experience?

A. *Not discussed.*

Q. How do you see the relation between innovation and Added Value?
A. *Not discussed.*

Benefits and limitations of "Added Value"

Q. What are the benefits of considering and talking about "Added value"?
A. Added value *makes you focus on the strategic aspects of FM and the clients', / customers' and end users' needs. It helps to speak the language that top management understands and gives a more constructive dialogue than focus on cost. It supports clear dialogues about each responsibilities and the benefits and costs of interventions. We like to start a dialogue with quality and not with costs in order to avoid too much focus much on costs which might result in lower quality and in the end higher costs.*

Q. What are the limitations of considering and talking about "Added value"?
A. *Added value is perceived differently by different people. Talking about added value on operational level will result in endless discussions without concrete results.*

Management of added value

Q. What are your top five of main values to be included in management of accommodations, facilities and services? Could you mention a few examples of concrete FM interventions to attain these added values and KPIs to measure them?
A. *See table below.*

Prioritized values	Concrete interventions	KPIs
1. Costs effectiveness		€ per m², per person, per service or activity; costs are partly charged to divisions. Benchmarking by National Facility Cost (NFC) index + dialogue in Platform 8 (sessions with people from 8 other firms).
2. Business alignment		NPS survey and MWO survey (about how customers experience end products and services) + management by walking around (observation, talking, getting a sense of what's going on).
3. Professionalism of FM/CREM staff	Education and training of cognitive skills, professional knowledge and skills, and leadership skills.	

Prioritized values	Concrete interventions	KPIs
4. Digitization/ logistics of document management	SENS (Samen Effectief Naar Succes - together working on success in an efficient way) using LEAN Six Sigma principles.	Based on analysis of whole supply chain.
5. Positive experience by customers	Managing of optimal service, e.g. if a client does travel but does not have a travel insurance they send a mail to remind them.	

The FM/CREM division follows a philosophy of "direction": make clear what you want and let providers decide how to cope with your requirements. We apply a matrix with seven main activities in connection to ownership per activity and each year we present a plan of approach with GO/NO GO moments and priorities. Not all you want can be attained. We try to formulate the goals SMART (specific, measurable, acceptable, realistic, and time specific i.e. including a clear planning when the target should be reached). We always search for the most efficient measures. An example of improving the supply chain is to replace current services by digital procedures and internet, which might reduce the costs with a factor 10 and support high speed responses. Regarding sustainability Achmea is no frontrunner but we do what we can; our sustainability level between basic and basic+. We also pay attention to architectural quality. For instance our location in Apeldoorn is well integrated in green area. However, its appraisal is very subjective. We try to operate CSR based and take care of where (potential) staff is living. We co-operate with educational organisations to get a win-win situation.

Q. Do you measure if you attain the aimed added values? If so, what Key Performance Indicators do you apply for each of the added values you mentioned before?

A. *Yes, see the table above. We also measure customer satisfaction and employee satisfaction, including satisfaction of the CEO. Productivity is being measured by SENS/LEAN; for instance we measure the time spent on particular service provision. All staff members have to record their time spent per main activity. Measuring of flexibility and adaptability is work in progress.*

Achmea uses both KPIs (Key performance Indicators) and KRIs (Key Risk Indicators), provided they can steer on it, and KOIs (Kritische Gedragsindicatoren / critical behaviour indicators). We pay much attention to compliance, and results on targets.

Q. Do you benchmark your data with data from other organisations?

A. *Yes, see table above. For instance we compare our cost ratios with average ratios and the range and discuss possible reasons for discrepancies.*

Q. What other methods do you use to document "Added value"?
A. *Not discussed.*

Last comment

Q. Are there other topics that you find important in connection to the concept of added value?
A. *Avoid to use added value as a defence mechanism. Only include this concept in dialogues when appropriate and on the right level. Adding value management of FM/CREM should also include HRM, IT and affecting customer delight.*

   An interesting question is how to cope with the greying staff? How to use FM/CREM and HRM to attract young talents ('YAP: young angry potentials'). We ourselves use traineeships. Another topic for further research is what happens with research reports and other reports (obliged by law or on companies' own initiative).

11 Innovation and creativity

Rianne Appel-Meulenbroek and Giulia Nardelli

Introduction

In today's hectic economies more than ever, innovation is not only recommended but also required for the survival and growth of organisations. A survey conducted by Corenet among 271 corporate real estate and facility managers worldwide (Gibler et al., 2010) asked respondents to rank seven possible real estate strategies on their importance. After reducing costs and increasing productivity and flexibility (mostly measured through efficiency/financial metrics), encouraging innovation of the client organisations was indicated as the most important real estate strategy to add value by improving the outcomes of the end user organisation.

In FM and CREM, various researchers have shown how innovation plays a significant role in adding value to the core business of organisations (e.g. Jensen et al., 2012). On the one hand, dedicated FM and CREM practices regarding workplace management, aimed at increasing knowledge sharing among employees and facilitating their creativity, were shown to contribute to the added value of FM and CREM by sustaining innovation across all layers of the served organisations (e.g. Appel-Meulenbroek, 2014; Kastelein, 2014; Martens, 2011; Dul and Ceylan, 2011). On the other hand, scholars outlined that innovation *of* the FM and CREM processes and/or services appears to contribute to its added value for the client/end users, as it increases the effectiveness and efficiency of FM and CREM practices (e.g. Goyal and Pitt, 2007; Jensen et al., 2012; Noor and Pitt, 2009a). The latter case describes the role of innovation as input and as a way to manage adding value (throughput) and, to avoid complexity, will in this chapter be termed innovation *of* FM and CREM practices. In the first case, the emphasis is on innovation and creativity of the organisation as a whole being *supported by* FM and CREM practices, i.e., on innovation as outcome of adding value through FM and CREM activities. The focus of this chapter is on supporting innovation of the client organisation, as one of the added values of CREM/FM discussed in this part of the book.

The term innovation has over the years become a buzzword, whose meaning can be interpreted in various ways. Innovation is here defined following Anderson et al. (2014, p.1298): *"Creativity and innovation at work are the process, outcomes and products of attempts to develop and introduce new and improved ways of doing things. The creativity stage of this process refers to idea generation, and innovation refers to the subsequent stage of implementing ideas toward better procedures, practices, or products. Creativity and innovation can occur at the level of the individual, work team, organization, or at more than one of these levels combined, but will invariably result in identifiable benefits at one or more of these levels of analysis."*

As Mozaffar et al. (2013) mentioned, innovation and creativity are results of processes that take place either within a person's mind or in interaction with others. Both can be influenced by CREM/FM, and are related to each other. So, the innovation process in client organisations needs the support of two main sub-processes: interaction and creativity (Oseland et al., 2011).

The chapter will be structured as follows. Having outlined the scope and purpose of the chapter in the introduction, a section dedicated to the state of the art will critically review existing theories and relevant literature on innovation as an added value of FM and CREM. The following section will present the core benefits and costs of typical interventions aimed at facilitating innovation and creativity *supported by* FM and CREM practices and interventions. We will then provide a list of KPIs associated with innovation and creativity supported by FM and CREM, and the key practices to manage and measure related benefits and sacrifices. This section first provides KPIs for innovation *supported by* CREM and FM and methodologies to measure them. Then, this section also presents the state of the art on innovation *of* FM and CREM practice itself, and discusses its potential as support for innovation of the client organisation, too. Finally, we will reflect upon the need for new knowledge and the research potential of topics related to innovation and added value of and supported by FM and CREM, and then identify an agenda for future research.

State of the art

The individual's creativity is generally defined as the production of novel and useful ideas. In addition, "a creativity-supporting work environment may not only increase the number of ideas, but also the quality of the ideas" (Dul and Ceylan, 2014, p.1256). A state of flow and the peace to stay in this flow are important (Martens, 2011), and are reached more easily outside the office while travelling or in the shower (Oseland et al., 2011). Nevertheless, such a state is achievable at the office too, for instance in the more private spaces and meeting rooms that support brainstorming and social interaction.

Besides creativity, knowledge sharing is also crucial for innovation, as it can ensure that these ideas are adopted, evaluated and diffused in the organisation. Also, new ideas might arise from these interactions between employees. In particular, the physical work environment can increase unplanned knowledge meetings between employees, which are typically frequent and often very short (Kastelein, 2014; Appel-Meulenbroek, 2014). In fact, communication and (social) networks have been shown to be a mediator between innovation and places (Sailer, 2011; Toker and Gray, 2008; Wineman et al., 2008). CREM/FM might therefore facilitate both creativity and knowledge sharing interactions to stimulate organisational innovation.

Several studies on layout aspects have looked at *how* to support the processes associated with knowledge sharing. First of all, the individual workplace layout is relevant, because a desk facing away from the entrance of a room decreases interaction with colleagues (Hatch, 1987). The floor layout and depth/width is also relevant, because most interactions are limited to colleagues sitting within twenty to thirty metres apart (Allen, 1977), with the majority of interactions taking place between colleagues seated within eight meters (Appel-Meulenbroek, 2014). Accessibility on the floor increases awareness of other people's need for help (Nonaka and Konno, 1998). Central spaces show more unplanned interactions with passers-by (Wineman et al., 2008). Oseland et al. (2011) specifically studied the design of meeting areas and mentioned their proximity, accessibility, privacy, legitimacy and functionality as relevant parameters. Interaction between floors, on the other hand, is limited (Allen and Henn, 2007; Becker, 2004) and influenced by vertical circulation. Finally, Slangen-De Kort (2001) mentions that the aesthetics of the work environment also stimulate people to help each other and find each other useful.

The innovativeness of organisations is also increased by knowledge sharing outside the home organisation, which leads to smaller risks of failure, a more effective learning process (Rogers, 2004) and access to complementary know-how (Nieminen and Kaukonen, 2001). A distinction is often made between horizontal relationships (with other companies, universities and public institutions) and vertical relationships within an organisation with end users, customers and clients (Solleiro and Castañón, 2005). Therefore, the proximity of suppliers, collaboration parties and related organisational departments seems relevant when choosing the right location. In addition, Becker et al. (2013) add facilities on campus, e.g., cafeterias and fitness centres, as contributors to inter-organisational interaction as well.

Dul and Ceylan (2006) wrote a first review on the few studies that existed at that time on the effect of the physical environment on creativity. The relevant building aspects they found are window view and sunlight penetration, plants and colours (for instance, green encourages creativity according to Lichtenfeld et al. (2012)), installations and control with regard to lighting and noise. In another paper, Ceylan et al. (2006) describe a case study among line managers, who rated pictures of work environments with the same aspects as the more creative ones. Additional aspects found relevant for stimulating creativity were materials used, furniture, and complexity of the individual workplace layout. Later on, Dul and Ceylan (2014) also added positive sounds (e.g. music), positive odours (e.g. fresh air) and further diversity of colours (relaxing versus stimulating) and views (nature versus other). Van Dijk (2014) showed that in practice, the focus of creativity supporting buildings changed from layout and routing in the early years to modern furniture and crazy objects more recently. From analysing forty creative buildings worldwide, he additionally identified aspects like openness within the building, the provision of inviting settings for meetings, and a nurturing environment. Several studies (e.g. Sailer, 2011; Martens, 2011) point to the four phases of creative processes, where the first (preparation) and the last (verification) phase need both communal and private spaces, while the middle two (incubation and illumination) should be more embedded in private spaces.

However, all these causal relations that CREM/FM could work with "should not be considered absolute but serve a pragmatic purpose to gain further insight in socially constructed reality" (Martens, 2011, p.64). In each organisation differences may emerge with regard to which causal relations are present and which are not. Previous studies have shown examples of how the context may influence the potential of CREM/FM in supporting innovation in the client organisation. For example, Sailer and Penn (2009) studied the effect of proximity on frequency of interaction and showed that each organisation had a unique behaviour pattern as a result of the workplace environment. Blakstad et al. (2009) even showed a different effect of similar office space within the same organisation. As Koch and Steen (2012) put it, the organisation itself and the space interact to form a certain spatial practice. Therefore it is necessary for CREM/FM to measure the spatial practice inside their client organisation. In that way the workings of different spatial mechanisms can be identified to select the most beneficial interventions.

With regard to knowledge sharing behaviour, six different mechanisms can be identified: visibility, placement in the room, exposure, centrality, proximity, and meeting areas (Appel-Meulenbroek, 2014). At the organisation studied by Appel-Meulenbroek, the most effective mechanism for stimulating knowledge sharing was visibility and placement in the room. She found that people on the outskirts of large rooms had more opportunity to share knowledge through evaluating together, while centrally located participants shared knowledge relatively more often through giving descriptions or making proposals. However, at another organisation, studied by Kastelein (2014), visibility and centrality had a minor role in generating additional interactions compared to close proximity. According to the employees in that study, common areas and facilities were the best facilitators of knowledge sharing, which indicates that context determines how to add value.

Sailer (2011) extracted a long list of mechanisms from a literature study, mentioning accessibility, density, proximity, privacy, layout, design (materials) and visual cues. Martens (2011) adds diversity, flexibility, visibility and the indoor climate (lighting and temperature). Mozaffar et al. (2013) actually tested several mechanisms and found a significant effect of three of them: privacy, beauty and spatial diversity/flexibility, with privacy being the strongest one and proximity/visibility not relevant. Context probably also has an impact here. Figure 11.1 provides an overview of mechanisms that might stimulate innovation in the client organisation. In reality, these mechanisms are also highly likely to trigger/inhibit each other.

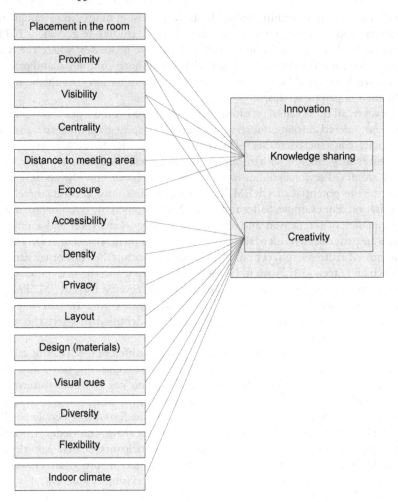

Figure 11.1 Mechanisms for stimulating innovation

Benefits and costs

Many interventions can be implemented to increase the mechanisms as described above; see Tables 11.1a and 11.1b for suggestions. Not all types of interventions have been the subject of research. Most studies have focused on changing the physical environment or the facility services. A third category of interventions in Table 11.1b deals with changing the interface with core business. Innovation is not achieved individually but as a process between different stakeholders, which requires knowledge and monitoring of each other's needs and thus trust and transparency.

Some mechanisms increasing knowledge sharing might decrease the creativity process, although they do support the interactive aspect of creative thinking.

Table 11.1a Benefits and sacrifices of FM/CREM intervention

Interventions	Management	Benefits	Sacrifices
Changing the physical environment			
Increase visibility and overhearing among employees and consider placement in the room	Evaluate the current layouts, looking at possibilities for: - Larger rooms - Fewer walls/obstacles - More compact layout types - Higher density Implement non-dedicated seating	More knowledge sharing and improvement of the interactive phases of creative thinking	Less visual and auditory privacy decreases 'flow' in individual phases of creative thinking
Provide opportunities for privacy	Evaluate the current layouts, looking at possibilities for: - Higher level of enclosure and less density	Promotes tranquillity and eases 'flow' in individual phases of creative thinking	Less visibility and exposure, so less knowledge sharing and less support of the interactive phases of creative thinking
Maximise exposure	Evaluate the current layouts, looking at possibilities for: - More open floor layouts - Enlarging entrances to rooms - 'Open door' policy	More knowledge sharing and improvement of the interactive phases of creative thinking	Less visual and auditory privacy decreases 'flow' in individual phases of creative thinking
Choose buildings with higher centrality + proximity for all employees	During site selection processes look for: - More compact building layouts ('square floors') - Limited number of floors	More knowledge sharing and improvement of the interactive phases of creative thinking	Fewer windows/views outside to inspire
Provide visual cues	Enhance the office décor by: - Expressing the organisation's creative identity - Adding design (plants, colours, furniture etc.)	Promotes tranquillity and a cooperative culture in the office	Additional costs

Table 11.1a continued

Interventions	Management	Benefits	Sacrifices
Maximise flexibility/diversity	Re-evaluate the office layouts on availability of different types of workspaces	More support of personal preferences for creative thinking and also promotes tranquillity	Possible additional costs for extra types of workspace and m^2
Optimise indoor climate	Improve the climate systems with: – Lighting with personal control – Temperature with personal control – Positive sounds and odours	More support of personal preferences for creative thinking and also promotes tranquillity	Additional costs
Reconsider the layout concept	Implement new ways of working/teleworking	More focus for creativity when able to withdraw and work at home	Danger of isolating employees and thus decreasing knowledge sharing

Table 11.1b Benefits and sacrifices of FM/CREM intervention

Interventions	Management	Benefits	Sacrifices
Changing the facilities services			
Provide different types of meeting areas	Manage meeting areas and their booking systems	More knowledge sharing and improvement of the interactive phases of creative thinking	Additional costs of these spaces and m^2
Provide informal areas (office leisure)	Re-evaluate the setup and location of restaurant, refreshments and communal areas	More knowledge sharing and improvement of the interactive phases of creative thinking	Additional costs of these facilities services
ICT support	Align ICT to the demand for virtual knowledge sharing	More knowledge sharing and improvement of the interactive phases of creative thinking	Additional costs of these facilities services
Changing the interface with core business			
Identify stakeholders	Map needs and expectations of stakeholders and involvement of those who can contribute to specific phases of the innovation process	May increase the rate of success of innovation process, as it allows customisation of innovation outcome to the specific needs and expectations of demand	Requires extended knowledge and continuous monitoring of needs and expectations of the stakeholders
Cooperate with stakeholders	Establish cooperative relationship with stakeholders of innovation process Track development of needs and expectations of stakeholders over time	May facilitate the management of the innovation process across different layers of the organisation and across the boundaries of the organisation as it ensures favourable relationship management	Requires transparency concerning needs, expectations and resources of each party, as well as trust between the involved individuals

This is the paradox of verbal communication, which up to date has not been solved in design studies or in practice. From this overview it becomes clear that this paradox exists both for visual matters (exposure versus privacy/feeling at ease) and auditory privacy (overhearing versus not wanting to be overheard). Although specific cost data are not yet available from the studies reviewed, it seems that additional costs are the main sacrifice to support the individual phases of creative thinking, in which 'flow' is necessary to come up with ideas and more knowledge sharing through better facilities services. These costs should be weighed against the benefits in the primary processes from being a more innovative company.

How to measure and manage

How to measure innovation supported by FM and CREM

Based on the interventions that are presented in Tables 11.1a and 11.1b, the following top five KPIs can be identified to measure the extent to which CREM/FM supports innovation among the end users of buildings:

1   Level of enclosure/openness of layout
2   Average walking distance between employees
3   Level of personal control of indoor climate
4   Diversity of available workspaces and meeting areas
5   Perceived quality of visual cues.

Other relevant KPIs are:

- Average distance to and visibility from room entrance of employees
- Number of employees/room (or number of visible workplaces/employee)
- m²/workplace
- Number of floors in the building
- Perceived quality of informal areas
- Frequency of meetings between FM/CREM and end user representatives.

The top five KPIs are all physical interventions that require evaluation of layouts, interior design and personal control. Spatial network analysis can be used to measure KPIs 1 and 2 quantitatively. See, for instance Appel-Meulenbroek (2014), who used this methodology combined with logbooks on knowledge sharing activities. The related perception can also be obtained directly from employees through employee surveys; see for instance Kastelein (2014), who combined spatial network analysis with employee perception surveys and a social network analysis, or Dul and Ceylan (2011), who used employee surveys to measure both creativity outcome and the physical work environment. KPIs 3 to 5 should be obtained directly from employees as well, as these are intended to support personal preferences.

To be absolutely sure that the level of innovation of the client organisation has risen due to interventions by CREM/FM, it would be necessary to also measure KPIs for the innovation outcome controlled by all other (not CREM/FM related) variables that influence innovation. Relevant methodologies are, among others, social network analysis, observation of innovative behaviour, and surveys on perceived innovation outcome, besides quantitative aspects of actual innovations of processes and products (like speed, costs etc.). As long lists of aspects influencing innovation have been identified (e.g. Dul and Ceylan, 2014), it seems an impossible task for CREM/FM to steer on these outcome KPIs, as many others will be influencing them simultaneously. More important is to cooperate with the end users in supporting their work processes in the best way by driving the input KPIs, because how to manage and measure the KPIs mentioned above depends on the focus of the organisation. As the tables on interventions showed, the benefits and sacrifices of increasing knowledge sharing versus creative individual processes seem to demand a precarious balance between privacy and exposure/overhearing. Also, interventions need to be weighed according to their additional costs versus the benefits.

How to manage innovation of CREM/FM processes and services

Besides managing the outcome of the KPIs mentioned, it is important to recognise the value of innovation *of* FM and CREM as well, given its potential as a driver for corporate innovation and consequently value-creation (Noor and Pitt, 2009a). Although the debate around innovation is still energetic, existing research has confirmed the value adding role played by innovation of FM and CREM practices, among other factors (Jensen, 2010; Scupola, 2012; De Vries et al., 2008; Jensen et al., 2012; Pitt et al., 2006; Lindkvist and Elmualim, 2010).

In practice, CREM/FM organisations typically manage their own innovation as a process and tend to have several projects being developed at the same time (Cardellino and Finch, 2006; Mudrak et al., 2005). However, they tend to lack the ability to establish progressive innovative routines that would enable a successful innovation management and thus better support of innovation of the client organisation. Also, FM innovation showed itself as being demand-led and less proactive within inhouse FM (Cardellino and Finch, 2006; Mudrak et al., 2005), whereas the current trend in the worldwide economy requests more radical approaches (Jensen, 2008). Encouragement of innovation culture within the CREM/FM organisation may increase creativity and knowledge sharing and therefore the number of generated ideas, as well as the amount of successful innovation outcomes in the CREM/FM organisation itself. More specifically, exploration, risk-taking and cooperation between several actors, as well as the responsibility of CRE and facilities managers to act as innovation managers, are crucial elements of FM innovation in a market that increasingly requires constant upgrades to meet growing expectations, needs and competitiveness (Cardellino and Finch, 2006; Goyal and Pitt, 2007; Mudrak et al., 2005; Moss, 2010; Nardelli, 2014). For instance, risk analysis and risk allocation across involved parties may reduce the risk of unsuccessful innovation and/or financial losses from the innovation process (Goyal and Pitt, 2007).

The need to establish cooperation between the different stakeholders in FM innovation is due to the complexity that characterises the management of interactions in FM. Jensen (2008), for instance, states the importance of the outsourced FM team being be given sufficient freedom to plan and implement activities aimed at optimising its productivity and use of competences. Noor and Pitt (2009b) similarly outline that innovative partnership approaches are crucial to bridge the demand and supply (inhouse or outsourced) of CREM/FM delivery by building an innovation network within the actors involved. Systematic measurement of innovation performance may increase transparency and facilitate communication between parties because it makes it easier to compare the invested resources with the obtained results (Goyal and Pitt, 2007; Nardelli, 2014).

Also, innovation management of CREM practices should pay dedicated attention to risk analysis and risk allocation, such as the drive for fast project completion, potential project cost escalation, encouragement for creativity and knowledge sharing aimed at innovation in project development, and appropriate management of maintenance costs (Akintoye et al., 2003). Standardisation of processes may reduce the complexity of innovation management of FM/CREM because it allows the clear mapping of the tasks and responsibilities of all involved parties, and their allocation over a pre-set sequence of phases and decision-making steps. On the other hand, this may reduce creativity because it constrains free thinking and increases the risk of low acceptance of failure (Goyal and Pitt, 2007). Identification of and support for FM innovation champions may raise awareness of the need to innovate, as well as support experimentation with new service concepts, processes and technologies (Goyal and Pitt, 2007; Leiringer and Cardellino, 2008).

In short, in FM and CREM, just as in other contexts, innovation needs to be approached as a mindset, rather than as a one-time event. The daily schedule of CREM and FM employees at all levels should incorporate innovation management principles, such as flexibility, open mindedness and acceptance of failure (Goyal and Pitt, 2007). The encouragement and management of innovation of FM and CREM practices is essential to maximise the added value for the client organisation (Pitt, 2008).

Perspectives

Although empirical research on the added value of CREM/FM is still scarce (Kastelein, 2014; Sailer and McCulloh, 2012; Steen and Markhede, 2010; Toker and Gray, 2008), this chapter has shown that many interventions are available for CREM/FM to increase their support of innovative processes in the client organisation. As there is little accordance among scholars who have studied the subject on exactly how the CRE/facilities-'organisation outcome' relationship exists (Sailer and Penn, 2009) and inconsistencies have been found (Koch and

Steen, 2012), the field would benefit from further research. Future study should focus on proving the effect of the CREM/FM interventions quantitatively in different contexts, and expressing this in money. Also it is necessary to identify relative benefits both within the CREM/FM range and as compared to the social work environment provided through other support functions. A recent study of creativity supporting work environments in practice (Dul and Ceylan, 2014) showed that the social work environment elements known to support creativity are present in many more cases than the known physical work environment elements that could also provide support. The physical aspects that were present in the work environments they studied focus on furniture and indoor climate, overlooking more tacit aspects like colours, nature, sound and smell. Therefore, these interventions especially need further effort and research. Also, studies on the balance between interaction and privacy at the office are still very much needed to be better able to add maximum value for the knowledge workers today and in the future.

Concerning the management of innovation of CREM and FM, future research should investigate the potential of open innovation (e.g. Chesbrough, 2003) and value co-creation (e.g. Vargo and Lusch, 2007) strategies and practices, as well as determine potential measures to evaluate the actual impact of CREM/FM innovation on the overall performance, productivity and innovativeness of the client organisation. Moreover, the potential of having CRE and FM managers act as innovation managers should be explored and measured, as it might offer a fruitful opportunity to add value to the core organisation through the innovation of CREM and FM processes, practices and services.

Overall, further research should uncover the costs and benefits of managing innovation in CREM and FM in relation to other added value parameters, and propose appropriate KPIs for the measurements of the related outcomes.

References

Akintoye, A., Hardcastle, C., Beck, M., Chinyio, E. and Asenova, D. (2003) 'Achieving best value in private finance initiative project procurement', *Construction Management and Economics*, 21 (5), pp. 461–470.

Allen, T.J. (1977) *Managing the Flow of Technology: Technology Transfer and the Dissemination of Technological Information Within the R&D Organization*, Cambridge, MA: The MIT Press.

Allen, T.J. and Henn, G.W. (2007) *The organization and architecture of innovation: Managing the flow of technology*, Burlington, MA: Butterworth-Heinemann.

Anderson, N., Potocnik, K. and Zhou, J. (2014) 'Innovation and creativity in organizations: A state-of-the-science review, prospective commentary, and guiding framework', *Journal of Management*, 40 (5), pp. 1297–1333.

Appel-Meulenbroek, R. (2014) *How to measure added value of CRE and building design: Knowledge sharing in research buildings*. PhD thesis. Eindhoven University of Technology.

Becker, F. (2004) *Offices at Work: Uncommon Workspace Strategies that Add Value and Improve Performance*. San Francisco, CA: Jossey-Bass.

Becker, F., Sims, W. and Schoss, J.H. (2013) 'Interaction, identity and collocation: What value is a corporate campus?', *Journal of Corporate Real Estate*, 5 (4), pp. 344–365.

Blakstad, S.H., Hatling, M. and Bygdås, A.L. (2009) 'The knowledge workplace – Searching for data on use of open plan offices', *Proceedings of the EFMC 2009 Research Symposium*, Amsterdam, 16–17 June. Amsterdam, the Netherlands: EuroFM.

Cardellino, P. and Finch, E. (2006) 'Evidence of systematic approaches to innovation in facilities management', *Journal of Facilities Management*, 4 (3), pp. 150–166.

Ceylan, C., Dul, J. and Aytac, S. (2006) 'Empirical evidence of the relationship between the physical work environment and creativity', *Proceedings of the 16th World Congress on Ergonomics*, Maastricht, 10–14 July, the Netherlands.

Chesbrough, H.W. (2003) *Open innovation: the new imperative for creating and profiting from technology*, Boston, MA: Harvard Business School Press.

De Vries, J.C., De Jonge, H. and Van der Voordt, T.J.M. (2008) 'Impact of real estate interventions on organisational performance', *Journal of Corporate Real Estate*, 10 (3), pp. 208–223.

Dul, J. and Ceylan, C. (2006) 'Enhancing organizational creativity from an ergonomics perspective: The Creativity Development model', *Proceedings of the 16th World Congress on Ergonomics*, Maastricht, 10–14 July, the Netherlands.

Dul, J. and Ceylan, C. (2011) 'Work environments for employee creativity', *Ergonomics*, 54 (1), pp. 12–20.

Dul, J. and Ceylan, C. (2014) 'The impact of a creativity-supporting work environment on a firm's product innovation performance', *Journal of Product Innovation Management*, 31 (6), pp. 1254–1267.

Gibler, K., Lindholm, A.-L. and Anderson, M. (2010) *Corporate Real Estate Strategy and office occupiers' preferences*, Atlanta, GA: Corenet Global.

Goyal, S. and Pitt, M. (2007) 'Determining the role of innovation management in facilities management', *Facilities*, 25 (1/2), pp. 48–60.

Hatch, M. (1987) 'Physical barriers, task characteristics, and interaction activity in research and development firms', *Administrative Science Quarterly*, 32 (3), pp. 387–399.

Jensen, P.A. (2008) *Facilities management for students and practitioners*, Technological University of Denmark, Centre for Facilities Management.

Jensen, P.A. (2010) 'The Facilities Management Value Map: a conceptual framework', *Facilities*, 28 (3/4), pp. 175–188.

Jensen, P.A., Van der Voordt, T., Coenen, C., Von Felten, D., Lindholm, A., Balslev Nielsen, S., Riratanaphong, C. and Pfenninger, M. (2012) 'In search for the added value of FM: What we know and what we need to learn', *Facilities*, 30 (5/6), pp. 199–217.

Kastelein, J.-P. (2014) *Space meets knowledge*, PhD thesis. Nyenrode Business University.

Koch, D. and Steen, J. (2012) 'Analysis of strongly programmed workplace environments: Architectural configuration and time-space properties of hospital work', *Proceedings of the 8th Space Syntax Symposium*. Santiago de Chile, 3–6 January, Chile, pp. 8146:1–16

Leiringer, R. and Cardellino, P. (2008) 'Tales of the expected: Investigating the rhetorical strategies of innovation champions', *Construction Management and Economics*, 26 (10), pp. 1043–1054.

Lichtenfeld, S., Elliot, A.J., Maier, M.A. and Pekrun, R. (2012) 'Fertile green: Green facilitates creative performance', *Personality and Social Psychology Bulletin*, 38 (6), pp. 784–797.

Lindkvist, C. and Elmualim, A. (2010) 'Innovation in facilities management: from trajectories to ownership', *Facilities*, 28 (9/10), pp. 405–415.

Martens, Y. (2011) 'Creative workplace: Instrumental and symbolic support for creativity', *Facilities*, 29 (1/2), pp. 63–79.

Moss, Q.Z. (2010) 'Facilities Manager = Innovation Manager?', *FM in the Experience Economy : CIB W070 International Conference in Facilities Management*, São Paulo, 13–15 September, Brazil: University of São Paulo, pp. 89–96.

Mozaffar, F., Hosseini, S.B. and Bisadi, M. (2013) 'Promotion of Researchers' Creativity and Innovation in an Architecture and Urban Design Research Center with Effective Spatial Aspects of Offices', *International Journal of Architectural Engineering and Urban Planning*, 23 (1), pp. 34–40.

Mudrak, T., Van Wagenberg, A. and Wubben, E. (2005) 'Innovation process and innovativeness of facility management organizations', *Facilities*, 23 (3/4), pp. 103–118.

Nardelli, G. (2014) *Innovation in services and stakeholder interactions: Cases from Facilities Management*, PhD thesis, Roskilde University.

Nieminen, M. and Kaukonen, E. (2001) *Universities and R&D networking in a knowledge-based economy: A glance at Finnish developments*, Helsinki: Sitra Reports series 11.

Nonaka, I. and Konno, N. (1998) 'The concept of "Ba": Building a foundation for knowledge creation', *California Management Review*, 40 (3), pp. 40–54.

Noor, M.N.M. and Pitt, M. (2009a) 'A critical review on innovation in facilities management service delivery', *Facilities*, 27(5/6), pp. 211–228.

Noor, M.N.M. and Pitt, M. (2009b) 'The application of supply chain management and collaborative innovation in the delivery of facilities management services', *Journal of Facilities Management*, 7 (4), pp. 283–297.

Oseland, N., Marmot, A., Swaffer, F. and Ceneda, S. (2011) 'Environments for successful interaction', *Facilities*, 29 (1/2), pp. 50–62.

Pitt, M. (2008) 'Innovation resuscitation and management in FM and operations management', *Journal of Retail and Leisure Property*, 7 (3), pp. 163–166.

Pitt, M., Goyal, S. and Sapri, M. (2006) 'Innovation in facilities maintenance management', *Building Services Engineering Research and Technology*, 27 (2), pp. 153–164.

Rogers, P.A. (2004) 'Performance matters: How the high performance business unit leverages facilities management effectiveness', *Journal of Facilities Management*, 2 (4), pp. 371–381.

Sailer, K. (2011) 'Creativity as social and spatial process', *Facilities*, 29 (1/2), pp. 6–18.

Sailer, K. and McCulloh, I. (2012) 'Social networks and spatial configuration – How office layouts drive social interaction', *Social Networks*, 34 (1), pp. 47–58.

Sailer, K. and Penn, A. (2009) 'Spatiality and transpatiality in workplace environments', *7th International Space Syntax Symposium*, Stockholm, 8–11 June, Sweden, TRITA-ARK, pp. 1–11.

Scupola, A. (2012) 'Managerial perception of service innovation in facility management organizations', *Journal of Facilities Management*, 10 (3), pp. 198–211.

Slangen-De Kort, Y.A.W. (2001) 'Imago van gebouwen – effecten op gebruikers', *De ontwerpmanager: vaktijdschrift voor ontwerpmanagement in de bouw*, 2 (1), pp. 4–7.

Solleiro, J. and Castañón, R. (2005) 'Competitiveness and innovation systems: The challenges for Mexico's insertion in the global context', *Technovation* 25 (9), pp. 1059–1070.

Steen, J. and Markhede, H. (2010) 'Spatial and social configurations in offices', *Journal of Space Syntax*, 1 (1), pp. 121–132.

Toker, U. and Gray, D. (2008) 'Innovation spaces: Workspace planning and innovation in US university research centers', *Research Policy*, 37 (2), pp. 309–329.

Van Dijk, J. (2014) *Creating creativity: A study into architectural means to stimulate the creative mind and enhance innovation*, Master's thesis, Delft University of Technology.

Vargo, S.L. and Lusch, R.F. (2007) 'Service-dominant logic: Continuing the evolution', *Journal of the Academy of Marketing Science*, 36 (1), 1–10.

Wineman, J.D., Kabo, F.W. and Davis, G.F. (2008) 'Spatial and Social Networks in Organizational Innovation', *Environment and Behavior*, 41 (3), pp. 427–442.

Interview 7: Jakob Moltsen, Sony Mobile, Sweden

Jakob Moltsen has ten years of experience in FM. His educational background is an MSc in Engineering with specialisation in Planning and Management. At the time of the interview he was global head of FM at Sony Mobile Communications based at the headquarters in Lund, Sweden. Before then he first worked for the international FM provider Johnson Controls, and after that in the FM department of Nordea – a large Scandinavian bank. He is now global head of FM at NKT, a Danish-based multinational technology company.

Added value in general

Q. What does the term "Added Value" mean to you?

A. *I use "Added Value" or similar terms like "Adding Value" or "Value Proposition" in my daily work. To explain it, I normally just give our vision of creating more value, including the value of having a global FM unit and the value of having an Integrated FM contract (I-FM).*

Q. How do you see the relation between "Added Value" and cost reduction?

A *It is part of the same. I prefer to talk about "cost optimisation" rather than "cost reduction", which has a negative feel. We also talk about savings and cost avoidance. Cost can never stand alone. All idiots can cut cost. We need to achieve increased user satisfaction and experience at the same time.*

Q. Is "Added Value" mostly treated at a strategic, tactical or operational level?

A. *We are a strategic function and have outsourced all tactical and operational functions, so we treat added value at the strategic level and our I-FM provider is responsible for the operational and tactical levels. We take the decisions at the strategic level together with our Governance Board. Our provider takes the daily decisions. In Sony, FM is part of Workplace Solutions together with Real Estate, Meeting and Travel, and Telecom. We are eight people working with FM globally.*

Q. In which context or dialogue is "Added Value" mostly being considered in your experience?

A. *It is used at all levels. We have started to work much more systematically with stakeholder management, and we use stakeholder dashboards in the dialogue with our internal partners. We experience that the partners have very different perspectives. Our provider has to prove himself at local, regional and global level.*

Q. How do you see the relation between innovation and "Added Value" in FM?

A. *Innovation is part of added value. I see it as one big box. Innovation does not have to be new to the market, but it has to be new to Sony. We also talk about improvements. We have an Innovation Board with our provider, where we decide and monitor innovations and improvements.*

Benefits and limitations of "Added Value"

Q. What are the benefits of considering and talking about "Added Value"?

A. *It is more fun to talk about value than cost as it is often perceived as something positive rather than just cost-cutting. When we talk with top management we prefer to discuss value propositions. We are in general moving towards a more value driven discussion. We want to be seen as a business partner. We are not just a cost centre, but we have also started to be a profit centre, which creates income by operating the Sony Shop in Lund. We are also trying to engage in co-creation of commercial products and services both with Sony's core business and with suppliers.*

Q. What are the limitations of considering and talking about "Added Value"?

A. *Added value is always relevant, but it is more fluffy than cost. For instance, managing legal exposure is an added value we need to deal with, but it is difficult to document.*

Management of "Added Value"

Q. What are your top five main values to be included in management of accommodations, facilities and services? Could you mention a few examples of concrete FM interventions to attain these added values and KPIs to measure them?

A. *See table below.*

Prioritised values in FM	Concrete FM interventions	KPIs
1. Cost optimisation	Cost reduction More for less Cost avoidance	Budget vs. spend at local, regional and global level Account payable (invoiced and paid) User satisfaction (related to optimisation)
2. Innovations	New and radical to Sony Access to new knowledge and networks Service Navigator is an example from the I-FM provider – a new way of organising FM services with a multi-skilled person servicing a floor of a building	Buyer satisfaction (FM and procurement) Supplier satisfaction (I-FM provider) User satisfaction
3. Flexibility	Scaling up and down	See examples in text

Prioritised values in FM	Concrete FM interventions	KPIs
4. Co-creation (new value)	Develop new services and apps together with suppliers to sell to customers Sony Map (positioning in buildings) is an example of a product developed with Sony's core business	See examples in text
5. Asset management	Reduce risk by life cycle focus	See examples in text

Q. Do you measure whether you attain the targeted added values?

A: *We use eight KPIs divided into four groups:*
   Quality measured as
   – *User satisfaction once a year, but we have also just started with instant feedback, i.e. smileys*
   – *Buyer and supplier satisfaction monthly – discussed at meetings with provider every month; this indicator is very important for alignment of expectations and to understand the importance of the services. It also forces the FM staff in Sony to be in contact with all sites. It forces people to talk.*
   Cost measured as
   – *Budget vs. invoiced BaseFM cost (services delivered as outlined in the Global Service Agreement) monthly*
   – *Budget vs. invoiced total FM cost (BaseFM, projects, optional service and assignments/moves) monthly*
   Innovations measured as
   – *Minimum of two innovations implemented per year, globally*
   – *Minimum of thirty improvements implemented per year, globally*
   Incidents measured as
   – *Response time to emergency actions*
   – *Reports on incidents*

Q. Which top three aspects do you find most important to include in your value adding management approach?

A. *1. Getting a definition – what is adding value*
   2. Relevancy of adding value – depends on who you are talking to
   3. Quantifying – what gets measured gets done

Q. Do you experience any dilemmas you have to cope with in relation to "Added Value"?

A. *Sony Mobile Communications is at the moment in a process of reducing FM costs by 30%, which is a major challenge as all focus goes towards this rather than adding value currently.*

Q.   Have you noticed a change in mindset regarding "Added Value"?

A.   *Earlier everything in FM was about cost. About five years ago facility managers started to focus on how to get access to the corporate management level. Now everybody is talking about adding value and value propositions. In Sony, FM changed from being under the CFO to becoming part of HR around 2010, which also reflects the increasing importance of FM in relation to user satisfaction.*

Q.   Are there other topics that you find important in connection to the concept of "Added Value"?

A.   *Important questions are how and where the value gets created.*

12 Risk

Per Anker Jensen and Alexander Redlein

Introduction

Risk Management (RM) is a managerial task concerned with continuously monitoring, evaluating and maintaining the risks levels that the company is or may be subject to, and to implementing suitable arrangements to prevent or limit the consequences of unacceptable risks. Such arrangements may typically consist of prevention in the form of security installations and guarding, preparedness in the form of disaster and emergency plans, and delegation of risks to service providers or insurances, which limit the financial consequences for the organisation.

RM is a generic discipline and can be applied to FM and CREM as well. RM has a longer history than FM, but there are clear parallels in the development of the two disciplines. The development in RM over time is illustrated in Figure 12.1.

RM has changed from a mostly technical and operational focus in the 1950s towards an increasingly broad and more strategic management focus. Today, enterprise or company-wide RM is an integral part of the governance structure of many large corporations. From 2008 strategic enterprise RM has been

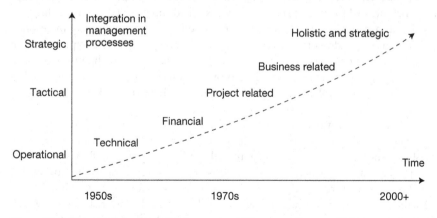

Figure 12.1 The development in RM over time (Ünver and Jensen, 2012)

obligatory for large companies in Europe due to the 8th EU Company Law Directive 2006/43/EC, providing strict requirements that entail mandatory auditing of annual and consolidated financial statements of companies with a public interest (Redlein and Giller, 2008).

Since the 1980s FM has also changed from a mainly operational focus towards an increasingly more tactical and strategic focus, as shown by Pathirage et al. (2008) in their identification of four generations of FM. In the first generation FM was merely considered as an overhead to the organisation and therefore was managed for minimum cost rather than optimal value. The second generation was characterised by an integration of processes between the FM organisation and the organisation's individual businesses. The third generation had a focus on resource and supply chain management, while the fourth generation includes an emphasis on alignment between organisational structure, work processes and the enabling physical environment in accordance with the strategic intent of the organisation.

The European FM standard on Taxonomy includes a so-called central (horizontal) function called Risk, described as "Evaluation and management of risks and threats to the (FM) organization" with a sub-function called Risk policy (CEN, 2011).

The view on risks in relation to the added value of FM/CREM varies a lot. Most Added Value frameworks presented in Chapter 1 include risk-related aspects in some form. The main exception is the CREM model by Anna-Liisa Sarasoja shown in Figure 1.2. The FM Value Map includes risk control in terms of reliability. Risk control was rated as the added value considered most important out of ten parameters in a questionnaire survey among staff in a Dutch bank (Gerritse et al., 2014). In the interview survey by Van der Voordt and Jensen (2014) risk-related aspects were not included in the top five prioritised added values from open questions, but when asked specifically the respondents' views on reliability varied a lot. One view was that reliability is at the lowest level of Maslow's pyramid of needs and therefore is not a motivation factor, which can add value. Another view was that the business continuity has become increasingly important, and for one of the interviewees it had top priority, e.g. regarding fire safety and data security. Another interviewee in a biotech company mentioned that preventing downtime is extremely important, and that compliance with legal requirements has top priority. The increasing focus on the importance of RM within FM practice is underlined by the international service provider ISS publishing a White Paper called "Managing and Mitigating risk within Strategic Facility Management" (Redlein et al., 2014).

When we look at risk in relation to added value of FM and CREM in terms of benefits versus sacrifices, risk control should definitely be seen as a benefit, while risks as such at least constitute potential sacrifices. However, RM should also help organisations to identify and take opportunities according to the Committee of Sponsoring Organizations of the Treadway Commission (COSO, 2004). Facilities managers are, according to Finch (1992), highly risk averse, which means that facility managers are inclined to avoid risk despite the favourable probability of success. This reflects the supportive nature of FM and contrasts with the risk propensity of a property investor, who sees acquisition and sales of property primarily as a financial opportunity and only secondarily as a risk.

Risks in FM/CREM can be divided into ongoing business risks (e.g. compliance, operations responsibility, business continuity), project-related risks (e.g. budget overrun, delays), and transaction and contract-related risks (due diligence, which will be explained later in this chapter). Several risk factors in relation to FM/CREM are covered by legislation and similar regulations, for instance fire hazards, health and safety, and the external environment. This chapter will focus on business continuity and general business risk factors. Project-related risks are a normal part of project management practice and will not be dealt with specifically. Risk factors related specifically to image, health and safety, asset value and environment are dealt with in other chapters. Due diligence will be mentioned briefly in the section discussing the state of the art, but no KPIs are included.

The chapter provides a state of the art discussion in the next section with the main concepts and definitions related to RM and a literature review of the research on RM in FM/CREM. The following section presents benefits and costs of risk interventions, followed by a section on KPIs and guidelines on how to manage risks in FM/CREM. The chapter finishes with a section about perspectives and the need for new knowledge and development in relation to risk in FM and CREM.

State of the art

Risk Management (RM)

RM can be defined as: "A process where an organisation adopts a proactive approach to the management of future uncertainty, allowing for identification of methods for handling risks which may endanger people, property, financial resources and credibility" (O'Donovan, 1997, cited by Lavy and Shohet, 2010). The International Organization for Standardization (ISO) has issued the 31000 series on RM and a guide with RM vocabulary (ISO/IEC, 2009).

In the general literature on RM it is common to distinguish between Risk Management at the strategic level, concerned with decision making about how to deal with identified risks, and Risk Assessment or Risk Analysis (RA) at a tactical and/or operational level. It is also common to divide the activities in RM and RA into a number of steps as shown in Figure 12.2.

RM has been an area of focus for FM since the early days. In 1987 there were three articles in the journal *Facilities* about RM. All three articles were short and very practice based. The first article in *Facilities* mentioning a research project on RM is by Alexander (1992). Since 1995 there have been a number of research-based articles in *Facilities* and later also some in the *Journal of Facilities Management* and the *International Journal of Facility Management*. A literature review on RM in the FM literature was conducted in 2011 by Ünver and Jensen

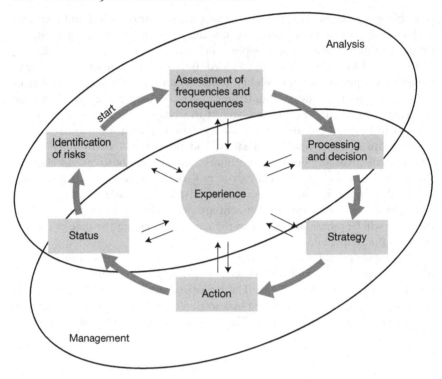

Figure 12.2 Framework for RM and RA activities (Ünver and Jensen, 2012)

(2012), revealed quite a limited number of research papers. The review concluded that the published research in general could be characterised as being very normative and prescriptive. There seemed to be a lack of more descriptive research investigating how facilities managers in practice work with RM in different companies. A new literature review conducted for this chapter shows that this has changed in recent years. A large number of research papers have been published recently, particularly in the *Journal of Facilities Management*, and many of these include descriptive empirical-based research and also the development of several conceptual models. Some of these are referred to in the following.

The treatment of risk in books on FM varies a lot. Among the first, the text book edited by Alexander (1995) includes a chapter on RM with a keynote paper (Dyton, 1995), while the text book by Barrett (1995) does not include the word "risk" in the index – not even in the third edition (Barrett and Finch, 2014). A more recent and practice oriented "desk reference" book by Wiggins (2010) includes specific chapters on Business Continuity, Fire Safety and Legislation, Electrical Supplies and Electrical Safety, First Aid at Work, Asbestos, and Security Management, as well as sections on risk in several other chapters.

RM strategies

The strategy for RM will typically be differentiated in relation to different types of risk handling. There are a number of generic RM strategies (Jensen, 2008):

- Risk avoidance
- Risk reduction
- Risk transfer
- Risk acceptance.

In the literature there is a distinction between "hazard" and "risk". Hazard is the potential for harm, while risk is a function of the probability of the harm actually occurring and the severity of its consequences (Wiggins, 2010).

Risk avoidance means that the hazard is eliminated. For instance, this can be done by changing a process, so that harmful products, substances or activities are replaced by less harmful ones. Risk reduction means that the hazard still exists but the likelihood of it occurring and/or the consequences are reduced. This can be done by preventive technical systems and by organisational preparedness like emergency planning. Risk transfer means that the hazard is unchanged, but the consequences for the company are reduced by transferring the risk (or some of it) to another party, for instance a service provider or an insurance company. More examples of such measures are provided later in the section on benefits and sacrifices of different types of interventions. Risk acceptance also means that the hazard is unchanged, but the risk is so low that the company is willing to accept the consequences, if the risk event occurs. The level of acceptance depends on the so-called "risk appetite" of the company.

Risk and legislation

There are in a number of areas of legislation related to safeguarding and safety. This concerns especially the following legislation (Jensen, 2008):

- *Construction legislation*, for instance including the building regulations with demands for house and building design, and demands of fire safeguarding, for instance
- *Preparedness legislation*, for instance with regulations for disaster planning, air-raid shelters, fire protection and preparedness in especially inflammable areas of companies, and also fire inspection in buildings with particularly large fire and/or person strain
- *Health and safety legislation*, for instance with regulations for health and safety organisation, machinery safety and design of workplaces
- *Environmental legislation*, for instance with regulations for the labelling of dangerous material and for especially polluting companies.

In relation to fire safeguarding the construction legislation aims to ensure the safety of persons and fire personnel, but not economic values, whereas the preparedness legislation, besides ensuring the safety of persons, also concerns valuable assets.

FM Risk Map

A study on RM in Austria (Redlein et al., 2014) based on the framework of "Company-wide Risk Management" (COSO, 2004) has resulted in the generic FM Risk Map shown in Figure 12.3.

The FM Risk Map builds on the best practice approach of Price Waterhouse Coopers. It defines three areas of potential risks/opportunities:

1 Strategy and goals with the sub-categories Procurement and Capital market
2 Members of staff and organisation with the sub-categories Sales market/ customers and Legislative requirements and internal directives
3 Business processes with the sub-categories Technology and Rival businesses.

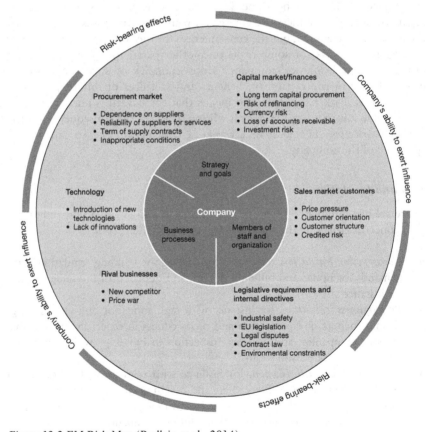

Figure 12.3 FM Risk Map (Redlein et al., 2014)

In several case studies examples of these risks/opportunities were identified together with facility managers. The facility managers stated dependency on suppliers and reliability as a major risk, especially if they want to use opportunities like economies of scale and scope and thereby reduce the number of suppliers. This example shows perfectly the trade-offs within RM. The benefits of the opportunity to attain lower prices and less coordination effort with one service provider contradicts with the sacrifice of higher threats, i.e. dependency on one supplier. The capital market involves investment and currency risks but also the opportunity to optimise assets and thereby ease the financing of the whole company (core business and real estate). The sales/customer market is especially important for the real estate business as it has a direct influence on income. The customer structure (one large tenant compared to several small ones) and orientation (offer of additional services to bind the customers) can have an important impact on risks as well. Customer orientation, including additional services to safeguard the customer, can lead to a surplus in income by up to 20% (expert interview with Deutsche Bank). The legislative requirements overlap significantly with the sub-section above on Risk and legislation. The technology sub-category also plays an important role, because it influences usability but also customer satisfaction. Too flexible and expensive technology use is a threat, but the right mix allows flexible usage and increases opportunities (Redlein et al., 2014).

Risk Assessment principles and methods

Risk Assessment (RA) involves identification of hazards and assessing the consequences and likelihood of risk events. It is an analysis of causes and effects. Dyton (1995) divides causes in natural forces, man-made forces and human acts, and the direct effects in liability for damages, property damages, loss of income and bodily injury. Besides these direct effects there can also be wider indirect effects. Dyton (1995) calls these "rippling effects", which start with victims and spread to company, industry and other technologies.

The results of RA are often presented in the form of a risk matrix showing a prioritisation of risks. An example of a risk matrix from the FM department in a large Danish industrial company is shown in Figure 12.4. It was developed in order to increase the transparency of maintenance activities and budget planning. An assessment of technical priority and operational priority according to the pre-defined values in combination determine the urgency/consequence.

Adedokun et al. (2013a and 2013b) carried out literature reviews concerning qualitative and quantitative risk analysis techniques, respectively. They identified a large number of techniques as listed in Table 12.1. For further information we refer to Adedokun et al. (2013a and 2013b).

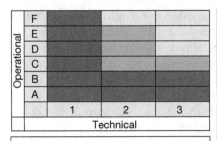

Technical priority:
1. High risk of consequential damage
2. Reduced functionality/minor consequential damage
3. Preventive maintenance

Legend:
Dark: Critical – as soon as possible
Medium: Serious, expected within 1–2 years
Light: To be performed, but can be posponed without consequential damage or operational impact

Operational priority:
A. Safety issues (avoid personal injury)
B. Regulatory/compliance
C. Risk for production
D. Influence on indoor climate/lower service level
E. Energy efficiency, payback <4 years
F. No operational risks

Figure 12.4 Example of a risk matrix for prioritisation of maintenance activities (reproduced with permission from interviewees Søren Andersen and Søren Samuel Prahl, Interview 8)

Table 12.1 Qualitative and quantitative risk analysis techniques

Qualitative risk analysis techniques	*Quantitative risk analysis techniques*
Assumption analysis	Decision trees
Cause and effect diagrams	Event and fault trees
Checklists	Expected value tables
Data precision ranking	Fuzzy logic
Event and fault trees	Monte Carlo and Latin hypercube simulation
Flowcharts	Multi-criteria decision-making support methods
Influence diagrams	(Analytic hierarchy process)
Probability and impact description	Probabilistic influence diagrams
Probability impact tables	Probability sums
	Process simulation
	Sensitivity analysis
	System dynamics

Business Continuity Management (BCM)

Business Continuity is the "capability of the organization to continue delivery of products or services at acceptable predefined levels following disruptive incident" (ISO, 2012). Business Continuity Management (BCM) is furthermore defined as a "holistic management process that identifies potential threats to an organization and the impacts to business operations these threats, if realized, might cause, and which provides a framework for building organizational resilience with the capability of an effective response that safeguards the interests of its key stakeholders, reputation, brand and value-creating activities" (ISO, 2012). A research paper about challenges faced by facilities managers in Australasian universities shows that emergency management and business continuity planning was the second most important challenge, with inadequate funding being the most important (Kamarazaly et al., 2013).

Until the terrorist attack in New York on 11 September 2001 BCM mostly resided within the IT department, but since then BCM has increasingly become an important area within FM and CREM. BCM involves a prioritisation of the criticality of both organisational functions and locations. The organisational prioritisation can be made with an "organisational onion" approach, identifying the most critical functions in relation to withstanding downtime, less critical but still important core functions, and non-critical functions. The locations can similarly be divided in "hot" sites with locations in continuous operation, "warm" sites with infrastructure in place for continuous operation at short notice, and "cold" sites that can be modified to suit a specific need (Gill, 2006).

Due Diligence (DD)

The European standard on FM agreements (CEN, 2006) defines Due Diligence (DD) as the "compilation, comprehensive appraisal and validation of information of an organisation at the appropriate stage of the Facility Management agreement required for assessing accuracy and integrity at the appropriate stage of the agreement process". DD is commonly used in major business transactions like mergers and acquisitions of companies and buying and selling real estate portfolios. The purpose of DD is to disclose major risks in relation to existing transactions and contracts. DD can include financial, legal, business, technical and environmental aspects. DD is often conducted by a team of experts from a group of specialised consulting companies.

Technical Due Diligence (TDD) is a special form of condition assessment of property (Jensen and Varano, 2011). TDD can be defined as "The process of systematic review, analysis and discovery in which a prospective purchaser, occupier or financier of property gathers information about the physical characteristics of the property in order to make an informed assessment of the risks associated with the transaction" (RICS, 2009). Varano (2009) investigated the use of TDD in two Danish FM supplier companies. The cases showed that a consultant can support the starting phase of a FM contract by conducting a survey of the property before signing a contract to determine the necessary costs to manage the building, or after signing a contract as guarantee of the physical condition of the property for the client of the FM supplier.

Benefits and costs

A number of typical types of interventions that FM/CREM can implement to add value in terms of RM and business continuity is shown in Table 12.2. A general management task for all interventions is to develop policies and ensure top management commitment. The management tasks mentioned in Table 12.2 are those specifically related to each intervention. The added value is specified as the benefits and the sacrifices for each type of intervention. Here sacrifices mainly regard the costs of interventions.

Table 12.2 Interventions, management, benefits and sacrifices of RM

Interventions	Management	Benefits	Sacrifices
Remove hazards	Change process/product Project Management	Risk avoidance	Cost of project, including disposal
Emergency planning	Engage the organisation Negotiate with authorities Arrange training and rehearsals	Risk reduction/ preparedness	Cost of planning and training etc.
Recovery planning	Establish and train teams Arrange rehearsals	Risk reduction/ preparedness	Cost of planning and training etc.
Security system installation	Ensure procurement, training and maintenance	Risk prevention/ reduction	Cost of installation, staff and monitoring
Back-up supply installation	Ensure procurement, training and maintenance	Risk and down-time reduction	Cost of installation, staff and monitoring
Insurance	Ensure procurement and administration	Risk transfer	Cost of insurance and procurement etc.
Outsourcing	Ensure procurement and administration/supervision	Risk transfer	Cost of contract and procurement etc.

Interventions that avoid risks can involve completely removing a hazard, for instance the renovation of buildings to remove harmful substances like asbestos or PCB in building components, or changing processes to substitute dangerous chemical products with less harmful products. Such interventions are typically implemented as individual projects or as programmes of projects rolled out over the property portfolio belonging to the corporation.

Risk reduction can take place by many different types of interventions. Such interventions can be organisational like emergency and recovery planning, but they can also be mostly technical like security systems, for instance the installation of video monitoring, and the installation of back-up supply systems, for instance Uninterrupted Power Systems (UPS) and diesel generators. These types of interventions are also typically implemented as projects, but unlike pure risk avoidance projects they will need ongoing monitoring, updating and maintenance.

Interventions with risk transfer can include insurance and outsourcing. Risk is in both cases transferred to a third party, but insurance is solely related to risks, while outsourcing involves many other factors than risk. This makes it difficult to make up the specific cost of the risk transfer in outsourcing. Both of these interventions include a procurement process but also ongoing administration and supervision. A limitation of risk transfer often is that even though the company can obtain an economical compensation for the direct cost of damage and maybe even for loss of production, the wider indirect "rippling" effects, for instance loss of market share, will not be avoided or compensated.

Table 12.3 Average length of business interruptions and their causes (Wiggins, 2014)

Cause of business interruption	Average length of interruption
Fire	28 days
Theft	26 days
Lightning	22 days
IT failure	10 days
Flooding	10 days
Power failure	1 day

Research shows, according to Wiggins (2014), that a company will experience a major disruption about once every four years. 57% of all business disasters are IT-related and 10% are due to power failures. The average lengths of business interruptions are shown for a number of different causes in Table 12.3.

How to measure and manage

Key Performance Indicators

Typical general KPIs for RM are cost related, for instance total risk expenses, insurance costs, emergency costs and the cost of RM activities. Such costs can be measured as ratios of total turnover or budget for an organisation or a project. In relation to business continuity KPIs often concern time, for instance uptime (the opposite of downtime) and recovery time.

We suggest the following top five KPIs for risks:

- Uptime of critical activities as percentage of total time
- Total risk expenses as percentage of company turnover
- Total insurance expenses as percentage of company turnover
- Total damage prevention expenses as percentage of company turnover
- Total actual damage expenses as percentage of company turnover.

How to manage risks

The main benefit of risk control is a high level of business continuity, which is best measured by uptime. The main purpose of BCM is precisely to ensure a high level of uptime. The most important interventions to ensure uptime are emergency and recovery planning and the installation of back-up supply systems, but also other interventions aimed at risk avoidance and risk reduction, like the internal control systems mentioned below, can contribute to increased uptime. Risk transfer by insurance has little effect on uptime, while risk transfer by outsourcing can have an effect, if the external party has special capabilities to ensure uptime.

The main sacrifice of risk control regards the required expenses. As a basis for the development of RM policies it can be useful to make up the total risk expenses. They include expenses for damage prevention, insurance, and recovery and repair after actual damages. Increased expenses for damage prevention will normally lower the expenses of the other areas. The typical correlation between risk expenses and safeguard level is illustrated in Figure 12.5.

It appears the total risk expenses have a typical minimum point. In an overall evaluation other conditions than the purely financial ones must be included. For instance, considerations for person risk may imply a higher safeguard level for specific hazards. The risk level in such cases is often based on the principle of ALARP – As Low As Reasonably Practicable – which means that the cost involved in reducing the risk further would be grossly disproportionate to the benefit gained (Faber and Stewart, 2003). Comprehensive damages can also imply considerable indirect consequences such as the delay of development projects and negative publicity. In order to make up total risk expenses it is necessary to continuously record and calculate the actual damages. This concerns both the insured and the non-insured damages. Such a recording is also an important basis to be able to prioritise damage prevention.

In spite of the correlation between expenses for damage prevention and insurance, which is shown in Figure 12.5, it can be difficult to gain a premium reduction on continuous insurances from increased investments in damage prevention. From experience it is not possible to get information from an insurance company about the discount on the premium one may gain from installation of a sprinkler system in a building, for instance. The insurance companies refer to the fact that the premium is determined on the basis of an overall risk assessment (Jensen, 2008).

On the other hand, large savings can be obtained from professional procurement of insurances in which the real and possible planned level of damage prevention is included as part of the basis for procurement. It can be recommended

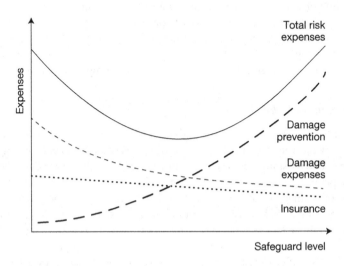

Figure 12.5 Correlation between risk expenses and safeguard level (Jensen, 2008)

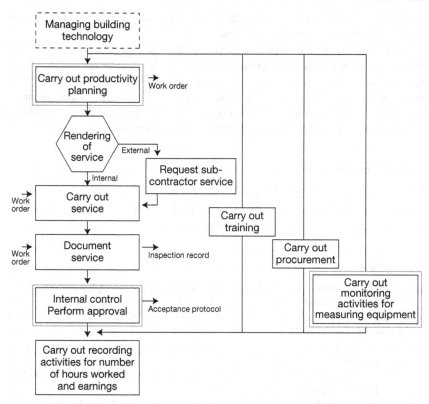

Figure 12.6 Building a technology management process with ICS activities (Redlein et al., 2014)

to engage an insurance broker to assist in preparation and implementation of insurance procurement, especially the first time it is done. Often considerable savings in the insurance expenses can be achieved by such a process (Jensen, 2008).

Internal Control System

The earlier mentioned study from Austria (Redlein et al., 2014) recommends that measures to reduce risk are integrated in a company's Internal Control System (ICS). An example of how this can be carried out in relation to a process concerning managing building technology is shown in Figure 12.6.

Perspectives

There is a need for further research to collect quantitative data about the costs and benefits of typical risk reducing or risk avoiding interventions in FM/ CREM, as well as exploring best practices (sound balance between costs and benefits) and worst cases (too little intervention resulting in huge hazards, or too costly interventions to reduce small risks).

In recent years awareness of climate change has increased the focus on RM in relation to the resilience of the built environment. This is an increasingly popular new research area, and it also forces property owners and FM/CREM organisations to take a more long-term perspective. Portfolio management of property is becoming more important due to this, and methods like forecasting, back-casting and scenario planning are becoming more commonly used.

Acknowledgement

The authors want to thank senior researcher Frank Markert, DTU Management Engineering, Technical University of Denmark, for his review of and useful comments on an early version of this chapter.

References

Adedokun, O.A., Ibironke, O.T., Dairo, D.O., Aje, I.O., Awodele, O.A., Opawole, A.D., Akinradewo, O.F. and Abiola-Falemu, J.O. (2013a) 'Evaluation of qualitative risk analysis techniques in selected large construction companies in Nigeria', *Journal of Facilities Management*. 11 (2), pp. 123–135.

Adedokun, O.A., Ibironke, O.T., Dairo, D.O., Aje, I.O., Awodele, O.A., Opawole, A.D., Akinradewo, O.F. and Abiola-Falemu, J.O. (2013b) 'Evaluation of quantitative risk analysis techniques in selected large construction companies in Nigeria', *Journal of Facilities Management*. 11 (4), pp. 354–368.

Alexander, K. (1992) 'Facilities Risk Management', *Facilities*. 10 (4), pp. 14–18.

Alexander, K. (Ed.) (1995) *Facilities Management – Theory and Practice*. London: E&FN Spon.

Barrett, P. (1995) *Facilities Management: Towards Best Practice*, Oxford: Blackwell.

Barrett, P. and Finch, E. (2014) *Facilities Management: The Dynamics of Excellence*, Third Edition. Oxford: Wiley Blackwell.

CEN (2006), *Facility Management – Part 2: Guidance on how to prepare Facility Management Agreements*. European Standard EN 15221-2. European Committee for Standardization

CEN (2011) *Facility Management – Part 4: Taxonomy, Classification and Structures in Facility Management*. European Standard EN 15221-4, European Committee for Standardization.

COSO (2004) *Enterprise Risk Management – Integrated Framework. Executive Summary*. Committee of Sponsoring Organizations of the Treadway Commission, September.

Dyton, R. (1995) 'Practical Risk Management', keynote paper in Alexander, K. (Ed.) *Facilities Management – Theory and Practice*, London: E&FN Spon.

Faber, M.H. and Stewart, M.G. (2003) 'Risk assessment for civil engineering facilities: Critical overview and discussion', *Reliability Engineering and System Safety*, 80 (2), pp. 173–184.

Finch, E. (1992) 'Risk and the Facilities Manager', *Facilities*. 10 (4), pp. 10–13.

Gerritse, D., Bergsma, F., and Groen, B. (2014) 'Exploration of added value concepts in facilities management practice: Learning from financial institutes', in Alexander, K.

(Ed.) *Promoting Innovation in FM. Research Papers. Advancing knowledge in FM. International Journal of Facilities Management, EuroFM Journal.* March.

Gill, T.J. (2006) 'Workplace continuity: How risk and technology will affect facilities strategy', *Journal of Facilities Management,* 4 (2), pp. 110–125.

ISO (2012) *Societal security – Business continuity management systems – Requirements.* International Standard IS0 22310. International Organization for Standardization.

ISO/IEC (2009) *Risk management – Vocabulary.* Guide 73. International Organization for Standardization and International Electrotechnical Commission.

Jensen, P.A. (2008) *Facilities Management for Practitioners and Students,* Centre for Facilities Management – Realdania Research, DTU Management Engineering. Lyngby, Denmark.

Jensen, P.A. and Varano, M. (2011) 'Technical Due Diligence: A study of current practice in assessment of buildings', *Journal of Performance of Constructed Facilities,* 25 (3), pp. 217–222.

Kamarazaly, M.A., Mbachu, J. and Phipps, R. (2013) 'Challenges faced by facilities managers in the Australasian universities', *Journal of Facilities Management,* 11 (2), pp. 136–151.

Lavy, S. and Shohet, I.M. (2010) 'Performance-Based Facility Management – An Integrated Approach', *International Journal of Facility Management,* 1 (1), pp. 1–14.

Pathirage, C., Haigh, R., Amaratunga, D. and Baldry, D (2008) 'Knowledge management practices in facility organisations: A case study', *Journal of Facilities Management,* 6 (1), pp. 5–22.

Redlein, A. and Giller B. (2008) 'Reference Processes and Internal Control Systems within Facility Management', in *European Facility Management Conference 2008 – Conference Proceedings,* pp. 181–191

Redlein, A., Redlein, B., Søborg, A. and Poglitsch, R.O. (2014) *Managing and Mitigating Risk within Strategic Facility Management,* ISS White Paper, August.

RICS (2009) *Technical Due Diligence of Commercial & Industrial Property.* London.

Ünver, K. and Jensen, P.A. (2012) 'Risk Management from Corporate and FM Perspectives: Two case studies', *Proceedings of the 11th EuroFM Research Symposium,* 24–25 May, Copenhagen, Denmark. Centre for Facilities Management – Realdania Research, DTU Management Engineering, and Polyteknisk Forlag, Lyngby, Denmark.

Van der Voordt, T. and Jensen, P.A. (2014) 'Adding Value by FM: Exploration of management practice in the Netherlands and Denmark', in Alexander, K. (Ed.) *Promoting Innovation in FM. Research Papers. Advancing knowledge in FM. International Journal of Facilities Management, EuroFM Journal.*

Varano, M. (2009) *Technical Due Diligence – A Comparison of Best and Current Practice.* Master's thesis, Technical University of Denmark, Lyngby, Denmark.

Wiggins, J.M. (2010) *Facilities Manager's Desk Reference.* Chichester: Wiley-Blackwell.

Interview 8: Søren Andersen and Søren Prahl, Denmark

Søren Andersen has twenty-two years of experience in sourcing, including serving as a head of FM since 2005, with an educational background in Economics and a Higher Diploma in Logistics. Søren Prahl has an educational background in building engineering and fifteen years of experience, starting with consulting and since 2008 as project manager in an inhouse FM unit. They both participated in the interview, and at the time they worked together in a Danish biotech company. Søren Andersen is today a senior project director and Søren Prahl is the head of portfolio management – both in large corporations in Denmark.

Added Value in general

Q. How do you understand the concept of "Added Value"?

A. *FM is very much cost driven in the company. We have a strong focus on creating transparency. Flexibility and scalability also have high priority. Creating time particularly by releasing managers of unnecessary responsibilities is important. We also emphasise when we can argue to avoid cost. The triple bottom line is used as a management measure.*

   FM in the company has four main drivers:

- *m^2: We have developed from having a strong focus on space management to a more strategic focus on developing site master plans*
- *Headcounts: We have a tool to make prognoses for developments in the number of staff*
- *Quality levels: We prepare standards for quality where there is a demand for it. We also participate in a Nordic benchmarking programme organised by Ernst & Young, where service levels and user satisfaction are compared. Workplace Assessments (APV) are also used. The contract with our service provider is based on open books*
- *Growth strategy: The company is expanding fast.*

Q. Is "Added Value" mostly treated at a strategic, tactical or operational level?

A. *All our operational activities have been outsourced. We have a strategic focus on developing site master plans and real estate strategy. Interior building maintenance has been centralised from the user departments to FM, and has been standardised to reduce cost and to create a transparent decision process.*

Q. In which context or dialogue is "Added Value" mostly considered from your experience?

A. *The dialogue with corporate management is related to decision making and the dialogue with other departments is mutual communication. In relation to decisions on maintenance and environmental projects we have developed a priority matrix (see Figure 12.4), which is used to create transparency in decision making.*

To visualise the scope and responsibilities of FM we have developed a "FM landscape" based on the Five Finger Model of a FM organisation from the Danish FM handbook. It makes use of "traffic lights":

- Green for areas where FM has the budget and decides
- Yellow for areas where FM has the budget and others decide
- Red for areas where others have the budget and decide.

The FM landscape is used internally in the FM organisation and in the dialogue with their customers.

Management of Added Value

Q. What are your top five main values to be included in the management of accommodations, facilities and services? Could you mention a few examples of concrete FM interventions to attain these added values and KPIs to measure them?

A. *See table below.*

Prioritised values in FM	Concrete FM interventions	KPIs
1. Transparency of cost and priorities	Priority matrix	Cost
2. Scalability	Site master plans/Space Management	Space utilisation
3. Release management resources, so that line of business can focus on their core-tasks	Outsourcing all operational services	Follow up on FM suppliers from Sourcing and the FM unit
4. User satisfaction	Friday morning brunch	People's Opinion (External measurements)
5. Satisfaction with service providers	Long-term relation/ partnership with key FM Suppliers and a "Management by Exceptions" concept: Reduce the need for detailed control of daily business	Key figures for price, quality and user satisfaction Monthly, quarterly and yearly follow up at different management levels

Q. Do you benchmark your data with data from other organisations?

A. *Yes, as part of the Ernst & Young programme, but also in projects like FM Deep Dive in 2011 with ten leading and globalised Danish companies.*

Q. What other methods do you use to document "Added Value"?

A. *For all staff in the company there are annually defined objectives related to business goals and personal goals, which are measured half-yearly. The objectives* are part of the salary bonus scheme, and the objectives are known to the manager's subordinates. Staff employability has high priority in the FM unit.

We are also working on developing a new model for a more holistic evaluation of investments.

Q. How is "Added Value" included in your communication with your stakeholders?

A. *It is included in terms of user satisfaction and in reporting on projects.*

Last comment

Q. Are there other topics that you find important in connection with the concept of "Added Value"?

A. *The borderline between inhouse and outsourced is interesting to discuss. What is the core of FM? Is it for instance best to have reception staff inhouse or outsourced? Or who should have the responsibility for IT systems for maintenance, to make sure that they are developed?*

13 Cost

Alexander Redlein and Per Anker Jensen

Introduction

Facility Management (FM) is a key function in managing facility services and working environments to support the core business of the organisation. A lot of companies have recognised FM as an important management strategy capable of reducing the costs of facilities (Chotipanich, 2004). A discussion is taking place about the added value of FM as a function within organisations and the services they provide. A clear expression of this is the large number of FM-related studies that have been conducted focusing on different aspects of FM and their added value for primary processes, quality, time, risk and relationship quality (Kok et al., 2011; Jensen and van der Voordt, 2015).

Since 2005 the Vienna University of Technology (TU Vienna) has analysed the demand side of FM on a yearly basis in different European countries such as Austria, Germany, Bulgaria, Italy, Romania, Spain, Turkey and the Netherlands (companies were selected randomly). The research has been based on a (standardised) questionnaire survey. One of the attempts to prove the profitability and efficiency of FM was performed by Susanne Hauk in her PhD study on "Wirtschaftlichkeit von Facility Management" (Hauk, 2007). Another research project at the TU Vienna analysed whether there are differences regarding efficiency and effectiveness according to whether a separate FM department has been established or not. The authors also defined additional parameters that influence the efficiency of FM. Examples of these parameters are: areas of cost saving, availability of cost and building data, and usage of CAFM (Computer Aided Facility Management). The study found that companies with their own FM department tend to achieve savings within more facility services areas (Redlein and Sustr, 2008).

This chapter presents further results of the actual surveys from the TU Vienna. The researchers used statistical models to find out whether there is a (significant) correlation between different variables/parameters. In the literature there are three major research paradigms for collecting the required data: quantitative research methods, qualitative research methods, and mixed method research. Quantitative and qualitative methods both have particular strengths and weaknesses. For this reason the authors used a mixed methods approach.

The chapter mainly focuses on cost savings, particularly concerning whether organisations with a FM department have more facility services with savings than organisations without a dedicated FM department, and whether outsourcing can be seen as a cost-saving approach.

State of the art

According to different publications (e.g. Scharer, 2002) it is possible to save between 10% and 30% of the costs of buildings through the (efficient) use of FM. In most cases, figures about the economic effects/benefits of FM are based on the study of a single company, or else the data presented is not specified in detail. In both cases, data cannot be used for a general proof of the economic efficiency/value added of FM (Zechel et al., 2005; Scharer, 2002). The figures are also subject to large variations. Therefore there is a need to determine the value added of the use of FM, particularly FM departments, and the parameters influencing the magnitude of this value added with the help of scientific models and methods. For this reason annual surveys have been conducted in various countries. The results were used to define the benefits and sacrifices of typical interventions.

The surveys from TU Vienna mainly focus on cost reduction and the increase of productivity of the FM organisation itself on the one side, and cost drivers on the other side (Mierl, 2012). Cost drivers require differentiated cost planning and cost control. They are measures of cost causation and resource use and output (Leidig, 2004).

The biggest cost drivers of the surveyed countries (measured by number of mentions) include areas such as energy, maintenance/repair, safety, cleaning and launching new software. The most relevant areas of cost savings (also by number of mentions) were areas such as energy, cleaning, maintenance/repair and personnel. Savings were mainly possibly through new types of contract, improved rates, technical upgrades, reorganisation and utilisation of synergies.

The most named areas in which an increase in productivity could be observed (in terms of number of answers) are administration, personnel, safety, and maintenance/repair. Reasons for an increase in productivity are: process optimisation, work utilisation, utilisation of synergies and personnel/employee workload optimisation.

Impact of an own FM department

Based on the data, several hypotheses concerning savings through the use of FM could be validated. The hypothesis is that companies with an own FM department tend to have a higher number of facility services with savings (areas of cost savings) than companies without an own FM department. An own FM department allows better management of facility services (e.g. cleaning, maintenance/repair) and guarantees the best realisation of optimal real estate management. As a result, economic optimisations in different facility services can be performed (Hauk, 2007). In relation to the typology of interventions in Chapter 2, this will often involve changing the interface with core business.

The number of facility services with savings (areas of cost savings) was analysed in detail. The Wilcoxon Test was used for comparing the average performance of two groups to verify whether there is a difference between two populations on the basis of random samples from these populations (Dodge, 2008). The data for Romania 2013 and Austria 2012 will be considered; see Table 13.1.

The results of the test show that there is a statistically significant difference (p < 0.05) between the two medians, which means that there is an effect. Companies with their own FM department tend to have more areas of cost saving than companies without their own FM department. A company's own FM department manages the different facility services better. In addition, synergies between the different services can be leveraged through the central management of facility services. This statistical model found that a company's own FM department allows better management of facility services, and therefore economic optimisation and cost savings in different facility services such as cleaning or maintenance/repair can be performed.

The proportion of companies with their own FM department is at a high level all over Europe. FM is a very important tool to achieve an increase in (annual) savings and productivity. According to a statistical analysis based on the data from the studies an inhouse FM department had positive effects on annual savings such as energy and cleaning. FM also leads to an increase in productivity. The most named areas were administration, personnel and maintenance/repair. The study proved that companies in Austria and Romania with their own FM department tend to achieve savings within more facility services in comparison with companies without their own FM department. An inhouse FM department also leads to higher annual savings. In both cases the Wilcoxon Test shows a significant result. That means that there is a (statistically) significant difference between the two groups (FM department yes/no) and the tested variables (annual savings, facility services with savings).

Table 13.1 Number of facility services with savings – FM department

	Austria 2012		Romania 2013	
	FM department	*Without FM department*	*FM department*	*Without FM department*
N	63	8	10	1
Mean*	1.86	.88	1.70	1.00
Median*	2.00	1.00	2.00	1.00
Std. Deviation*	.998	.641	.674	–
p – value	= 0.000000025466		P = 0.011412	

Outsourcing as cost-saving approach

A commonly used method to achieve cost reductions is outsourcing. In relation to the typology of interventions in Chapter 2 this concerns changing the supply chain.

The International Facility Management Association (IFMA) conducted surveys on the practice of outsourcing in the FM field in 1993, 1999 and 2006. The results reveal that over the years the use of out-tasking (hiring individual, specialised vendors to provide one or more FM functions) has decreased from 91 per cent of the responding companies in 1993 to 77 per cent in 2006. The steepest decline was from 1999 to 2006, with a corresponding increase in the number of companies that are outsourcing (hiring full-service, single vendor organisations to provide many services bundled together). The most commonly outsourced/out-tasked services are housekeeping, architectural design, trash and waste removal and landscape maintenance. The most important criteria when deciding whether or not to outsource are financial in nature: controlling costs, freeing capital funds, improving ROI, and reducing turnover and training costs. Over half of the companies surveyed saved money through outsourcing/out-tasking, and a third saw a quality improvement. Two out of five companies brought services back inhouse after outsourcing the service. Typically the reasons for this were to regain control of the service, either in terms of costs, quality or response time. One half of the companies consolidated their vendor base to use fewer service providers (IFMA, 2006).

The fundamental argument for introducing outsourcing and market competition to management services is that such a delivery approach can save costs by reducing bureaucratic inefficiencies, allowing large organisations and governments to access economies of scale, bypassing costly labour and generating competition among service providers. Competitive tendering can, according to the literature findings, yield 10 per cent to 30 per cent in cost savings, with no adverse effect and sometimes an improvement in service quality (Lam, 2011).

Several studies on the risk factors associated with outsourcing functions have been reported. Kremic et al. (2006) carried out a survey of risk factors for outsourcing IT functions. These risk factors include: unrealised savings with a potential for increased costs; employee morale problems; over-dependence on a supplier; loss of corporate knowledge and future opportunities; and inadequate requirement definitions. Ikediashi et al. (2012) analysed the risks associated with outsourcing FM services in Nigeria. Findings from the study reveal that poor quality of services was rated the most critical, security issues was rated second, followed by the inexperience of the client.

The areas of outsourced facility services in surveys from Austria, Spain and Germany in 2014 are shown in Figure 13.1. The degrees of outsourcing of cleaning and technical maintenance are among the highest in all three countries with levels of well over 80%. Outsourcing of heating/ventilation/air conditioning is also high in all three countries, while the results for other services vary considerably between the countries or are generally much lower.

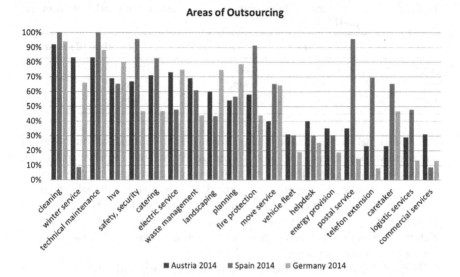

Figure 13.1 Outsourced FM Services in Austria, Spain and Germany 2014

The theoretical background underlines the assumption that organisations who outsource their facility services gain more added value than organisations that provide their facility services inhouse. That is why the researched organisations were asked how they control their FM functions and which percentage they source out (Smit, 2008). To see if the degree in outsourcing of facility services has an effect on the added value of an organisation, the degree of outsourcing is compared with the parameter "annual savings". A regression analysis was used to make quantitative estimates of the relationship between these two variables/ parameters. The dependent variable was the annual savings (in %) and the independent variable was the degree of outsourcing.

The hypothesis that a larger degree of outsourcing result in more added value, i.e. a higher degree of perceived annual savings, is only partially confirmed. An increasing degree of outsourcing does indeed lead to a slight increase in annual savings, but there is no statistically significant correlation between the degree of outsourcing and the annual savings, i.e. the degree of outsourcing has only a weak effect on the annual cost savings. Only 0.9% of the variance in annual savings is explained by the outsourcing degree. It can be concluded that there is no significant relation between the degree of outsourcing in an organisation and the way respondents perceive the added value of their FM organisation.

As mentioned before, many facility services such as cleaning, safety, winter servicing and catering are outsourced (see Figure 13.1). Therefore, outsourcing is still an important strategy for companies. The number of external service providers in the years from 2010 to 2012 is shown in Table 13.2.

The results show a change from the majority having more than ten external providers in 2010 towards most of the surveyed companies having between three

Table 13.2 Number of external service providers in Austria 2010–2012 (Redlein and Zobl, 2014)

	1–2 providers	*3–10 providers*	*More than 10 providers*
2012	6%	63%	31%
2011	6%	68%	26%
2010	12%	41%	47%

and ten external service providers. This reduction in the number of commissioned service providers shows the tendency to engage external service providers with an integrative service offer. The statistical analysis of the relation between annual savings and number of external service providers shows that the annual savings for companies with three to ten external service providers are highest, and that they decrease with an increasing number of external service providers. An increasing or high number of external service providers does not automatically generate more cost savings (annual savings).

The surveys also showed that the share of companies with only one or two external service providers under contract remains at a low level. In 2010 the share of companies with only one or two external service providers under contract was 12%. This share of companies under contract decreased in 2011 and 2012 to 6%. Thus, at least in Austria and the EU countries that have been analysed, there is no trend towards only one external service provider with an integrative service offer.

It might be assumed that the less external service providers a company has to commission, the less complex is the internal administration and coordination of contracts in connection with external service providers. Indirect costs may include contract monitoring and oversight, contract generation and procurement, intangibles and transition costs. These costs may increase with an increasing number of external service providers and reduce the annual savings. Another view is that external service providers with an integrative service offer cannot provide the full range of services required of companies that outsource. They offer a range of disparate services and fail to do anything well (Drion et al., 2012). If important functions are being outsourced, an organisation is heavily dependent on the external service provider. Risks such as bankruptcy and financial loss cannot be controlled. These risks increase with a decreasing number of external service providers. This may also reduce the annual savings of the demanders. Companies have to find a trade-off between the costs of complex administration and coordination of contracts for a high number of external service providers and the dependence on the external service provider.

Benefits and sacrifices

Based on the results of the surveys, the statistical models and expert interviews, practical examples of interventions in the area of cost saving and increase of productivity are:

- *Establish own FM department:* As an own FM department manages the different facility services better and leverages through the central management synergies between the different services, as a first step companies should establish their own internal FM departments.
- *Reorganisation:* Most companies concentrate on process optimisation mainly for the core processes. Secondary processes are not the focus of optimisation projects. To use methods like business process optimisation for secondary processes as well can reduce costs and increase productivity.
- *New type of contract:* The use of results-orientated contracts instead of task-oriented ones makes it possible for the service provider to use innovative production methods. This type of contracts makes investments in new technology more reasonable.
- *Rates:* The use of benchmarking pools and the comparison and evaluation of hourly and m² rates offered makes it possible to save costs. However, it is also necessary to validate the offers. That means analysing whether the tasks can be carried out within the given cost framework. Collective agreements provide information about hourly wage rates and help to validate the offers.
- *Technical upgrade:* Investments in energy saving technology (LED, free cooling chillers, thermal isolation) lead to lower energy consumption, reduction of CO_2 emissions and cost reduction. Also, investment in automation (electronic entrance systems) reduces personnel costs. Due to high wages especially in Western Europe, the payback period is rather short.
- *Utilisation of synergies:* The reduction of service providers can lead not only to economies of scale but also to economies of scope. According to Redlein and Pichlmüller (2007) 80% of all routine inspection can be done by additionally trained personnel. Therefore job enlargement in the area of security etc. can be used to leverage economy of scope and thereby increase productivity. Persons carrying out security patrols can also perform routine inspections during their tours. In addition, costs can be saved due to reduced preproduction costs, for example. The time to travel to the premises of a technician carrying out the routine inspection can be saved as a person already in the facility can do the tasks.

Such interventions can result in cost savings, which can be seen as benefits, while costs and expenses can be seen as sacrifices. However, cost savings can have related sacrifices as well. For instance, cost savings that involve lower service levels might cause dissatisfaction and resistance among staff and problems with attracting and retaining the best employees.

How to measure and manage

Key Performance Indicators

Proper interventions and benchmarking as described in the EN15221-7 (CEN, 2012) can be used. Strategic benchmarking can be used for resource allocation decisions and budget planning. Process benchmarking can be used to identify the prioritisation of process optimisation projects. Performance benchmarking supports the assessment of property performance and cost effectiveness.

Examples of benchmarks to identify relevant areas for cost savings and increase of productivity are given in Chapter 5.2 of EN15221-7. Commonly used financial ratios are:

1   FM cost per m², workstation or full time equivalent (fte)
2   Space and infrastructure cost per m², workstation or fte or more detailed
3   People and organisation cost per m², workstation or fte or more detailed
4   Space cost per fte etc.
5   Workplace cost per fte etc.

Out of the comparison of these ratios internally or with external parties, relevant areas for cost savings and/or potential increase of FM productivity can be identified. A prerequisite is to structure costs adequately. A proper structure can be found in EN15221-7 (Table C.1 in the standard). Companies can start with the top level of operational facility products shown in Table 2.3 in Chapter 2 of this book. Later on companies can use the more detailed specification to identify problems at a detailed level.

Perspectives

The significant results of the surveys show that there are several areas for cost savings, and that FM can help to leverage them. Companies with their own FM department tend to have higher annual savings than companies without their own FM department. An inhouse FM department guarantees the best realisation of optimal real estate management. The expert knowledge of optimal management makes it possible to achieve savings through the use of FM. For example, clearly arranged real estate documents, contracts and floor plans in one central place help to identify cost saving potentials (Redlein et al., 2007). The surveys identified potential areas and interventions. The benchmarking process according to EN15221-7 can help to identify the relevant areas in each company, as well as proper intervention to leverage the potential.

Even though FM does not equal outsourcing, it is an important method within FM. The most outsourced facility services in Austria, Spain and Germany are cleaning, technical maintenance and heating/ventilation/air conditioning. Most of the companies had between three and ten external service providers, and the annual savings are highest for these companies, but the correlation between the number of external service providers and the degree of outsourcing and the annual savings is weak. Although organisations may outsource for cost-related reasons, there are no guarantees that expected savings will be realised. The literature also warns that there is an initial tendency to overstate the benefits of outsourcing, and that suppliers are likely to perform better at the beginning of a contract to make good first impressions. If outsourcing is to be fully integrated as a valid and respectable management tool, it must be pursued with a clear sense of where, when and why it leads to enhanced value and cost savings (Alexander and Young, 1996).

The effects of outsourcing on an organisation's costs are not yet fully understood and perhaps the variables and their relationships are more complex than expected. Further studies should also include an investigation of "soft factors". More detailed analyses of different industries would be useful to gather more information and expand the data within this research field, in order to get more valid and reliable cost information and a better understanding of cause-effect relationships between interventions, cost effects, additional effects including sacrifices, and the effects of intermediating variables. In addition, a comparison of these studies with similar ones from other countries would help to gather more information about this research field.

References

Alexander, M. and Young, D. (1996) 'Outsourcing: Where's the value?' *Long Range Planning*, 29 (5), pp. 728–730.

CEN (2012) *Facility Management – Part 7: Guidelines for Performance Benchmarking*. European Standard EN 15221-7, European Committee for Standardization.

Chotipanich, S. (2004) 'Positioning facility management', *Facilities*, 22 (13/14), pp. 364–372.

Dodge, Y. (2008) *The Concise Encyclopedia of Statistics*, Berlin, Heidelberg: Springer Verlag,

Drion, B., Melissen, F. and Wood, R. (2012) 'Facilities management: Lost, or regained?' *Facilities*, 30 (5/6), pp. 254–261.

Hauk, S. (2007) *Wirtschaftlichkeit von Facility Management*, PhD dissertation, TU Wien.

IFMA (2006) *An inside look at FM Outsourcing*. Summary available at: http://www.ifma.org/publications/books-reports/an-inside-look-at-fm-outsourcing (accessed 12 January 2014).

Ikediashi, D.I., Ogunlana, O., Boateng, P. and Okwuashi, O. (2012) 'Analysis of risk associated with facilities management outsourcing: A multivariate approach', *Journal of Facilities Management*, 10 (4), pp. 301–316.

Jensen, P.A. and Van der Voordt, T. (2015) *The added value of FM: How can FM create value to organisations. A critical review of papers from EuroFM Research Symposia 2013–2015*. Research Report. Baarn: EuroFM publication.

Kok, H.B., Mobach, M.P. and Omta, O.S.W.F. (2011) 'The added value of facility management in the educational environment', *Journal of Facilities Management*, 9 (4), pp. 249–265.

Kremic, T., Tukel, O.I. and Rom, W.O. (2006) 'Outsourcing decision support: A survey of benefits, risks and decision factors', *Supply Chain Management: An International Journal*, 11 (6), pp. 467–482.

Lam, T.Y.M. (2011) 'Outsourcing of housing services in Hong Kong: Optimization of transaction value', *Journal of Facilities Management*, 9 (2), pp. 114–126.

Leidig, G. (2004) *Prozesskosten-Management*. Arbeitskreis Klein- und Mittelbetriebe, Eschborn. http://www.rkwkompetenzzentrum.de/fileadmin/media/Dokumente/Publikationen/2004_FB_Prozesskosten-Management.pdf (accessed on 07.01.2014).

Mierl, M. (2012) *Trends und Status Quo 2011 im Facility Management im Vergleich mit 2010*, Master's thesis, TU Wien.

Redlein, A. and Pichlmüller, H. (2007) 'Value Engineering im Bereich Facility Services', in Burtscher, D. (Ed.) *Value Engineering, Partnering, PPP – Neuer Wein in alten Schläuchen? Beiträge aus Theorie und Praxis, Tagungsband ICC 2007.* Innsbruck University Press, 1. Auflage, Innsbruck, ISBN: 978-3-902571-27-4; S. 35 - 46.

Redlein, A. and Sustr, F. (2008) 'Economic Effective Implementation of FM: A Research Based Upon Financial Data', *Proceedings of 7th EuroFM Research Symposium*, EFMC2008, Manchester, pp. 221–229.

Redlein, A., Schauerhuber, M. and Hauk, S. (2007) 'Parameters for an economic efficient implementation of FM'. *6th EuroFM Research Symposium, Zurich, Switzerland, 26 and 27 June 2007*, Conference Papers, pp. 109–116.

Redlein, A. and Zobl, M. (2014) 'Outsourcing: A Cost-Saving Approach in FM?' in Jensen, P.A. (Ed.) *Proceedings of CIB Facilities Management Conference: Using Facilities in an Open World – Creating Value for All Stakeholders. Joint CIB W070, W111 & W118 Conference. Technical University of Denmark, Copenhagen, 21-23 May 2014.* Centre for Facilities Management – Realdania Research, DTU Management Engineering, and Polyteknisk Forlag, May.

Scharer, M. (2002) *Wirtschaftlichkeitsanalyse von CAFM Systemen*, Diplomarbeit, WU Wien.

Smit, I. (2008) *A study on the added value of Facility Management*, Thesis report, Wageningen UR.

Zechel, P., Bächle, A., Balck, H., Felix, P., Flecker, G., Friedrichs, K., Geertsma, C., Henzelmann, T., Hovestadt, L., Hovestadt, V., Janecek, M., Mende, W. and Neumann, G. (2000) *Facility Management in der Praxis: Herausforderung in Gegenwart und Zukunft*, Renningen: Expert Verlag.

Interview 9: Victor Collado, Sodexo, Spain

Victor Collado has twenty-four years of experience in FM. His educational background is business administration. For most of his career he has worked in client organisations, but for the last four years he has worked as a service provider at Sodexo. His responsibilities concern creating solutions to sell FM globally, as well as managing processes locally. In addition, he is responsible for research in IFMA-Spain and active in EuroFM.

Added value in general

Q. What does the term "Added Value" mean to you?

A. *I use the term in my daily practice – also as a marketing tool. It is related to the contribution for the achievement of the objective of an organization. This is the explanation I use to open the conversation. It is valid for any kind of organisation in both the private and the public sector. However, the value proposition depends on the purpose of the organisation and the character of the buildings and facilities considered. For example, for manufacturing clients, the main priorities are usually business continuity and Health, Safety, Security, Environment and Quality (HSSEQ), in labs quality used to be the primary driver, and in offices corporate image and employee satisfaction are normally more important. Of course, cost efficiency is always there.*

Q. How do you see the relation between "Added Value" and cost reduction?

A. *The usual conversation between client and supplier is about "cost reduction" – people usually think about reducing the price. It is not common practice to have a conversation about Total Cost of Ownership, because it is a more complex topic. I try to focus the conversation on "How do you manage your facilities?" There is a limit to price reduction, but with the help of process analysis and technology there are always new ways of improving efficiency.*

Q. Is "Added Value" mostly treated at a strategic, tactical or operational level?

A. *It is mostly treated at the tactical and strategic levels. During the sales process it is a very clear argument with central and high level clients. At this level the topic is more conceptual and generic. When you design the local solution, added value becomes more tangible. At the operational level it is normally a more complex conversation, sometimes because of lack of knowledge and other times due to resistance to change or cost pressure.*

Q. In which context or dialogue is "Added Value" mostly being considered in your experience?

A. *At user and customer level, people are not always clear about how FM can add value, because FM is not their main activity or they see it as a commodity. At the provider side it is also a challenge to communicate internally with the operational level. It is important to train colleagues to be more convincing about the added value of FM. As a service provider I have added value conversations in three different stages in relationships with clients:*

1   *Contract negotiation with central client. Added value is treated in a conceptual way*
2   *Contract deployment with local client. Focus is more on savings than other factors*
3   *Contract operation – with a large difference between years one and two*
     a   *During year one: Focus on building trust and delivering savings.*
     b   *From year two: When the collaboration starts to consolidate, a real partnership between local operational teams (client and provider) can develop.*

Q.   How do you see the relation between innovation and "Added Value" in FM?
A.   *As part of contract commitments, the service provider must come up with a certain number of suggestions for innovation every year. Maybe innovation is not a prerequisite for added value, but it is always included in any conversation about added value.*

Benefits and limitations of "Added Value"

Q.   What are the benefits of considering and talking about "Added Value"?
A.   *From a service provider point of view, the main benefit is to avoid FM being seen as a commodity and to be able to make a differentiation compared with single service competitors. We focus on the benefit of integration of different singular services within one service provider and talk about people and experiences. Sodexo's delivery strategy is to self-deliver most of the services. The outcome of this strategy is a better alignment between different teams (cleaning, maintenance, security, food, reception, etc.).*

Q.   What are the limitations of considering and talking about "Added Value"?
A.   *Added value depends on the vision, background and objectives of people, and their level in the client organisation. They usually adopt a different position depending on their level of maturity within the outsourcing process. There are three different types of clients:*

1   *Those who tell you how to do things: No possibility to discuss added value*
2   *Those who tell you what to do: Some possibility to discuss added value*
3   *Those who share the why of their needs with you: Good possibility to discuss added value.*

Management of added value

Q.   What are your top five main values to be included in management of accommodations, facilities and services? Could you mention a few examples of concrete FM interventions to attain these added values and KPIs to measure them?

A.   *See table below.*

Prioritised values in FM	Concrete FM interventions	KPIs
1. Financial value: Cost Reduction	Ensuring cost reduction, traceability, accuracy and standardisation of reporting.	Budget accuracy. Savings materialised. Invoicing and reporting accuracy.
2. Customer satisfaction: To ensure the right level of satisfaction and identify improvement areas.	Clients: By a consistent governance process. User: By fluent feedback on their perception of services.	% of satisfaction in different and periodical surveys for users, customers and client leadership.
3. HSSEQ: To ensure compliance and the best level of HSSEQ.	Producing an integrated HSSEQ and business continuity plan for all services provided; adapted to the specificities of the client core activities on site.	Number and severity of incidents. Level of outputs from regulatory audits.
4. Innovation: To demonstrate know-how, best practices implementation and continuous improvement.	Sharing benchmarking and best practices across sites and contracts.	Number and impact of innovations proposed.
5. Service and asset performance: To ensure compliance with the level of quality agreed for each service.	Produce an operating manual with the key elements to integrate and manage the services in scope.	Outputs from monthly service audits. Availability of assets. Service continuity.

Q.   Do you measure whether you attain the targeted added values?

A: *User satisfaction is measured by surveys at least once a year in order to explore user perception of services (accessibility, time of response, quality etc.). Customer satisfaction is measured by leadership surveys and interviews usually conducted on a quarterly basis. The points considered are more related to the management of the services provided and the achievement of the agreed objectives. When the contract is global, regional or multi-site, KPIs must be designed in a standard way, but with adjustments of the priority weighting and the acceptable performance criteria for each service line. This is the only way to track and control of the service performance.*

Q.   Do you benchmark your data with data from other organisations?

A.   *We use internal benchmarking to identify areas for improvement, implement best practices and achieve savings. In addition our operational experts use data coming from different sources, depending on the services provided.*

Q. What other methods do you use to document "Added Value"?

A. *We produce a business case for each proposed innovation. All relevant processes are documented in a "Site Operating Manual", which is shared with the client and updated at least once a year. We need to demonstrate that all processes, teams and facilities are properly managed. This manual compiles all relevant service operation processes, governance and reporting procedures, and plans for business continuity and HSSEQ, etc. Every HSSEQ incident must be investigated and the corresponding corrective and preventive actions are documented. The learning is shared with the client.*

Q. How is "Added Value" included in your communication with your stakeholder?

A. *We use the Balanced Scorecard to demonstrate compliance with added value commitments. It usually contains the same elements but with a different approach depending on the site function and industry segment (offices, manufacturing, R&D, warehouse etc.).*

Q. Which top three aspects do you find most important to include in your value adding management approach?

A. 1   *Ensure the right level of service*
   2   *Resources optimisation*
   3   *Continuous improvement.*

Q. Have you noticed a change in mindset regarding added value?

A. *In Spain the FM discipline is relatively new. The concept of integration and management of facilities in a more strategic way is now starting to grow. I think that promotion of value added in FM is the next step on our maturity journey. Except for cost reduction, user satisfaction and risk transference, it is difficult to introduce any other added value topic today.*

Last comment

Q. Are there other topics that you find important in connection with the concept of added value?

A. *I believe that there is a strong relationship between value added and the alignment of the client's FM department with the strategy of its client organisation. If this link is effectively created, it is possible to create a real partnership with FM providers.*

14 Value of assets

Hilde Remøy, Aart Hordijk and
Rianne Appel-Meulenbroek

Introduction

This chapter will focus on the financial side of Corporate Real Estate (CRE). The effect of ownership or leasing on the balance sheet will be discussed. In addition, the lifecycle effects of ownership will be looked at from the point of view of aspects such as renovation, restructuring or alternative use. The importance of regularly valuing CRE at market value will be highlighted, as well as the financial risks of not valuing and not strategically managing CRE. CRE is not always easy to value: it might have specific characteristics, which will not have a market value, or only for similar enterprises and specific use. The value might also be influenced by industry trends or labour costs followed by shifts of the company's activities to other locations or even other countries. Consequently, active CRE financial management should be given high priority. Involvement of CREM in business plans and decisions is essential to fulfil that role.

State of the art

Typically, one or all of the following three approaches might be used to value real estate: a sales comparison approach, a cost approach or an income capitalisation approach (Lusht, 2012). There are interconnections between the three approaches. The direct capitalisation method is the most common approach for investment purposes, and is based on an estimate of the annual potential gross income, taking vacancy and rent collection losses into consideration when determining the effective gross income. Next, the annual operating expenses are deducted and the annual net operating income is calculated. Finally, the capitalisation rate is estimated and is applied to the property's annual net operating income to estimate the value of the real estate. The applied capitalisation rate will be derived from market evidence, and that is where the sales comparison approach comes in: transactions of comparable real estate will be looked for and adjusted for differences in the characteristics and the time of the transaction. In the case of development opportunities, the cost approach (costs of new development with the same functions and appearance) is also applied to justify the choice between investment in existing real estate or a new build.

The three approaches might also show up individually. For home-owner occupiers the sales comparison approach is the most common approach. In case of a comparison between alternatives of renovation or building new, the cost approach might be most appropriate. The cost approach might also be the most appropriate approach for CRE, especially when a building with specific characteristics or a special location is needed. In other cases, CREM might also have a look at existing real estate – for instance, if the company wants to expand and a next-door building might be for sale.

The frequency at which valuations should be made in the case of CRE is arguable. It will depend on business plans or possible changes in the location and/or the size of the premises. In that case, it might be part of a scenario analysis to create a better view of the consequences of the CEO/CFO/COO's management decisions and to support those decisions. However, it might be wise for the CREM department to update their perception of market values by valuations on a regular basis.

The ability of a property to yield income depends on its physical characteristics. When relocating, office organisations consider properties within geographically defined markets that optimally facilitate their main processes as well as supporting image and financial yield. An often-heard statement is that the rent levels and the asset prices of office properties are determined by "location, location and location". The chosen location is assumed to influence the profitability of the office organisation, and this should result in increased willingness to pay, expressed in rent levels and asset prices, for locations with preferred characteristics (Koppels et al., 2009). On a building scale, sustainable and durable office space is important for image, status and possible expansion (Remøy and Van der Voordt, 2014). Research has revealed value, rental, sale price and occupancy premiums for green office buildings. Moreover, discounts in value were revealed for lower energy rating categories (Newell et al., 2011).

According to the lifecycle perspective, real estate management follows a cyclical process. During the initial phases (initiative, briefing and design), the use value of CRE for the corporation is central. During the property's lifespan, use and operation alternate with adaptations. At certain stages the property will reach a situation where its future usability and value will have to be assessed, and obsolescence may occur if the use value is no longer aligned to the value of the property or the rent paid. This can happen because of deterioration of the building's technical or functional characteristics, or because the costs of use exceed the benefits of occupation. Then the current function of the property is no longer the highest and best use. At this point the property may face major adaptation or be demolished (Blakstad, 2001). Even at this stage real estate always has a value, though it might be limited to the land value minus the costs of demolition, or might even be negative. This residual value of CRE should always be included in strategies for CRE.

CRE on the balance sheet and its impact on the value of assets

Since CRE is seen as a means of production, the underlying market value is normally not looked at as it should be, and regularly monitoring of the market value is not specified or seen as part of the core business. However, the total value of CRE is huge. Brueggeman et al. (1990) mention that CRE is estimated to control as much as 75% of all commercial real estate, and on a book value basis (original investment + additional investments -/- cumulative depreciation) roughly one third of the total assets of Fortune 500 companies is estimated to be real estate. For a corporation, real estate will show up in a balance sheet in different forms; see Table 14.1:

a   Real estate as an asset
When owned, real estate will show up on the balance sheet at book value at the asset side. The corporation is allowed to depreciate real estate according to tax regulations, which vary from country to country. However, according to the International Accounting Standards (IAS), the book value is not allowed to be higher than the market value. If so, according to IAS 17, impairment has to take place, which means a write-off of the book value to market value; this will show up in the profit and loss account. When the book value arrives at market value, the following year depreciation might continue to be based on the market-adjusted book value.

b   Rent and lease obligations (liabilities)
At the liabilities side of the balance sheet the rent obligations show up first. This reflects future rent obligations until the contract expires. The International Financial Reporting Standards (IFRS) 16 requires this, but not all companies and countries have to follow that rule. If not, only the rent payment will show up on the profit and loss account until the rent contract expires. On the liabilities side lease obligations are mentioned as well. These might be in a slightly different form compared to rent obligations, since in some cases the lessee might have an option to buy the real estate on or before the expiry date according to stipulated conditions. If the market value gets higher, the option to buy might be profitable for the lessee.

The best solution for the accommodation of a given company – own, rent, lease, or sale and lease back – will depend on the above mentioned circumstances.

Table 14.1 Real estate on the balance sheet of corporations

Assets	*Liabilities*
tangible	equity
intangible	debts
real estate	**rent obligations**
	lease obligations

Additionally, the company has several options for managing their real estate, including management to maintain the current situation, adaptation and upgrade, adaptive reuse, disposal, or demolition and new build. Brueggeman et al. (1990) conclude that in the case of ownership of CRE the company will enjoy the same benefits as a pure investor. They mention lease income, tax benefits because of depreciation and the right to sell the CRE in the future. On the other hand, if CRE is not the core business, according to Brueggeman et al. the company should include other factors in CREM decision making, especially the opportunity costs of capital invested in CRE.

CRE risks

If CRE is not the core business, one of the risks is that the company will not pay sufficient attention to CRE. In addition, the company will probably be able to earn more money with the capital invested in CRE by using it for its core business. For example, in the supermarket business operating profit from sales will usually be much higher than the costs to rent a facility.

Heywood et al. (2009) stated legislative evolution and broader risk management as two of the seven most important CREM practice issues. Also Brueggeman et al. (1990) established the importance of Corporate Real Estate Risk Management (CRERM). Gibson and Louargand (2002) stated: "Organizations insure themselves against many risks but it is the corporate real estate manager who must develop contingency plans for re-housing the critical operations". To be able to innovate and keep up with economic growth in prosperous economic times, CRE and facilities are needed to facilitate this growth. Business continuity is not the only risk (for a more extended discussion of business continuity see Chapter 12 on risk). With regard to the added value discussed in this chapter, the global financial crisis showed to be an especially important risk factor in terms of being tied to long-lasting expensive rent contracts or having to sell beneath the book value.

Gibson and Louargand (2002) suggested four benefits of CRERM that can help to achieve the goal of maximising shareholder value. CRERM can help to: 1) identify CRE risks and make them visible and understandable for others; 2) implement plans for risks that require a response; 3) create procedures to manage them; and finally 4) help managers focus their time on the most important CRE risks. Using an extensive review of the CRE risks literature combined with expert interviews, six risk categories were determined (Bartelink, 2015):

- Development risks (e.g. zoning plan, land acquisition, tender, planning, budget)
- Financial policy risks (e.g. liquidity, solvability, cost of capital, budget cuts, book value)
- Operational and business policy risks (e.g. FM, malfunctioning installations, health and safety, real estate flexibility, occupancy rate, relocation)
- Location risks (e.g. preferred location, uptime production, accessibility)

- Appearance risks (e.g. design, maintenance)
- External and regulation risks (e.g. natural disasters, terrorism, economy, exchange rate, property market, contracts).

With regard to asset value, some of the financial policy and the external risks are particularly relevant in terms of the acquisition or disposal of CRE, so only these aspects are discussed in more detail here.

CRE choices based on financial policy can influence the asset value in terms of necessary changes. Liquidity risk is the possibility that the money stuck in real estate is suddenly needed for other purposes. Huffman (2002) and Rasila and Nenonen (2008) add that this liquidity risk is not necessarily reduced by leasing, since accounting-wise this money is also spent. This could go even further and become a solvency risk, if the organisation is not able to fulfil its long-term financial obligations due to the amount of debt equity invested in CRE. If the cost of the necessary capital to buy the CRE increases, this can also be a financial risk. The next financial policy risk regards the book value. If the value of the real estate is in reality not the same as the value accounted for on the balance sheet, this becomes a problem when it is disposed of. Systematic valuations of the portfolio could prevent or reduce this risk. Bartelink (2015) showed that when the risks were ranked according to likelihood x impact, it seemed that CRE managers who operate at a strategic level rank financial policy risks higher than respondents working at the tactical or operational level.

External risks are the risks of economic shocks and recession (CBRE, 2012). As Gibson and Louargand (2002) stated, the value of assets is driven by both the property market and its regional and national economy. The CRE performance can be affected by both market developments and the economic situation in general (Bartelink, 2015). In times of economic/market growth, material and labour costs increase and likewise the CRE costs, but possibly also the market value of assets. During a recession vacancy rates increase and rent prices are likely to drop, which can also be a benefit. However, lower rents mean that the CRE generates less turnover and thus its book value should decrease too. From an international perspective, changing exchange rates can cause additional risks. A market effect is that owner occupiers usually want to dispose of their surplus real estate assets in an economic downturn. The available space on the property market increases, leading to reduced CRE value. CRE rents tend to fluctuate too, making the timing of rent negotiations very important (Huffman, 2002).

Prevent or reduce loss of value due to vacancy

Financial and real estate crises and "new ways of working" reduce the need for office space, and office markets become replacement markets without a quantitative need for new office buildings (Lokhorst et al., 2013). Users face an oversupply of square metres that will be disposed of once leases end (Jongsma, 2011). Current space consumption reflects past demand, which depends on past rental rates and past expectations of required space during the course of the contract. It does not reflect the current need for space (Gunnelin, 2005). The implementation of new ways of working by sharing a variety of activity-based workplaces started as early as the early 1990s and is expected to continue. New ways of working is an umbrella concept covering flexible ways of working, including office types like e.g. the home office, working in cafés or other public spaces, flex offices and more (Van Meel, 2015). When lease contracts end and the use of office space is assessed, unused office space will be disposed of. Lokhorst et al. (2013) found that the trend of using less m² per employee seems irreversible. Besides this, demographics show that the (working) population is shrinking.

When a company uses less m², (parts of) CRE will become vacant. The vacancy will have consequences for the financial value of the asset. The market value of a vacant building is much less than the value of an occupied building, as it is based on the potential rental income. Moreover, assessments of the market value become less reliable the higher the vacancy level and the longer the vacancy lasts (Schiltz, 2006; Mirzaei, 2015). This is due to lack of transparency, information asymmetry, infrequency of transactions etc. Hence, investment decision making procedures and investment risk assessing procedures (to determine the property value) are prone to some uncertainty. These procedures are mostly based on a qualitative assessment. Difficulties in identifying the risk sources in real estate investment, and the complexity of measuring it, result in uncertainty and inaccuracy.

A common reaction to vacancy is to do nothing, waiting for better times. However, there is little reason to believe that obsolete properties will be taken up in a market characterised by oversupply and structural vacancy. Possible strategies to cope with vacancy are adaptive reuse, demolition and new building, adaptation or upgrade for the same function, or disposal (Remøy, 2010; Remøy and Van der Voordt, 2014). Which strategies can be chosen also depend on the ownership of CRE.

Benefits and sacrifices

It is clear that CRE managers as well as facility managers are faced with a much more dynamic environment than earlier on. It requires close cooperation with the CEO, CFO and COO functions to try to achieve the most efficient financial solutions under rapidly changing circumstances. CREM includes all decisions regarding real estate assets, properties and facilities. Typical CREM interventions are as follows.

To give up ownership by the sale of CRE

Whether or not this is a realistic option will first of all be a question of market circumstances: sufficient demand from developers and institutional or private investors is essential. Unfortunately, those circumstances are hard to predict (Hordijk et al., 2008). If favourable market conditions coincide with the right moment to sell from a business point of view, it will probably be more a matter of luck rather than a carefully planned decision (Hordijk et al., 2004). Selling CRE can be done in different ways:

Disposal: the CRE does not have a function for the core business anymore and will be sold. It is essential to anticipate the disposal as much as possible to avoid or reduce value loss. Because the market value of vacant real estate is lower than for occupied real estate, renovating or adapting the property to accommodate a new use and finding a new tenant should be considered, even though this may not be part of the corporation's core business.

Sale and lease back: in this case the company sells one or more CRE and leases it back from the new owner. Tipping et al. (2007) describe a number of ways to realise sale and lease back: inclusion of all operational property within the structure of a single trading entity, a separate property and trading division, a straightforward sale and lease back, sale and manage back, and property outsourcing. Especially when a company anticipates obsolescence, sale and lease back might be a financially efficient option. In the past a rental period of ten years was quite common, but that has changed. The reasons for this are that companies that have to obey the IAS/IFRS regimes have to take up the rent obligations in the financial statements, which affects the company's financial position. Another argument is the unpredictability of the achievable rent and the required yield. Hordijk et al. (2010) analysed 275 sale and leaseback transactions in the Netherlands between 2000 and 2008. They came to a conservative conclusion that from the available information about sale and lease back transactions the contract rent is most of the time higher than the market rent, and the contract yield is most of the time higher than the market yield.

To renegotiate the rental contract on or before the lease expiry

CREM should be aware of its position in negotiations and financial opportunities. During a period of economic downturn and with lot of vacancy, CREM might well start to negotiate a lower rent and a longer contract in exchange, if that will fit with the company's long-term business plan. As another example, just recently the largest department store in the Netherlands negotiated a discount on the rent payments with 95% of its property owners, threatening that without such a discount they would go bankrupt.

To buy the leased property

This option is the reverse of sale and lease back. Most likely this option will only be considered by the company if the value of the property is higher than the price to be paid to the lessor, in most cases a bank.

To improve owned CRE

It might be that part of the CRE will not serve the company's core business in the most optimal way. CREM should anticipate that and come up with solutions to improve CRE that the company owns. The composition and management of CRE depends on non-financial and financial objectives, such as accommodation strategies, corporate social responsibility, and accommodation costs. At certain moments the CRE will reach a situation where its future usability and value will have to be assessed, and obsolescence may occur. Once it is clear that a building is obsolete, the CRE manager bases strategy and future use of building on its value and fitness for use, and distinguishes four types of future use: consolidation, adaptation and upgrade for continued use, demolition (and new build), and adaptive reuse. Next to accommodation strategies, the best option depends on location and building characteristics and is schematically showed in Figure 14.1. The four types of future use are discussed further below.

Consolidation

Consolidation includes actions like subletting and searching for new tenants, and disposal of (or selling) the property. The choice is based on forecasting the potential future fitness for use and use value. The cash flow of an occupied office building comprises a diverse range of costs, including operational, maintenance, and capital expenditure on both the building and infrastructure (Borst, 2014). However, vacant office buildings often generate a negative cash flow. When more offices are vacant this can even lead to a negative return on investment of the whole portfolio.

Adaptation or upgrade of the property for continued use

Though smaller renovations are performed every five years (Vijverberg, 2001, Douglas, 2006), at some point the building might be functionally obsolete and

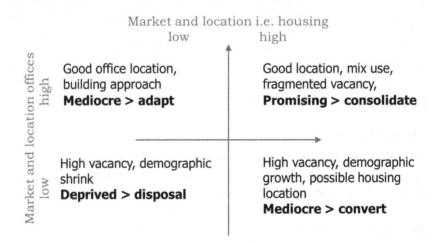

Figure 14.1 The possibilities for dealing with obsolete properties depend largely on market, location and building factors

a more radical intervention is needed to retain the use value. However, in markets with high levels of vacancy there is a risk that the positive effect of adaptation and renovation will be less than the costs of other interventions, like moving and disposal of the building. If the building is still in use, adaptation or renovation may cause disruption for the user. If a thorough renovation is needed, this may be the reason for office organisations to move (Van der Voordt et al., 2012).

Demolishment and new construction

This intervention creates possibilities to better fit to the users' needs. However, redevelopment of CRE takes time and leads to a delay of income, the necessity to move, and if the building is technically in a good state, demolition and new building are a waste of resources.

Adaptive reuse

Adaptive reuse means a physical adaptation and conversion to new use. Adaptive reuse usually requires major changes to the building. Adaptive reuse may be expensive and disrupt the use and the income from the building. The future market value of the new function must be higher than for offices. However, if it works out successfully, adaptive reuse sustains the beneficial and durable use of the location and building, implies less income disruption than demolition and new build, and has high social and financial benefits. For an owner-user adaptive reuse is an interesting option for dealing with a property that is no longer suited for office use.

Adaptive reuse for the owner-user almost certainly includes disposal of the building. Remøy(2010) identified the six most important drivers for adaptive reuse of office buildings:

- Demand for the new function
- Segregation of real estate markets; actors' roles
- Purchasing price of office buildings for adaptive reuse
- Location characteristics
- Building characteristics
- Transformation costs.

Office buildings with cultural-historical, architectural, symbolic, intrinsic values or experience values are often reused successfully. The adaptive reuse potential of newer office buildings depends more on financial/economic, functional, technical and legal aspects influencing feasibility. Office over-supply, sustainability aims, obsolete office buildings and a tight housing market are the most important adaptive reuse drivers. To define the highest and best future use of obsolete CRE, the highest and best use approach is an approach to analyse possible future use and to compare it with the current state of the building. The use or function with the highest probability of use determines the value of the property (Lusht, 2012). The highest and best use analysis is a step-by-step approach, defined by the following steps:

- Physically possible
- Legally permissible
- Financially feasible
- Appropriately justified.

A market analysis provides the value of the property for the four options described above: consolidation, adaptation and upgrade for continued use, demolition (and new build), and adaptive reuse. Furthermore, the functional requirements of new use are analysed and tested by the physical constraints of the existing property. New ways of working have functional requirements that might not always fit within the existing building, and likewise, adaptive reuse might comprise functional requirements that are difficult to fit in, or that lead to a 'waste' of space in the existing building. In the adaptive reuse of existing properties, physical constraints are translated into construction costs, which can be crucial for the financial feasibility of restoration and adaptive reuse. The next step is to study legal possibilities. Environmental laws, central and local government regulations are legal issues that could stop the adaptive reuse of a property (Geraedts and Van der Voordt, 2007). Finally, deciding on the best option for future use depends on which function generates the higher income, the construction costs and rental/selling incomes and costs, and the perceived risk of the different possibilities.

Table 14.2 summarises the benefits and sacrifices of a number of typical CRE interventions.

Table 14.2 Benefits and sacrifices of CRE interventions

Typical interventions	Benefits	Sacrifices	Risks
Disposal of CRE	Unused CRE has no value to the owner-user	The market value of unused office space might be (too) low, or lower than the book value. Partial vacancy in a building might be difficult to lease because of security or lay-out	Timing
Sale and lease back	Real Estate is not core business. Free resources for core business. Anticipate obsolescence	Long leases are common, which may affect the company's financial position negatively. Contract rent is often higher than market rent	Long-lasting obligations

Typical interventions	Benefits	Sacrifices	Risks
Renegotiate rental contract	Possibility to lower the rent, especially during economic downturn and periods with high vacancy	Less flexibility	Market conditions
Buy the leased property	Beneficial if the market value is higher than the asking price	'Freezes' resources in CRE	Less money for core business
Improve owned CRE by adaptive reuse	Reduce operation costs of (unused) CRE Improve the value and return from real estate Reduce vacancy in the portfolio Add value before disposal	Adaptive reuse of properties tend not to be the core business of owner-users	Project management Insufficient market insight

How to measure and manage

Key Performance Indicators

First of all from the benefits and sacrifices referred to in Table 14.2, the different CRE areas of influence are presented in Table 14.3. From that table KPIs have been identified and described.

Prioritised management activities

From Table 14.3 with a long list of areas of influence as described above, the following 5 KPIs are crucial to CREM and should be involved in the business developments and decisions:

In the case of ownership

1   Operational: if CREM wants to sell, achieving the highest price will be highly dependent upon market circumstances.
    CRE intervention/action: regular market valuations to identify the right time to sell.
2   Location development: rather than wait until the premises are out of date, CREM should anticipate obsolescence by timely actions.
    CRE intervention/action: regular reporting to CEO/COO/CFO about the development of the location, like new regulation/zoning plans, new development and/or renovation.

Table 14.3 CRE influence depending on position as owner, tenant or lessee

CRE position	Possible interventions	Areas of influence					
		Operational consequences	RE market exposure	Business changes	Regulation restrictions	Location development	Duration
Position as tenant	Contract is binding maybe renegotiate	Occupancy and operational costs	Adjustable rent	Depends on RE and the owner	Restrictions in use of RE	Higher or lower market rent	Contract expires or renegotiations
Position as lessee	Contract is binding maybe sublease	Occupancy and operational costs	Fixed lease amounts	Depends on RE and the owner	Restrictions in use of RE	Option to buy?	Until contract expires, long lasting
Owner-ship	Consolidate/maintenance (hold)	Occupancy and operational costs	Low	Low	Restrictions in use of RE	Obsolescence risk	Long or until to be sold
	Adapt and upgrade	High one-off costs, lower operational costs	Low, if corporation stays; high, if relocation required	Possible change of way of working	Restrictions in use of RE	Contribute to value of location	Long or until sold
	Demolish and new build	High one-off costs, lower operational costs	New market segment	Make changes to keep business model	Planning regulations	Contribute to value of location	Long or until sold
	Adaptive reuse (conversion)	High costs, sales benefits	High (different market)	Different business model	Planning regulations	Change of use, contribute to improve liveability	Temporary up to ten years or sold after conversion
	Disposal	Sales benefits	Low or high value?	End of business	Tax restrictions?	None	Required time to sell

3   Business changes: Adapt and upgrade, demolish and new build or adaptive use are options to deal with CRE that will become obsolete. All three options require an assessment of temporary or permanent relocation. The first two options mean physical changes that could sustain change in the way of working of the corporation.

CRE intervention/action: regular cooperation with CEO/COO/CFO to identify the business needs, the possible adjustments of the premises and the financial justification.

CRE position as tenant

4   Business changes: depends on whether the owner is willing to adapt or renovate when business developments require it.

CRE intervention/action: involve the owner, communicate the possible changes in the business and as a consequence the usefulness of the premises for the company. Negotiate possible adjustments/renovations as a condition to stay in the premises for a longer period.

CRE position as lessee

5   Duration: lease obligations are usually long lasting, eventually ending in ownership, but at that time the CRE might not be suitable anymore. So it is important to anticipate which of the options mentioned in Table 14.3 are most appropriate.

CRE intervention/action: involve the lessor, communicate the possible changes in the business and as a consequence the usefulness of the premises for the company. Negotiate possible sublease if premises have potential to become obsolete.

Perspectives

This chapter has shown the benefits and sacrifices of having CRE on the balance sheet, and the influence of CRE on the value of assets. The CRE risks and the benefits and sacrifices of typical interventions in CRE were described.

Future studies on CRE risks need to provide more insight on the actual impact of certain events on the asset values of owned properties. Second, more insight is needed on the impact of property attributes on the depreciation rate of real estate. In addition, research is needed on whether or how CRE, property, asset and facility managers could act to prevent value loss or even increase the value of owned properties. There is still a lack of theory building on the risks of CRE and how to manage them and thus provide more added value, both for asset value as well as for the other added values discussed in this book.

References

Bartelink, R. (2015) *Corporate Real Estate Risk Management*, Master's thesis, Eindhoven University of Technology.

Blakstad, S.H. (2001) *A strategic approach to adaptability in office buildings*, Trondheim: Norwegian University of Science and Technology.

Borst, S. (2014) *Now hiring: Vacancy management at portfolio level*, Master's thesis, Delft University of Technology, Delft.

Brueggeman, W.B., Fisher, J.D. and Porter, D.M. (1990) 'Rethinking Corporate Real Estate', *Journal of Applied Corporate Finance*, 3 (1), pp. 39–50.

CBRE (2012) 'Risk management in corporate real estate within the global banking and finance sector', *Global ViewPoint*, pp. 1–7.

Douglas, J. (2006) *Building Adaptation*, Oxford: Butterworth-Heinemann,.

Geraedts, R.P. and Van der Voordt, D.J.M. (2007) 'The New Transformation Meter: A New Instrument for Matching the Market Supply of Vacant Office Buildings and the Market Demand for New Homes', paper presented at the *International Conference Building Stock Activation*, Tokyo, November.

Gibson, V.A. and Louargand, M. (2002) 'Risk Management and the Corporate Real Estate Portfolio', *Proceedings American Real Estate Society Annual Meeting*, Naples, USA: The American Real Estate Society (ARES).

Gunnelin, A., England, P., Hendershott, P., Soderberg, B. (2005) *Adjustment in Property Space Markets: Estimates Form the Stockholm Office Market*, Cambridge: National Bureau of Economic Research.

Heywood, C., Kenley, R., and Waddell, D. (2009) 'The CRE toolbox: Addressing persistent issues in corporate real estate management', *Proceedings of 15th Pacific Rim Real Estate Society (PRRES) Conference*, Sydney, January.

Hordijk, A.C., de Kroon, H.M. and Theebe, M.A.J. (2004) 'Long-run return series for the European continent: 25 years of Dutch commercial real estate', *Journal of Real Estate Portfolio Management*, 10 (3), pp. 217–230.

Hordijk, A.C. and Teuben, B. (2008) 'The liquidity of direct real estate in institutional investors' portfolios: The case of the Netherlands', *Journal of Property Investment and Finance*, 26 (1), pp. 38–58.

Hordijk, A.C., Koerhuis, L. and Rompelman, D. (2010) 'Ten Years of Sale and Leaseback transactions in the Netherlands', *Journal of Corporate Real Estate*, 12 (1), pp. 26–32.

Huffman, F.E. (2002) 'Corporate real estate risk management and assessment', *Journal of Corporate Real Estate*, 5 (1), pp. 31–41.

Jongsma, M. (2011) 'Verborgen Leegstand Bij Kantoorhuurders', *Financieel Dagblad*, 27 May 2011.

Koppels, P.W., Remøy, H. and de Jonge, H. (2009) 'Economic value of image, location and building', *Real Estate Research Quarterly*, 8 (3), pp. 31–38.

Lokhorst, J., Remøy, H. and Koppels, P.W. (2013) 'Hidden Vacancy; the Occurrence, Causes and Consequences', paper presented at *ERES*, Vienna..

Lusht, K.M. (2012) *Real Estate Valuation: Principles and applications*, USA: KML Publishing.

Mirzaei, S. (2015) *Valuation accuracy in vacant office properties: A comparison between appraised cap rates ad transaction cap rate*, MSc thesis, Delft University of Technology, Delft.

Newell, G., MacFarlene, J. and Kok, N. (2011) *Building better returns*, Sydney: Australian Property Institute (API).

Rasila, H.M. and Nenonen, S. (2008) 'Intra-firm decision-maker perceptions of relocation risks', *Journal of Corporate Real Estate*, 10 (4), pp. 262–272.

Remøy, H. (2010): *Out of Office: A Study of the Cause of Office Vacancy and Transformation as a Means to Cope and Prevent*, Amsterdam: IOS Press.

Remøy, H. and van der Voordt, D.J.M. (2014) 'Priorities in accommodating office user preferences: impact on office users decision to stay or go', *Journal of Corporate Real Estate*, 16 (2), pp. 140–154.

Remøy, H.T. and van der Voordt, D.J.M. (2014) 'Adaptive reuse of office buildings: Opportunities and risks of conversion into housing', *Building Research and Information*, 42 (3), 381–390.

Schiltz, S. (2006) 'Valuation of vacant properties', *Real Estate*, Amsterdam, Amsterdam School of Real Estate

Tipping, M. and Bullard, R. (2007) 'Sale-and-leaseback as a British real estate model', *Journal of Corporate Real Estate*, 9 (4), pp. 205–217

Van der Voordt, D.J.M., Remøy, H., and Hendrikx, T. (2012) 'Renoveren of Verhuizen? Duurzaamheid Speelt Grote Rol Bij Besluitvorming', *Facility Management Magazine*, (25) 202, pp. 50 –52.

Van Meel, J. (2015) *Workplaces Today*, Rotterdam: ICOP.

Vijverberg, G. (2001) *Renovatie Van Kantoorgebouwen Literatuurverkenning En Enquete-Onderzoek Opdrachtgevers, Ontwikkelaars En Architecten*. Kantorenmarkt 5. Delft: DUP Science.

Interview 10: John Suyker, Suyker Invest BV, the Netherlands

John Suyker has thirty-five years of experience in CREM. He is a co-founder and past President of IDRC Europe, the first European CRE platform established in 1990 which later converted to Corenet Global. John came to the Netherlands to study Mechanical Engineering. He formerly worked at Sealand Shipping, Tandem Computers, ABN AMRO, AMS-Cgi and Johnson Controls. For the last six years he has owned a private consultancy company, Suyker Invest BV. He lectures at Hanze University of Applied Sciences in the post-graduate MRE course. His main topics of interest include CREM and FM Integration, Integrated Global Workplace Solutions, CRE-masterplans, CREM modelling, change management and re-accommodation plans, global accounting, and technological and functional performance. John is also involved in organisational transition and change management with the Kgottla Company, supporting various transition projects.

Added value in general

Q. What does the term "Added Value" mean to you?

A. *Being conscious about costs but also benefits for the company's products and services. Added Value for the internal and external client can be tangible, but may also be less tangible or soft – e.g. a higher speed of delivery, increased wellbeing of employees, extra workplace comfort and flexibility and making complex things simpler, easier and faster for management to make decisions. Clients trust you based on your authority or objective advice and experience, not always necessarily based on hard data. In the case of the "trusted advisor" clients will be open to more than just giving a quick solution to a specific problem. For instance, once our target and task was to reduce the European occupancy costs by 5–10%. We succeeded in doing so, but after benchmarking with an industry peer group we concluded that 30% more cost reduction should be possible by the introduction of stricter space policies and eventually New Ways of Working. By exceeding management expectations your position is safeguarded and value-addition proven.*

Q. Are there other related terms that you prefer to use rather than "Added Value"?

A. *The use of the Added Value concept has changed over time. In the 1990s the focus evolved from bottom line Earnings Per Share (EPS) to Shareholder Value and eventually Economic Value Added (EVA). Nowadays AV has a wider scope and encompasses a variety of financial, economic and softer factors affecting the image and sustainability of the company's future. Whatever terminology we use depends on the person we talk with, e.g. stakeholder position, or their level in the organisation, e.g. CEO or operational manager.*

Q. How do you see the relation between "Added Value" and cost reduction?

A. *Cost reduction can be an added value in itself, but it should be assessed in connection with effects on other factors and their values. Too many cost reduction programmes have proven to be short-sighted, poorly communicated and seen as a quick easy fix to an immediate problem.*

Q. Is "Added Value" mostly treated at a strategic, tactical or operational level?

A. *At all levels, but in different ways: a) Strategic: focus on long-term decision support, risk avoidance, shareholder satisfaction (CEO/CFO level); b) Tactical: focus on speed of delivery and achieving what is asked or expected; and c) Operational: focus on cost reduction and budget alignment, employee satisfaction and customer satisfaction.*

Q. In which context or dialogue is "Added Value" mostly considered in your experience?

A. *Internally, within and between the CREM and FM organisation (to discuss the roles and responsibilities and who delivers what and when), and with service provider companies on contract negotiations and Service Level Agreements. Externally, the AV concept is used in selling our own values and contribution: your competencies, the value of CREM and FM for the company in connection to being seen as a cost centre ('value for money'). It is always about balancing between tangible costs and softer benefits, i.e. better performance in communicating your value added to the organisation.*

Q. How do you see the relation between innovation and "Added Value"?

A. *FM/CREM can support innovation and innovation can result in added value by developing better products, faster delivery, an interdisciplinary approach etc. Innovation is a continuous "thinking out of the box" process. The implementation must be achievable or tangible. Innovative thinking is a must and needs to be stimulated for CREM/FM to be successful. You also have to use knowledge and approaches from different disciplines and learn from these, understanding their relevance to CREM and FM.*

Benefits and limitations of "Added Value"

Q. What are the benefits of considering and talking about "Added Value"?

A. *The AV concept helps to focus on the impact of FM and CREM on the client organisation and the needs and expectations of the client, their customers and end users. The AV concept elevates you to the more strategic aspects of FM and CREM. It also lets you speak the language that top management understands. It is important to understand which value is most important for your client or customer and to raise the right questions to understand what the client really needs. This can often be more than just simply solving the current problem.*

Q. What are the limitations of considering and talking about "Added Value"?

A. *Added Value is perceived differently by different people at different levels or perspectives. It can be difficult to make AV concise and concrete or operationally achievable. The term requires a clear story and explanation. Story telling with examples of value added can convince clients better than explaining theory or presenting a load of data with tables and graphs. Talking about AV at the hands-on operational level can be confusing to some individuals, or too complex and counterproductive.*

Management of "Added Value"

Q. What are your top five main values to be included in management of accommodations, facilities and services? Could you mention a few examples of concrete FM interventions to attain these added values and KPIs to measure them?

A. *See table below.*

Prioritised values	Concrete interventions	KPIs
1. Cost reduction	Less m², e.g. by New Ways of Working (new workplace-sharing ratio policy) Reduction of occupancy costs Length of rental contracts and flexibility Space management is the key driver. Cost reduction by FM (e.g.by less cleaning, less catering) is more difficult: slower, more painful, less impact	Rental + service charge/m² Total cost owned or leased/m² Total occupancy/ workplace Note: avoid €/fte or €/ headcount due to desk sharing concept
2. Affordability	The ratio of total occupancy cost as percentage of total income generated. Can you afford this building portfolio given the organisation's performance and productivity (revenue per person)?	Occupancy cost/revenue Revenue/m² (retail related) Revenue/person Example: benchmark deviation of 3% is OK; > 6% means you have a problem
3. Speed of delivery and problem solving	Consistent and simple A4 templates for Business Case approvals, Dashboard Metrics, Progress Reports	Quarterly progress reporting Business Case statistics % completed Capital Investment projects Deviation from targets on programmes
4. Increase clarity and risk avoidance	Make complex databases, tools and policies/procedures simpler and easier to manage. Increase lease flexibility	Three year listing lease terminations/actions Average lease duration Lease database accuracy %

We look continuously at maximising flexibility as well as optimising total cost of occupancy and risk avoidance in connection with investment costs. Branding and identity are important, but keeping in mind that exclusive-looking buildings are perceived by customers as too expensive and a waste of (our) money. Sustainability is becoming more important but is not prioritised clearly at this stage of its evolution, and is mainly focused on industrial products rather than on buildings. Taking one company as an example, a new CEO was more aware of sustainability and wanted to improve their buildings from level 6 to 8 and if possible to 9. Level 10 was seen currently as not necessary. BREEAM is used as an indicator in Europe and Japan; in the USA usually the LEED accreditation is used.

An example of proven added value is the increased employee satisfaction and productivity in a French organisation, from far below average to highly above average due to an organisational change allowing a relocation to a much more effective (one third the number of floors) and 50% more efficient building space. Instrumental was the implementation of a worldwide IT-mobility programme and changes in the workplace concept and sharing. However, the direct cause-effect relationships of the workplace effect compared to the IT tools and organisation change factors are difficult to trace and allocate to the three factors. Effects are then the result of an integrated approach.

Q.   Do you measure whether you attain the targeted "Added Values"? If so, what Key Performance Indicators do you apply for each of the added values you mentioned before?

A.   *See table above. In addition, satisfaction of employees is measured by annual satisfaction surveys and employee sick leave percentage. Customer satisfaction is also measured by a survey. Security and safety are important as well, in themselves and in connection with health and wellbeing. Safety is measured by the number of accidents and sick leave due to accidents.*

Q.   Do you benchmark your data with data from other organisations?

A.   *Yes, for instance by comparing our use of space with the best practice standards: 9 to 11 average net m^2 rentable floor space per workspace, 15 to 16 net m^2 rentable floor space per workspace when including central circulation support space such as corridors, and archives and indirect support space such as auditoria; and 18 to 19 net m^2 rentable floor space per workspace in our head offices. Comparisons of staff turnover percentage are used as an indicator to leverage a change process and need to start actions towards a new workplace concept. Industry sector key data comparisons have always created a strong leverage for management to support a change process to comply with best-in-class target settings.*

Last comment

Q. Are there other topics that you find important in connection with the concept of "Added Value"?

A. *Pay attention to relations between CREM and FM. Be aware of four fields to create added value:*

Business strategy	**CREM**
	Portfolio management
	Transactions (acquisition, disposal, rental contracts)
	Ownership, renting, sale and lease back choices
FM	**Project management**
Workplace management	Task of architects, consultants
Maintenance management	

We need a book or course on "Mastering CREM in 10 Steps or Days" to better support the CEO's perspective and tell people how to sell value added. We also need case studies in connection with theory, including practical examples and best practices.

15 Sustainability

*Susanne Balslev Nielsen, Antje Junghans and
Keith Jones*

Introduction

This chapter examines sustainability as a value parameter within Facilities
Management (FM) and Corporate Real Estate Management (CREM). The
overall objective is to provide a common approach for measuring and managing
sustainability impacts throughout a building's life-cycle. Public and private
sector organisations, as building owners, facilities managers and users are
addressed as primary internal stakeholders in the value chain; external
stakeholders are considered as the wider beneficiaries of sustainability
improvements at the community, regional, national and global levels. The
framework outlined in this chapter will enable organisations to release latent
value through the transition to more sustainable FM and CREM operations,
while simultaneously helping society address its wider sustainability goals.
However, the key to realising this latent value is the ability to identify the links
between value parameters, sustainability and organisational goals, which will be
addressed in this chapter.

The built assets owned or occupied by an organisation provide the greatest
opportunity to embrace the sustainability agenda. Through effective management
of buildings over their life cycle, building owners and occupiers can reduce the
negative impact that their buildings have on the environment, improve the impact
that their buildings have on social wellbeing, and provide economic benefits to
the business through reduced maintenance and refurbishment costs. Such an
approach provides value not only to the organisation but also to the wider society
of which they are an integral part. However, for many organisations these
opportunities are missed or fail to have the level of impact they could (Kwane et
al., 2009, Baharum and Pitt, 2009, Elmualim et al., 2010; Durmus-Pedini and
Ashini, 2010) due, in the authors' opinion, to the difficulty many organisations
have in identifying and measuring the sustainable performance of their built
assets. The aim of this chapter is to provide a practical framework based around
key performance indicators that facilities managers and corporate real estate
managers can use to develop built asset management plans that improve the
sustainable performance of their built assets and deliver value to core business by
aligning these plans with the organisation's strategic sustainability goals.

Hodges and Sekula (2013), Falkenbach et al. (2010) and Brown et al. (2010) identified the benefits of sustainability offered through FM. These can be summarised as:

1　*Increased productivity:* healthier buildings and better designed workplaces result in increased employee satisfaction and hence increased productivity.
2　*Lower operation and maintenance costs:* reduced demand for resources and reduced waste production results in lower annual costs.
3　*Enhanced competitive edge:* people increasingly want to purchase products and services that are sustainable, as sustainability initiatives are important to attract future customers.
4　*Improved company image:* positive company stories about sustainability aspirations and achievements can enhance company profile. Since more and more people want to work for companies that have a sustainability agenda, this helps in attracting the best people.

In their study of the benefits and risks of greening existing buildings, Durmus-Pedini et al. (2010) identified financial risks, market risks, industry risks, performance risks and legislative risks as potential risks that need to be countered in FM risk management.

In considering how sustainability could be explicitly linked to organisational value, Jones and Cooper (2007) suggested that value in the context of built asset maintenance decision making goes beyond a consideration of building technology issues, to a perspective that acknowledges the impact of the built asset on the long-term viability of an organisation.

Readers are recommended to consult additional book chapters in this volume, and in particular the closely related Chapter 16 on Corporate Social Responsibility (CSR), which emphasises the social aspects of added value creation and complements this chapter's emphasis on the building management perspective and the balance that needs to be achieved between the environmental, economic and social dimensions of sustainability.

State of the art

Definition of sustainability

Sustainability is a broad but contested term that has been used and misused in various ways to describe mankind's symbiotic and/or parasitic relationship with the planet. Indeed, the wide-ranging definitions developed to describe the concept of sustainability provide a continual challenge when it comes to measuring and managing the added value of FM and CREM (e.g. Laedre et al., 2014; Bond and Morrison-Saunders, 2011).

Sustainability as ideal has a long history. This concept has been associated with slightly different meanings over time; but the vision of a sustainable planet is still in use and our current understanding is based on the accumulation of the various meanings. The first definitions are found in sources dating back to the eighteenth century (Kloepffer, 2008; Kummert et al., 2013; Von Carlowitz and von Rohr, 1732). Von Carlowitz and von Rohr (1732) focus on the protection of natural resources from an organisation's perspective. Von Carlowitz developed ideas and concepts about how timber production could be achieved in a way which ensures the availability of wood as needed and protects the forest as a whole system. As superintendent of silver mines in Germany, he expressed that the mining and silver production processes were dependent on good practices for cultivating the forest. Likewise today, mining and other production processes rely on good practices which sustain the resources that feed into their production.

The second meaning which evolved around the 1960s focused on the world's economy and its dependence on limited resources. The concern about exponential growth in a finite and complex system was identified in the research report "The Limits to Growth" (Meadows et al., 1972). An international and interdisciplinary team of researchers developed a model to gain insight into the limits of the world system. Meadows et al. (1972) identified five key factors that determine and limit growth on Earth: population, agricultural production, natural resources, industrial production, and pollution. They studied the limits of the world system by considering different scenarios of human numbers and activities development, referring to exponential growth as the main problem (Meadows et al., 1972).

The third meaning of sustainability is associated with the consideration of the interdependence of society, environment and economy in complex sustainability thinking. The World Commission on Environment and Development (WCED) published the "Brundtland report" in 1987, named after the commission's chairman, the former Norwegian Prime Minister Gro Harlem Brundtland. In this report sustainability was defined as: "…development that meets the needs of the present without compromising the ability of future generations to meet their own needs" (Brundtland et al., 1987, p.9). The United Nations (UN) Conference on Environment and Development (UNCED), commonly referred to as the Rio Conference or Earth Summit, is often cited as referring to the three sustainability dimensions: social, economic, and environment, leading to three sub-parameters of sustainability and a hierarchy of sustainability strategies at global, national and regional levels (UNCSD, 2012). Figure 15.1 shows the three dimensions in a FM and CREM model. These dimensions are also known as the three-P triplet of People-Planet-Profit or People-Planet-Prosperity.

Despite the fact that sustainability is not a new concept, it is here argued that sustainability is still a vision and an ideal more than a reality especially on a global scale, whereas the UN proclaims that Sustainable Development, which balances current needs with the needs of future generations, will be at the core of the UN's development agenda after 2015 (UN, 2015). It should also be noted that massive initiatives to reduce environmental problems have to some extent changed the overall picture of environmental threats and problems. For instance, local air pollution from heat and electricity production that was common twenty years ago hardly exists today. However, we also see new problems arising, for example due to the increasing use of chemical substances in various products,

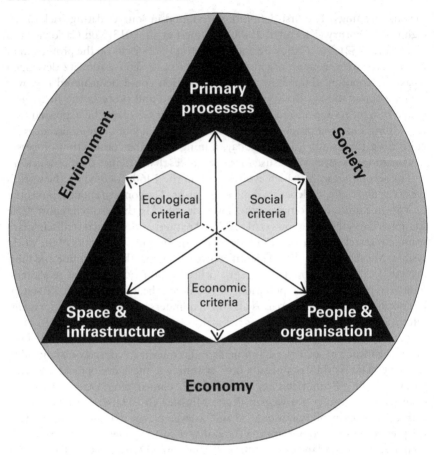

Figure 15.1 Sustainability in FM and CREM model (adapted from Junghans, 2011)

which complicates both waste treatment and sewage treatment and where we do not know the long-term impact on nature. Sustainability in FM seems to be an ongoing quest rather that an absolute, as stated by Fennimore (2014).

Life cycle thinking and environmental impacts

The lifecycle thinking is essential when considering the sustainability of facilities and services. Figure 15.2 shows a simple structure for screening an artefact like a cup, a chair or a concert hall to identify the largest sustainability problems: what kind of materials (resources) are used, and how the artefact is produced, transported, used, and finally disposed of. The structure also identifies the sustainability issues in each of the life-phases, e.g. the use of scarce and non-renewable materials, energy usage, the use of chemical or other sustainability issues such as health and safety issues, or hazards to vegetation and wildlife.

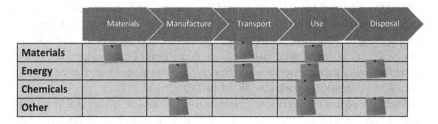

	Materials	Manufacture	Transport	Use	Disposal
Materials					
Energy					
Chemicals					
Other					

Figure 15.2 Structure for screening an artefact's lifecycle and environmental impact (materials, energy and chemicals) and other sustainability issues (adapted from McAloone and Bay, 2009)

As FM and CREM are highly related to buildings and the built environment, a building focused approach to measuring and managing the added value of sustainability efforts is an obvious strategy. A variety of sustainability certifications are already available such as the "Building Research Establishment's Environmental Assessment Method" (BREEAM), "Leadership in Energy and Environmental Design" (LEED) and the "Deutsche Gesellschaft für Nachhaltiges Bauen" (DGNB) assessment methods. The common approach is to break down the sustainability parameter into a hierarchy of indicators and sub-indicators. For example, the Danish adaptation of the German DGNB system claims to take a holistic approach to sustainability and operates with five main themes: Environmental Quality, Economic Quality, Socio-cultural and Functional Quality, Technical Quality, and Process Quality. These five themes are divided into forty-five main criteria, which are themselves divided into 188 sub-criteria, too many to mention in this chapter. Other certifications are available too, and they vary slightly in scope (number of indicators included), how they award points to each evaluation criteria, and how these individual evaluations add up to a specific certification class. DGNB uses the classes Gold, Silver and Bronze. For a facilities manager the ideal certification system might seem to include as many sub-criteria as possible. However, the typical problem is lack of data (e.g. Jensen et al., 2009). As a consequence a full certification process can be time demanding. An alternative is to select a number of indicators/criteria from the certifications which are particularly meaningful and operational in the specific context, and use these for one's own balanced scorecard.

Stakeholders

Different stakeholders may experience different sacrifices and benefits from sustainable transitions into sustainable products, services and practices in FM and CREM. The British Institute of Facilities Management (BIFM) has identified employees, government and clients/customers as the three main stakeholder groups, which organisations report their sustainability activities to (BIFM, 2013). A similar survey, which was conducted in Norway, confirmed the same stakeholder groups in slightly different order: the government, clients/customers, and employees (Junghans et al., 2014).

Hodges and Sekula (2013) provide a longer and more general list of stakeholders who will be impacted by sustainability initiatives, and highlight that the facilities manager needs to understand all stakeholders involved. Who is impacted, and how are they impacted? A plan should be developed for how to meet their sustainability needs while leveraging their varied interests at the same time. The stakeholder groups are listed in Table 15.1.

When the senior management commits to a sustainable FM/CREM plan it is more likely to get the resources needed to implement it. However, other stakeholders still have a say in the implementation process, perhaps because they have formal authority, like a real estate department or in the case of certifying buildings. Other stakeholders might have little or no formal authority over the decision, but may still desire to influence decisions and their outcome. Depending on their motives and agendas, this might not be evident on the surface (Hodges and Sekula, 2013). This makes stakeholder management and involvement one of the core disciplines within sustainable FM/CREM.

Sustainability in European FM standards

From a generic perspective, sustainability is already integrated in the European standards for FM. To the extent that the standards are used to educate future generations of facilities managers, there is a basis to assume that sustainability will be integrated gradually as a value parameter in FM. Sustainability in the meaning of environmental impact is addressed in the standard EN15221-7 on Performance Benchmarking (CEN, 2012), in which the indicators for primary environmental ratios (e.g. total CO_2 emissions in tonnes per annum), primary energy ratios (e.g. total energy consumption in kWh per annum), primary water ratios (e.g. m^3 per annum), primary waste ratios (e.g. total waste production in tonnes per annum), and other environmental scores are used. However, our review of all seven European FM standards (EN15221-1-7), particularly EN15221-4 on Taxonomy (CEN, 2011), shows that sustainability in FM is only explicitly addressed in twelve out of more than one hundred facilities services categories and products, representing less than 10% of the overall scope of FM.

Table 15.1 Stakeholders impacted by sustainability initiatives (with inspiration from Hodges and Sekula, 2013, pp. 70–71)

Internal stakeholders	External stakeholders
• Facility Management	• External building owners (landlords)
• Real estate	• Tenants
• Procurement	• External service providers and vendors
• Legal	• Governing authorities
• Human resources	• Utility providers
• Finance and accounting	• Neighbouring businesses and residents
• ICT support	• The community at large
• Marketing and sales	• Nature, as a non-human stakeholder
• Senior management	

The main focus of sustainability assessment is on the "Space and Infrastructure" service group, and the least attention is given to the "People and Organisation" service group. The twelve facilities service and product categories which are addressed for the collection of qualitative environmental data are: 'Building Initial Performance', 'Property Administration', 'Maintenance and Operation', 'Land, Site, Lot', 'Occupier Fit out and Adaptations', 'Health and Safety', 'Environmental Protection', 'Mobility', and 'Procurement', and at product level 'Energy', 'Water' and 'Waste' (CEN, 2012).

Organisational integration of sustainability

The authors of this chapter claim that the generic perspective is not sufficient to ensure a fast voluntary commitment to sustainability, as the organisational context is so important for the level of commitment (Nielsen, 2012). We argue that sustainability in CREM and FM requires translations of the general sustainability definitions into context-specific definitions referring to organisations' value norms. So how can one align the sustainability strategy of organisations' primary activities with their strategies for supporting facilities and services management? The challenge is to develop a scalable approach which helps to break down the overall ambitions of sustainable development at an organisational level, but is still comparable with the overall understanding of sustainability. The categorisation of societal, environmental, and economic impacts should be done by the organisation itself. This approach from inside to outside is visualised in Figure 15.3. However, it does not mean that triple bottom line indicators have to be balanced symmetrically in each organisation. Looking back to the forestry example from the eighteenth century, it is more important to focus on those aspects which are most important for each organisation's sustainability impact (Junghans, 2011).

In considering how sustainability could be linked to organisational value, Jones and Cooper (2007) suggested that value (in the context of built asset maintenance decision making) goes beyond a consideration of building technology issues to one that acknowledges the impact of the built asset on the

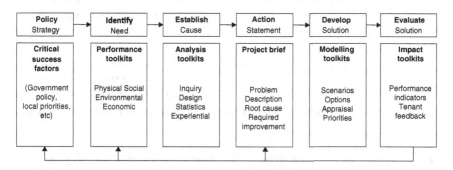

Figure 15.3 Sustainable based built asset management process model (adapted from Jones and Cooper, 2007)

long-term viability of an organisation. Jones and Cooper argued that 'value' should be explicitly linked to the ability of the built asset to support organisational performance, and that built asset management should be viewed as a strategic issue managed within the broader context of an organisation's strategic planning framework. In the context of sustainability Jones and Cooper identified the need to define value in the context of the organisation's sustainability goals, aspirations and critical success factors. These goals and aspirations will normally be informed by the organisation's Corporate Social Responsibility Strategy, its Sustainability Strategy, its HR Strategy, its Environmental Management Strategy, its Quality Management Strategy and others. These documents will provide the direction for the organisation in terms of its environmental, social and economic targets and set timescales by which the targets need to be met. These targets provide benchmarks against which the performance of buildings can be judged. Although benchmarks will be contextually dependent upon each organisation's strategic aspirations and goals, the benchmarks are likely to be informed by national policy underpinned by sector-wide key performance indicators and benchmarks, codes of practice/policy guidance, or international references such as UN sustainability goals (UN, 2015).

Jones and Cooper (2007) applied these principles to the development of a sustainable built asset management model (Figure 15.3) for UK social housing. In this model the long-term strategic goals of an organisation are used to set the context within which FM needs to deliver value to that organisation. FM then can use this context to develop a series of toolkits (using Key Performance Indicators – KPIs) to measure the current performance of a building or service, and to devise a series of analytical processes in order to interrogate the KPIs to establish the root cause of any underperformance or identify process changes that are needed to improve performance. Such performance improvements can then be expressed as an action statement that will provide benchmarks (for the KPIs) against which potential solutions can be evaluated. As many KPIs used to measure performance will be inter-dependent, a multi-criteria approach is needed to model the impact that various solutions have under different operating and future scenarios. Once a solution has been identified and implemented the original KPIs and benchmarks can be used to evaluate the actual performance of the solution in use and inform long-term strategic plans. In many ways this is a Plan-Do-Check-Act framework approach (see also Chapter 17).

Cooper and Jones (2009) further developed this approach in an action research project with a UK social housing landlord. During this project Cooper (2015) identified the issues that facilities managers need to consider when operationalising the sustainability framework. In particular facilities managers have to:

• Consult organisational strategy documents to identify strategic goals and, through discussions with senior managers, express these goals in terms of critical success factors

- Map the critical success factors against a range of possible KPIs taken from existing (generic) toolkits and develop bespoke solutions to specific organisational challenges that were not covered by the generic approaches
- Develop a survey methodology to apply the KPIs to a service or built asset. While the generic KPI toolkits have survey instruments included as part of the toolkit, the bespoke KPIs require survey instruments to be developed
- Develop analytics to analyse levels of performance. The analytics draw on quantitative assessments (e.g. operational research techniques) supplemented with qualitative assessments (e.g. focus groups) to establish the root cause of any performance issues
- Use the action statement as the basis for the design brief or service specification for commissioning new, more sustainable, solutions
- Develop a multi-criteria scoring method to rank the sustainability indicators. This requires consultation with all key internal stakeholders and a consideration of the importance that the organisation places on the wider implications of their sustainability agenda (e.g. social responsibility, image etc.)
- Develop a scoring mechanism to prioritise individual solutions against the sustainability agenda. This involves developing complex multi-criteria assessment models that could accommodate future business and environmental scenarios (e.g. future climate change impacts on internal building temperatures and the link this could have on productivity) as part of the decision making process and support detailed options appraisal against each scenario to identify the most business critical solutions over any given timescale
- Develop and implement a continuous review process that provides actual data on the performance of a solution in use.

While it was clear from the study that facilities managers could operationalise the sustainable built asset management framework, the amount of work required to do so was significant and required the collection of a large amount of data that was not routinely gathered by the organisation through its stock condition survey process. As such, facilities managers need to be aware of the scope of the challenge they face when addressing the sustainability agenda.

How to measure and manage

Key Performance Indicators

In this section the overall perspective is broken down to a smaller scale of single organisations or business corporation units. We look at FM/CREM with a focus on an organisation's demand for sustainable buildings and facilities services in the total building life-cycle. This management perspective includes all life-cycle phases, such as the development, production, management and redevelopment of buildings. The management phase includes operation, maintenance and service provision.

This section will provide a practical framework based around KPIs which facilities managers can use to develop built asset management plans, to improve the sustainable performance of their built assets while simultaneously improving the value of these assets to core business. Once the value criteria have been established they can be used to develop organisation-specific KPIs that can be used to assess the current performance of a building (performance toolkits). The KPIs need to cover the range of value criteria (notionally physical, environmental, social and economic performance) and should be informed by existing KPIs where available (e.g. DGNB, standards etc.). However, when developing (or selecting) KPIs the facilities manager needs to be aware of potential inter-relationships between indicators. In any multi-criteria decision making framework those factors or variables that influence the final decision should only be measured once and should be independent of each other. When considering sustainability this condition is rarely satisfied due to significant overlap and interference between the factors.

One of the largest challenges in measuring and managing sustainability in FM is to establish the scope and alignment with the organisation's strategic goals. It is useful to see how others are dealing with reducing complexity and defining a focus. One example is Laedre et al. (2014) who used the Sustainability Impact Assessment (SIA) methodology (OECD, 2010) as an analytic assessment tool in the context of project management. They applied SIA to the assessment of an infrastructure project in the early phases of a construction process. As part of the findings they suggest a prequalification approach to identify the most relevant assessment categories. In this study sustainability impact assessment is structured at a strategic, tactical and operational level and according to the categories of economy, society and environment; see Figure 15.4. The nine-category structure aims to capture the economic, social and environmental effects of a construction project, which seems less technical and process oriented than building certifications like BREAM, LEED and DGNB, but adds an interesting consideration of time perspectives and risk management to the process of setting goals and value parameters. The time and risk perspectives can be helpful for identifying strategic goals and guiding the eventual simplification of lifecycle evaluations of intervention options.

Häkkinen et al. (2012) provide an overview of indicators and issues which are covered by most of the buildings sustainability assessment tools, like BREEAM, LEED, DGNB and others. Based on the review of existing tools and interviews the following Key Performance Indicators are presented (Häkkinen et al., 2012, p.23):

- Environmental core indicators:
  - Primary energy consumption
  - Water management, materials (rational use and low impact)
  - Waste (construction and operation)
  - Global warming potential (CO_2 emissions)
  - Land use and ecological value of the site

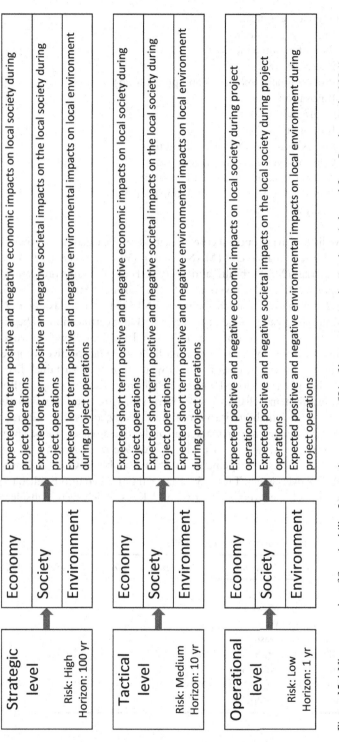

Figure 15.4 Nine categories of Sustainability Impact Assessment – SIA of large construction projects (adapted from Laedre et al., 2014)

- Economic core indicators:
  - Building adaptability
  - Ease of maintenance
  - Life cycle costs
  - Process quality (planning and preparation)
  - Innovation
- Social core indicators:
  - Indoor air quality
  - Access to transport (for building users)
  - Comfort (visual, thermal, acoustic)
  - Access to public services and amenities
  - Access for users with physical impairments
  - Safety and security.

The above listed environmental core indicators can be organised in the categories energy (ENE), management (MAN), water (WAT), waste (WAS), land use (LUE), materials (MAT), health and safety (HEA), pollutants protection (POL), and transport (TRA), which were used in the European Standard (EN15221-7). Table 15.2 shows the alignment between EN15221-7 and Häkkinen (2012). A similar comparison can be made to specific building certifications. The difference is likely to be the inclusion of non-building-related indicators like work-related transportation, or the extent to which a full life-cycle perspective is applied.

A slightly different approach on how to structure sustainability in FM assessment has been developed within the German research project "Return on Sustainability System" (ROSS). KPIs were structured according to the triple bottom line of sustainability into economy, ecology, and socio-cultural aspects, and in addition divided into management and process-related KPIs. The research findings were published in the book *Nachhaltiges Facility Management* (Sustainable Facilities Management) which is available in German (Kummert et al., 2013). The overview in Table 15.3 has been translated into English to provide an impression of the overall structure and KPIs which were considered most relevant.

Table 15.2 Short list of environmental KPIs

No	Short name	Indicator	Häkkinen et al. (2012)	EN 15221-7 (CEN, 2012)
1	ENE	Energy	X	X
2	MAT	Materials	X	X
3	WAT	Water	X	X
4	WAS	Waste	X	X
5	LUE	Land use	X	X
6	POL	Pollution	X	X
7	TRA	Travel	X	X
8	HEA	Health	X	X

Table 15.3 KPIs overview (adapted and translated from Kummert et al., 2013)

	Management KPIs	Process KPIs
Economy	1. Earnings before interest and taxes (Ebit) (EUR) 2. Equity ratio (%) 3. Customer complaint (%) 4. Customer relation duration (month)	1. Land/area consumption quota (%) 2. Process additional expenditure (%) 3. Process rework (%)
Ecology	5. Water consumption (m^3/employee) 6. Heating energy consumption (kWh/m^2 GFA) 7. Electricity consumption (kWh/m^2 GFA) 8. Fuel consumption (l/km) 9. Waste generation (kg/employee)	4. Green suppliers ratio (%) 5. Green operating equipment and material (%)
Socio-cultural	10. Health rate (%) 11. Labour turnover (%) 12. Training effort (EUR/employee)	6. Accident rate (%) 7. Personal contribution quota (%)

In addition to the above mentioned suggestion for qualitative data collection, the focus on energy, waste and water is used to provide the following examples for quantitative environmental data collection (CEN, 2012):

1 Primary environmental ratios: total CO_2 emissions (tonnes per annum), and CO_2 emissions per annum per full time equivalent (fte) and per m^2 Net Floor Area (NFA)
2 Primary energy ratios: total energy consumption (kWh per annum), and kWh per annum per fte, and per m^2 NFA
3 Primary water ratios: total water usage (m^3 per annum), and water usage per fte
4 Primary waste ratios: total waste production (tonnes per annum), and waste per fte and per m^2 NFA.

The quantification of environmental impact, as described in EN15221-7 (CEN, 2012), is based on calculation of CO_2 emissions. First, quantities in each category are recorded based on typical units like kWh, litres and tonnes. Second, impact factors are used to convert these quantities into CO_2 emissions.

Reflections on using these KPIs

Three questions or issues for discussion regarding this approach are: 1) availability of qualitative and quantitative data; 2) demand of appropriate factors for sustainability impact calculation; and 3) workload and qualification for sustainability assessment and management.

The first question concerns how to acquire exact quantities in often very abstract categories, which might not be documented in the FM registrations. Concerning energy management, for example, how can analysts differentiate the measurement of the non-renewable energy consumption and the renewable energy consumption? Concerning waste management, how can they identify different types of waste within regular, recyclable, and biodegradable waste categories, and how can they measure the amounts? Can all relevant information about the basic processes be considered? For example, how is the renewable energy supply managed? Is it done in a sustainable way? For instance, zero carbon electricity can be produced with hydropower, which has a strong impact on the ecosystem and biodiversity etc., and its extensions are limited. Other examples are biofuel or wood pellets etc.

Second, how can analysts ensure the availability of correct impact factors for conversation of quantities into CO_2 emissions? How does changing technology affect the converting processes and national varieties of production etc., and how should this be considered? Within the European standard it is suggested: "In calculating CO_2 emissions from energy related data, please ensure you use the right CO_2 conversion factors ... CO_2 emission factors differ from country to country, especially for electricity. Therefore, it is important to verify the CO_2 conversion factor provided with the relevant authorities or your energy provider" (CEN, 2012, Annex E).

Third, how can analysts evaluate the additional workload for data collection and the calculation and communication of results, and all reporting efforts? Who does this, and who has the necessary qualifications? Ideally, a data collection template should be completed by an environmental expert on the facility management team or the appropriate building manager, and data should be collected on a building by building basis (CEN, 2012, Annex E). Some organisations have implemented sustainability offices. However, a challenge might also be presented by the internal and external services provision and the availability of complete documentation.

Perspectives

Sustainable FM and CREM in practice is not a simple matter. It includes complex challenges with numerous dilemmas, such as how to prioritise energy savings in comparison with quality, economy, wellbeing and health. This is a part of the everyday life of FM and CREM professionals. When it comes to measuring and managing sustainability in FM/CREM it is important to:

1   Be context specific in the formulation of strategic goals and KPIs, as sustainability challenges as well as implementation possibilities and barriers vary between locations, buildings, businesses and organisations
2   Apply a lifecycle perspective when you plan your next FM and CREM task
3   Use a balanced scorecard for structuring an evaluation and comparing between the current situation and intervention options. Expect that possible

benefits and disadvantages might be estimates to begin with, and build up your knowledge base to improve validity.

Strategic Sustainable FM and CREM calls for a leadership perspective on the integrated whole, consisting of a building/facility, processes (operation and use), and management practice in its specific context. It calls for FM change agents who acknowledge that however FM is performed, FM has an impact beyond the internal organisation, and also has to take part of the responsibility for contributing to sustainable development beyond the organisation. Facilities managers needs to learn new concepts which reflect the sustainability values, and they need to learn how to best implement, operate and manage the new and updated building assets as well as the FM services.

Future research should investigate whether and how the FM and CREM sector is developing into potential change agents for sustainable development on a societal scale to qualify policies and regulation in the field. This includes research on building performance and studies of the application of building certifications, CEN standards and other enablers for making the built environment more sustainable. Another important role is to support the change management of FM practices at an organisational level. This can be done by evaluating the outcome of more sustainable practices in front-runner organisations, and disseminating learning points of general interest. Finally, research should lead to the development of new management tools and FM services, which may be very general tools for wider application, with a recognition that adaptation is needed to fit the tools to every unique context.

References

Baharum, M.R. and Pitt, M. (2009) 'Determining a conceptual framework for green FM intellectual capita', *Journal of Facilities Management*, 7 (4), pp. 267–282.

BIFM (2013) *Sustainability in Facilities Management Report 2013*, United Kingdom.

Bond, A.J. and Morrison-Saunders, A. (2011) 'Re-evaluating sustainability assessment: Aligning the vision and the practice', *Environmental Impact Assessment Review*, 31, pp. 1–7.

Brown, Z., Cole, R.J., Robinson, J. and Dowlatabadi, H. (2010) 'Evaluating user experience in green buildings in relation to workplace culture and context', *Facilities*, 28 (3/4), pp. 225–238.

Brundtland, G.H., Khalid, M., Agnelli, S., Al-Athel, S.A., Chidzero, B., Fadika, M.L., Hauff, V. and Al, E. (1987) *Our Common Future – Report of the World Commission on Environments and Development*, The World Commission on Environments and Development.

CEN (2011) *Facility Management – Part 4: Taxonomy of Facility Management – Classification and Structures*, European Standard EN 15221-4. European Committee for Standardization.

CEN (2012) *Facility Management – Part 7: Guidelines for Performance Benchmarking*. European Standard EN 15221-7. European Committee for Standardization.

Cooper, J. (2015) *Sustainable building maintenance within social housing*, PhD thesis, University of Greenwich.

Cooper, J. and Jones, K. (2009) 'Measuring performance in-use in UK social housing', in RICS 2009: *Proceedings of the Construction and Building Research Conference of the Royal Institution of Chartered Surveyors*, London: Royal Institution of Chartered Surveyors (RICS), pp. 916–928.

Durmus-Pedini, A. and Ashuri, B. (2010) 'An Overview of the Benefits and Risk Factors of Going Green in Existing Buildings', *International Journal of Facility Management*, 1 (1), pp. 1–15.

Elmualim, A., Shockley, D., Valle, R., Ludlow, G. and Shah, S. (2010) 'Barriers and commitment of facilities management profession to the sustainability agenda', *Building and Environment*, 45 (1), pp. 58–64.

Falkenbach, H., Lindholm, A.L. and Schleich, H (2010) 'Environmental Sustainability: Drivers for the Real Estate Investor', *Journal of Real Estate Literature*, 18 (2), pp. 203–221.

Fennimore, J. (2014) *Sustainable Facility Management: Operational Strategies for Today*, Pearson, USA.

Häkkinen, T., Antuña, C., Mäkeläinen, T., Lützkendorf, T., Balouktsi, M. and Al., E. (2012) 'Sustainability and Performance Assessment and Benchmarking of Buildings', in Häkkinen, T. (Ed.) Final report. Espoo: VTT Technical Research Centre of Finland.

Hodges, C. and Sekula, M. (2013) *Sustainable Facilities Management: The facility manager's guide to optimizing building performance*, USA: VisonSpotsPublishing.

Jensen, J.O, Kjærulf, A., Wilhelmsen, C. and Nielsen, S.B. (2009) *Bæredygtighedsprofiler for bydele i København*. Aalborg University: SBi Danish Building Research Institute.

Jones, K and Cooper, J. (2007) 'The Role of Routine Maintenance in Improving the Sustainability of Social Housing', in *Proceedings of the European Network for Housing Research*, Rotterdam.

Junghans, A. (2011) 'State of the art in sustainable Facility Management', in *Proceedings of the 6th Nordic Conference on Construction Economics and Organisation – Shaping the Construction/Society Nexus. Volume 2: Transforming Practices*, Aalborg University: SBi Danish Building Research Institute, CIB Publication.

Junghans, A., Elmualim, A. and Skålnes, I. (2014) International Trends for Sustainability in FM: Evaluation of Policy and FM Competence in Norway and UK, in *Proceedings of CIB Facilities Management Conference – Using Facilities in an open world, creating value for all stakeholders, Joint CIB W070, W11 & W118 conference, Technical University of Denmark, Copenhagen, 21–23 May 2014*, Polyteknisk Boghandel og Forlag, pp. 305–316.

Kloepffer, W. (2008) 'Life cycle sustainability assessment of products', *The International Journal of Life Cycle Assessment*, 13, pp. 89–95.

Kummert, K., May, M. and Pelzeter, A. (2013) *Nachhaltiges Facility Management*, Berlin: Springer.

Kwane, A.-D., Liow, K.H. and Neo Yen Shi, S. (2009) 'Sustainability of Sustainable Real Property Development', *Journal of Sustainable Real Estate*, 1 (1), pp. 203–225.

Laedre, O., Haavaldsen, T., Bohne, R.A., Kallaos, J. and Lohne, J. (2014) 'Determining sustainability impact assessment indicators', *Impact Assessment and Project Appraisal*, 33 (2), pp. 99–107.

McAloone, T. and Bey, N. (2009) *Environmental improvement through product development: A guide*, Copenhagen: The Danish Environmental Protection Agency.

Meadows, D.H., Meadows, D.L., Randers, J. and Behrens III, W.W. (1972) *The Limits to Growth: A report for the Club of Rome's Project on the Predicament of Mankind*, New York: Universe Books.

Nielsen, S.B. (2012) 'Claims of Sustainable FM: Exploring current practices', in Jensen, J.O. and Nielsen, S.B. (Eds.) *Facilities Management Research in the Nordic Countries: Past, Present and Future*, Copenhagen: Polyteknisk Forlag.

Nielsen, S.B. and Jensen, J.O. (2010) 'Translating sustainable development to the domain of a local authority: The case of urban districts in Copenhagen', in *Proceedings from EASST 010: Practicing Science and Technology, Performing the Social. Track 38: Towards Zero Emission Buildings, Settlements and Cities.* 1-2 September, Trento, Italy.

OECD (2010) *Guidance on Sustainability Impact Assessment (SIA)*, OECD.

Shah, S. (2007) *Sustainable Practice for the Facilities Manager*, Oxford: Blackwell Publishing.

UN (2015) *Transforming our World: The 2030 Agenda for Sustainable Development*, A/RES/70/1, New York: United Nations.

UNCSD (2012) Institutional framework for sustainable development. Available at: http://www.uncsd2012.org/isfd.html/ [accessed 4 August 2015].

Von Carlowitz, H.C. and Von Rohr, J.B. (1732) *Sylvicultura oeconomica.*

Interview 11: Olav Egil Sæbøe, Pro-FM Consulting, Norway

Olav Egil Sæbøe has 21 years of experience in CREM/FM. His educational background is from the Royal Military Academy of Norway and various university courses. He initially served in the Norwegian military and was later Senior Vice President in a Norwegian bank and CEO in a hotel chain. In 1994 he became CRE/FM Director of a Norwegian bank, from 2000 to 2003 he was FM Director at Aberdeen Property Investors, and since 2003 he has been management consultant and owner of his one-man company Pro-FM Consulting. Olav has for many years been active in EuroFM and NordicFM.

Added Value in general

Q. What does the term "Added Value" mean to you?

A. *There is an increasing recognition of added value as strategic objective among my clients. The meaning of added value depends on the nature and the objectives of the business. For investors the concern is to add value both for the owner and the tenants, and in healthcare both for the hospital and the patients. In my communication with portfolio investors I emphasise the importance of professional FM as value driver, e.g. by providing space and work environments and well adapted services which are competitive to tenants' high expectations, resulting in higher rents and longer tenancies. Tenancies like this will most likely create higher values for the tenant and consequently also for the investor.*

Q. Are there other related terms that you prefer to use rather than "Added Value"?

A. *The concept of "right cost" is also used. In some cases, when analysis of needs and expectations conclude that providing assets or services to a specific quantity and quality level in order to secure the achievement of a desired objective requires a cost level higher than average, his cost may be referred to as the "right cost". "Right cost" is then justified by satisfying needs to achieve the added value.*

Q. How do you see the relation between "Added Value" and cost reduction?

A. *It is a challenge. It depends on the nature of needs and the description of required quality. In the Norwegian Facility Management Network the focus in benchmarking has shifted from a "lowest cost" focus towards comparing specific needs and requirements and required quality with realised cost in order to identify "right cost".*

Q. Is "Added Value" mostly treated at a strategic, tactical or operational level?

A. *It is mostly considered at a strategic level with a focus on the relation between cost and value. The overall objectives and policies should be formulated at strategic level and the implementations and effects are made transparent at the tactical level.*

Q. In which context or dialogue is "Added Value" mostly considered in your experience?

A. *It is mostly used between clients and providers, particularly in contract negotiations, and between clients and consultants. For property investors the focus is on the relation between cost and value and increasingly on the potentials of creating added value for clients and tenants.*

Q. How do you see the relation between innovation and "Added Value" in FM?

A. *The aspiration of creating added value can be an inspiration for innovation. There is for instance a renowned Norwegian property investor and owner of a large restaurant located in natural surroundings in Oslo, who created a unique sculpture park around the restaurant to make it more exiting and attractive to visit the area and the restaurant. This is to me being very innovative and adding value at the same time.*

Benefits and limitations of "Added Value"

Q. What are the benefits of considering and talking about "Added Value"?

A. *It makes you focus on the strategic aspects of FM and gives a more constructive dialogue than focus on cost. Added value is most important on strategic level and in relation to different aspects of CSR. For instance, the property investment company Norwegian Property sees FM as one of the five important drivers for added value. The Norwegian hospital A-hus is also a good example of an organisation that focuses on the added value of FM, e.g. in terms of shorter recovery time and better turnover of "bed days".*

Q. What are the limitations of considering and talking about "Added Value"?

A. *There are definitely limitations but I see few downsides. One downside might be that differences in environmental attitudes might have as results that what one person sees as successful "added value" might be considered as a deplorable environmental sacrifice by another.*

Management of Added Value

Q. What are your top five main values to be included in management of accommodations, facilities and services? Could you mention a few examples of concrete FM interventions to attain these added values and KPIs to measure them?

A. *See table on page 278.*

Prioritised values in FM	Concrete FM interventions	KPIs
1. Business' strategic acknowledgement of market demand and expectations which may positively affect value	Analysis and high level exchange of timely information on market needs and expectations relevant to successful application and operations of business accommodation and facilities	Investments and cost budget targets Shareholder value
2. Business' strategic focus on space application and user/tenant perspectives	Analyse and inform management of practical consequences of user/tenants trends and expectations for accommodation and service provision Recommendations	Vacant/hired ratios. Hire turnover Rent budgets targets
3. Flexible usability of property and space for different primary processes	Long-term/lifecycle planning and sustainable operations, maintenance and space developments, based on analysis and confirmed trends Availability and adaptability of service provision to changing needs	Scope and frequency of property condition analysis Hire turnover
4. High reputational profile on ethics and social responsibility	Reliable sourcing and procurement processes Environmental sustainability	Reputation survey scores Different energy scores (selection/configuration of energy sources, efficiency etc.)
5. Reputation for high quality tenant/user satisfaction	Close tenant/user follow-up based on firm information and understanding of their primary activities and associated accommodation and service needs	Satisfaction indexes (client, tenant, user) Quality/quantity scores on estimated/agreed requirements/targets in each relevant service

Q.   Do you measure whether you attain the targeted "Added Values"?

A: *A number of examples are given in the table. However, one should be careful not to jump to conclusions on satisfaction indexes. Responses to satisfaction surveys tend to be too general and dependant on the formulation of questions and the situation in which responses are given. Indexes may give an inaccurate response indication. It is recommended to compare responses with other KPIs on each of the related operations and services incorporated in the satisfaction index. I don't use benchmarking myself, but I recommend my clients to do so and use the aforementioned KPIs.*

Q.   How is "Added Value" included in your communication with your stakeholder?

A.   *It is included more and more both as part of verbal communication and in business cases. This mainly concerns discussions of relevant KPIs, which indicate targeted achievements at a strategic and to some extent at a tactical level, where possible added values to the main business objectives are most likely targeted and identified.*

Q. Which aspects do you find most important to include in your value adding management approach?

A. *I find analyses at a strategic level and from a long-term perspective most important. However, it is hardly possible in general terms to prioritise as the aspects, which are most important may vary according to which business you are referring to and also between private and public activities.*

Q. Do you experience any dilemmas in relation to "Added Value"?

A. *The most important dilemma I experience is the understanding of the term "added value" itself. It is important to distinguish between creation of values and values added, and I am not sure this is consistently dealt with in many cases, e.g. what is in the short term categorised as "added value" may in the longer term become part of a value creation scheme.*

Q. Have you noticed a change in mind-shift regarding "Added Value"?

A. *It seems to be becoming more and more recognised as being important, but I am not sure everyone has a clear understanding of "what is in it".*

16 Corporate Social Responsibility

Brenda Groen, Martine Vonk, Frans Melissen and Arrien Termaat

Introduction

Over the last decades, organisations have become more aware of the importance of Corporate Social Responsibility (CSR) (Visser and Tolhurst, 2010). They recognise the social, ecological and economic consequences of their activities and seek ways to incorporate their responsibilities in their governance and be transparent about it. This change is driven by several factors. An economic driving factor is the scarcity of resources, leading to innovations that focus on lean processes and cradle-to-cradle principles. Likewise, to remain competitive companies need to innovate, and CSR functions are triggers for new business concepts (Loew et al., 2009). This chapter will give some examples of such innovations in the fields of Facility Management (FM) and Corporate Real Estate Management (CREM). An example is the Job Development Center developed by ISS, which enables unemployed people to get the qualifications and confidence needed to acquire a job (Nordic Innovation Center, 2010).

Besides economic drivers, morality has become an important factor as well, both internally, i.e. inside organisations, and externally, from society. Due to social media, public opinion has gained influence by revealing corporate activities and denouncing misconduct. In order to keep the trust of citizens, organisations need to be transparent about their policies and practices, whether this concerns safety issues, the way they treat their employees or the ecological impact of their activities. However, the choice for CSR cannot be based on public morality alone. For an organisation to truly embrace CSR, it needs to become part of corporate ethics and personal leadership: an organisation can only be trustworthy when it backs up its words by actions.

What exactly does CSR mean? The United Nations Global Compact (2000), the OECD Guidelines for Multinational Enterprises (2008) and the ISO 26000 Guidance on Social Responsibility (2010) are internationally recognised principles and guidelines for CSR. These guidelines cover a broad range of ways to translate Corporate Responsibility (CR) into policy and ethical behaviour that contributes to sustainable development, including the health and welfare of individuals and society. The United Nations Global Compact consists of ten principles within the fields of human rights, labour, the environment and anti-corruption. The OECD guidelines mention additional fields regarding consumer interests, the contribution of science and technology to national development and economy, fair competition and tax compliance. The ISO 26000 acknowledges seven principles of social responsibility, including accountability, transparency, ethical behaviour, respect for stakeholder interests, respect for the rule of law, respect for international norms of behaviour and respect for human rights. With these principles as a starting point, an organisation may identify its stakeholders and analyse the core subjects of social responsibility policy. As core subjects, ISO 26000 recognises human rights, labour practices, the environment, fair operating practices, consumer issues and community involvement and development. The seven principles form an ethical direction for how to deal with CSR. Many of these issues are directly related to the responsibilities and practices of facility managers and real estate managers. In this chapter we will focus on specific CSR guidelines and practices that relate to FM and CREM.

CSR helps companies to take responsibility and encourages sustainable development. It can also bring benefits regarding risk management, customer relationships, cost savings, human resource management and innovation (European Commission, 2011). In addition, working from a CSR perspective may lead to more satisfaction and engagement among the employees (see for example Bauman and Skitka, 2012; RICS, 2013). CSR offers opportunities to connect with values that employees and stakeholders adhere to. In society we can recognise tendencies or shifts towards values such as co-operation, decentralising, engagement with local communities, meaning, and a growing emphasis on access to spaces and products rather than ownership (Rotmans, 2014). These tendencies give way to innovations in the fields of FM and CREM in adhering to CSR and the welfare and wellbeing of employees and stakeholders, such as possibilities for flex-work in alternative (more local) workplaces, web conferencing, and the shift to a more circular economy in which products are no longer bought but hired from the producer to whom the products (and thus the resources) can be returned for recycling.

State of the art

The evolution of the concept of CSR, and especially how it is dealt with both in literature and in practice, reveals a clear and undeniable trend. Only five decades ago the American economist and Nobel Prize winner Milton Friedman stated the following:

> Few trends could so thoroughly undermine the very foundations of our free society as the acceptance by corporate officials of a social responsibility other than to make as much money for their shareholders as possible. If managers used corporate resources for any cause other than profit maximization, it would constitute a form of theft.
>
> (Friedman, 1962)

Today, such a statement by the CEO of any company would probably be considered short-sighted by the vast majority of academics and would definitely lead to his or her strategic and marketing experts pointing out the high likelihood of negative consequences, such as other companies no longer wanting to do business or engage in partnerships and consumers turning their backs to the company's products. In other words, values in our society have shifted and the need to address CSR is more or less a given for any company wanting to be successful in today's market place. Whereas in 1977 less than half of the Fortune 500 firms mentioned CSR in their annual reports, this number had grown to almost 90% at the turn of the millennium (Lee, 2008). Today many companies acknowledge this responsibility. The following excerpt from TNT's 2009 annual report provides a typical example of how this can be worded:

> Given TNT's view that the interests of all stakeholders must need to be managed in a balanced way, TNT's annual report should report on both financial and non-financial performance (...) this is the first time TNT has integrated CR in the actual annual report and as such CSR performance and strategic performance must be the outcome of improved actions taken in day to day management of TNT's core business.
>
> (TNT Holding B.V., 2010, p. 17)

However, the way in which companies address and fulfil this responsibility is anything but a given. Yam (2013) claims that even today, one can distinguish two schools of thought with respect to CSR: 1) those that claim that companies ought to focus on maximising profits within the boundaries of the law and certain minimal ethical considerations; and 2) those that claim that companies have a (much) wider range of responsibilities and should act accordingly. The first school of thought might not take matters to the extreme as Friedman did, but adherents still stress that ultimately companies need to focus first and foremost on their financial results. This approach is often referred to as the Anglo-Saxon model and is linked to e.g. the USA, the UK and Singapore. The second school of thought takes a different approach to describing the role of companies in our societies and usually relies on applying a systems perspective to point out that companies are part of a wider social, economic and environmental system and thus have at least a shared responsibility in the functioning of those systems (see e.g. Evans, 2009). This vision is also referred to as the Rhine model, named after countries such as Switzerland, Germany, France and the Benelux countries, all located near the river Rhine (Michel, 1993).

Simultaneously, other authors have moved beyond simply distinguishing two schools of thought and have introduced (various) levels of CSR or the alignment of companies with sustainable development. For instance, Masalskyte et al. (2014) developed a generic sustainability maturity level model for CREM. Based on acknowledging the need for a sustainability agenda for companies to be able to continue to meet the needs of all relevant stakeholders and thereby maintain a license to operate, they focused on five specific maturity levels that link to compliance (the lowest level) up to innovation (the highest level). The focus areas distinguished by Masalskyte et al. (2014) include resources, processes, governance, communication, finance and strategy.

Similarly, Baumgartner and Ebner (2010) established four generic sustainability strategies for companies in all sectors, ranging from introverted to visionary, which can be linked to appropriate aspect profiles. They differentiate between more than twenty aspects that can be grouped based on contributing to either the environmental performance of the firm, the (long-term) economic performance of the firm, or the social performance of the firm. An illustration of an aspect from the environmental group would be emissions to water, air or ground coming from the company's processes. The economic group of aspects includes initiatives such as reporting on environmental performance and active collaboration with business partners, which are both crucial to long-term survival in today's market place. Typical examples from the social group would be health and safety (within the company) and corporate citizenship (initiatives to promote the wellbeing of people outside the company). For each of these aspects, Baumgartner and Ebner distinguish four maturity levels. The lowest level would, for instance, involve limiting one's emissions to what is allowed by law, but not reducing levels any further, whereas the highest level would involve trying to avoid any emissions and even helping other companies, such as supply chain partners, to reduce their emissions as well.

Obviously, the more ambitious the specific sustainability strategy of a company is, the higher the maturity level they would need to strive for with respect to these aspects. However, this does not apply to all aspects equally. For instance, a company that focuses on communicating its sustainability commitment to the outside world for different reasons, not mainly because of an intrinsic motivation regarding sustainable development, will probably aim at higher maturity levels for aspects such as sustainability reporting, corporate citizenship (to promote a positive image) and health and safety (to prevent a negative image) than for aspects such as safeguarding biodiversity and human capital development (which do not directly affect the public image of the company). Therefore, sustainability strategies can regard specific aspect profiles and different maturity levels for the various sustainability aspects.

The above portrays just a fraction of all the choices available to companies in aligning their business activities with the full range of environmental, economic and social CSR aspects. Given that this responsibility is not governed by law, but rather represents a form of self-regulation (Moon, 2007), this implies that, in principle, all different options are available to companies in shaping their efforts, regardless of whether this is referred to as Corporate Social Responsibility (CSR), Corporate Responsibility (CR), Strategic CSR/CR, Creating Shared Value, Business Ethics or any other term that has popped up in recent years.

This also implies that the relationship between FM and CREM on the one hand and CSR on the other is not fixed; the actual contribution of FM and CREM to CSR depends on these choices at the corporate level. If a company decides to pursue a visionary sustainability strategy and aims for the highest level of maturity for most aspects, reaching this level requires a significant contribution by FM and CREM, for instance regarding health and safety and purchasing. At first glance a less obvious aspect would be corporate citizenship. This aspect usually involves initiatives and activities that move beyond the core business of the company and relates to the company's role as a 'citizen' and how it contributes (often through dedicated projects) to the local, regional and global community. However, if the company's employees are involved in these initiatives using company resources and during office hours, then this aspect could rely heavily on the contribution of FM and CREM. In contrast, if a company chooses to go for an introverted or conservative sustainability strategy, it is likely that some aspects will be dealt with by simply abiding legal constraints. In those circumstances, the contribution of FM and CREM is probably more directed towards cost savings and consistency, and supporting primary processes in an efficient way, rather than trying to enhance employee satisfaction and building a company's image of being a leading partner within the wider social-environmental-economic system through facilitating non-core business activities.

Consequently, choices made by a company with respect to how it wants to fulfil its CSR not only influence the focus and objectives of specific FM and CREM processes, but also the overall status and relevance of FM and CREM within the wider context of the company's ability to reach its objectives. Whereas more mature interpretations of an organisation's CSR policy could be argued to support an emphasis on the potential added value of FM and CREM in creating shared value from a society perspective, less mature strategies, at least from a sustainable development perspective, could be seen as forcing FM and CREM back to their former role as a cost centre.

Benefits and costs

Aguinis and Glavas (2012) have developed a comprehensive theoretical framework of CSR, based on an extensive literature review. The framework describes drivers for CSR, outcomes, factors that moderate the relation between drivers and outcomes, and underlying mechanisms (mediators). Drivers for firms to choose to implement CSR actions and policies are affected by stakeholder pressure as well as regulations, standards and certifications. Provided that these actions are not merely greenwashing or "window dressing", outcomes at societal level may include improved reputation, support of consumer preferences and loyalty, and improved stakeholder relations. At company level instrumental reasons such as expected financial outcomes as well as normative reasons based on the firm's values ('doing the right thing') predict CSR, especially in large firms with high public visibility.

Indeed, there is a small but positive relationship between CSR actions and policies and financial outcomes. Furthermore, there are several non-financial outcomes, like improved firm capabilities such as increased product quality and operational efficiencies, as well as a more diverse workforce. An underlying mechanism is managerial interpretations of CSR as an opportunity.

Working for socially responsible companies may have a number of positive effects, such as increased organisational identification, engagement, retention, organisational citizenship behaviour and commitment. The strength of the effect depends on individual factors (awareness, values), supervisory commitment, a strong organisational identity and visionary leadership.

How to measure and manage

Key Performance Indicators

Based on an extensive literature review as well as consultation of FM professionals about facility performance measurement, Lavy et al. (2010) distinguish between financial, functional, physical and survey-based indicators related to users' opinions of a facility. They have developed a holistic view of performance by identifying major KPIs suitable for a broad range of buildings and facilities for short- and long-term goals. A number of these indicators refer to the People-related aspect of the Triple-P: employee productivity, community and customer satisfaction, health and safety, and diversity-related aspects. Other indicators can be connected to the Environmental (Planet) aspect of the Triple-P as a mixture of use of space – with indicators such as adequacy, utilisation and quality of space – and sustainability, with indicators such as waste management and use of resources (energy, water and materials). Financial indicators are related to Profit, and refer to costs and expenditures associated with operation and maintenance, utilities, building functions and real estate. These indicators may be used for appraisal of financial performance to support short- and long-term decision-making by various management levels. Some CSR indicators are not included in Lavy et al. (2010), for instance no use of child labour.

According to CFP (2014), based on research into the importance and benefits of CSR in the Netherlands, the main barriers for facility managers regarding CSR are budget and time (40%), not enough support within the organisation (approx. 30%), and lack of knowledge (25%). This is supported by Twijnstra Gudde (2012), who stated that CSR needs support from higher management, and needs to be incorporated in the mission and vision of the organisation. Furthermore, a long-term strategy is essential. FM is unanimous regarding the advantages: improves image (85%), decreases costs (65%), improves competitive power (34%), higher turnover and profit (less than 10%). Therefore, profit-based KPIs (Lavy et al., 2010) as well as image-related KPIs (Aguinis and Glavas, 2012) seem to be most appropriate to steer CSR. Table 16.1 shows the KPIs for FM (based on Lavy et al., 2010), linked to the outcome variables for CSR in Aguinis and Glavas (2012) and classified according to the Triple-P framework.

Any positive outcome with respect to these KPIs is not necessarily caused by CSR initiatives; the underlying intention of measures is crucial. For example, improving operational efficiency may be very suitable to increase the level of sustainability, but the underlying motive may be purely financial and not related to CSR at all. Nappi-Choulet and Décamps (2013) have demonstrated that

Table 16.1 KPIs for FM (based on Lavy et al., 2010), linked to outcome variables for CSR in Aguinis and Glavas (2012), in the Triple-P framework

Category	Facility performance indicators	Relationship to CSR outcomes
People	Productivity (occupant turnover rate, absenteeism, occupants' satisfaction and self-rated productivity)	Firm capabilities: product quality
	Space utilisation (adequate, sufficient, over/under used, space management)	Demographic diversity
	Employee/occupant turnover rate	
	Employees' health and safety (accidents, lost work hours, compliance with codes)	
	Accessibility for disabled employees	
	Customer/occupants' satisfaction with products or services	Employee engagement
		Attractiveness to (new) employees
		Consumer choice and loyalty
		Organisational identification
	Community satisfaction and participation	Stakeholder relations
		Reputation
	Quality of indoor environment	Firm capability: product quality
	Physical building condition (quantitative and qualitative)	Firm capabilities: product quality
Planet	Total waste produced	Firm capabilities: operational efficiency
	Resource consumption (water, energy, raw materials etc.)	Firm capabilities: product quality, operational efficiency
	Security, site and location	Reduced risk
Profit	Operating costs	Financial performance
	Occupancy costs	Firm capabilities: operational efficiency
	Custodial and janitorial costs	
	Churn rate and churn cost	
	Maintenance costs (including costs for deferred maintenance)	
	Maintenance efficiency	
	Facility condition index	
	Groundskeeping costs	
	Utility costs	

energy efficiency – which could be part of environmental CSR goals – has a positive impact on asset value and rent. On the other hand, measures taken by an organisation that are clearly part of CSR might benefit the operational efficiency of FM, even though this was not the primary goal. A Swedish housing association implemented a number of people-oriented projects, like youth projects (e.g. a graffiti group and summer jobs), tenant mobilisation (stairway hosts, safety walks organised by parents at weekends in the evenings/at nights). These are people-oriented projects aimed at improving the relationship between the housing association and its tenants. However, the result was that a block of apartments with CSR had 4.5% lower annual operating and maintenance costs, especially maintenance of the outdoor environment, compared to a similar block of apartments without CSR. This is a clear example of the influence of a CSR project on FM KPIs.

Management of CSR interventions

CFP (2014) has shown that approximately two thirds of facility managers have incorporated CSR in their FM policies, with environment, social awareness and labour conditions being the most important aspects. They perceive improved image and decreased costs as most important benefits. According to Jones Lang LaSalle (2013), the health and productivity of their employees are the main reasons for seeking sustainable office accommodation. Mentioned aspects include daylight, visual comfort and improved internal air quality. Energy takes second place, and transport third. The top four CSR aspects among members of CoreNet Global (2012) are quality of working environments, energy conservation, building retrofitting and reliable performance measurements for CSR. Organisations do indeed mention (many) aspects of CSR in their annual CSR reports. As these reports do not specify the contribution of FM or CREM, the following examples are primarily taken from service providers.

People

If one of the CSR goals of an organisation is to improve labour conditions for its own staff, in order to increase e.g. employee engagement and organisational citizenship behaviour, FM can play a role by providing excellent workspace that optimally support the employees, thereby increasing employee satisfaction and productivity. According to CoreNet Global (2012) this is an important aim of CSR. Aspects would be e.g. adequate and sufficient workplaces and a healthy indoor environment (Miller et al., 2009; Feige et al., 2013). These aspects influence employee related survey-based outcomes like job satisfaction. These aspects are dealt with in Chapter 5 on satisfaction and Chapter 8 on health and safety.

If one of the CSR objectives of an organisation is to increase the diversity of its workforce, FM can contribute by optimising the accessibility of the building and adjusting workplaces for employees with physical impairments. Furthermore, FM may influence human resource management and employ people with a mental or physical impairment. By offering flexible working hours FM may support the work-life balance of the employees. These kinds of measures may make the organisation an interesting employer for a diverse workforce. On a wider scale this item includes respect for human rights, such as no forced labour and no child labour. These aspects require adequate supply chain management (see *Profit* below).

Examples:

- Sodexo UK/Ireland employs a diverse workforce, and provides supported internships to disabled young people (Sodexo, 2014)
- ISS' first priority is a safe workplace without fatalities; furthermore it emphasises its diverse workforce, and aims to improve female representation in management positions (ISS, 2014)
- RICS supports employees' wellbeing and health, e.g. by its Wellbeing Month, inviting employees to get involved in a number of wellbeing promoting activities (RICS, 2013)
- Facilicom offers social return and employs challenged workers. New ways of working are offered as well. Facilicom promotes a recognisable and pleasant working atmosphere and at the same time efficient use of space (Facilicom, 2014)
- ISS is committed to respect human rights, in compliance with the ISS Code of Conduct. For example, this applies to working in environments with children and vulnerable people. Treating people with respect is a fundamental leadership principle in ISS and this is reflected in ISS' Code of Conduct (ISS, 2014).

Charity and community involvement are also part of CSR. Examples are:

- Blycolin donates used linen to children's homes (Blycolin, 2012)
- Sodexo donates to charity, and lets employees volunteer in projects run by charitable partners (Sodexo, 2014).

Planet

If one of the CSR goals of an organisation is to decrease its impact on the environment by using fewer resources and producing less waste, FM may support this CSR initiative, as FM is responsible for operations and maintenance. FM can influence the use of utilities (energy, water) by striving for operational efficiency. Certification (LEED, BREEAM, etc.) is both a means and an end in this process. Furthermore, CREM may contribute by choosing a location near to public transport and employees' residences.

Examples:

- Sodexo aims to cut emissions from their fleet of vehicles (Sodexo, 2014)
- RICS promotes public transport (RICS, 2013)

- The ING Group is shifting from a linear to a circular purchasing model, by moving from purchasing goods that are typically wasted after usage to purchasing services through results-based contracts such as renting or leasing (ING Group, 2013)
- AirFrance-KLM is increasing the sustainability of its in-flight catering (AirFrance-KLM, 2014).

This topic is also covered in Chapter 15 on sustainability.

Profit

The most obvious aspect would be that both measures regarding people and planet may lead to cost reduction, as fewer resources are used and staff may be more productive. Furthermore, the profit aspects cover supply chain management, fair competition and anti-corruption measures. Peloza (2009) has performed a literature review on thirty-one practitioner reports and 128 academic studies, to ascertain how companies measure the financial impacts of their CSR strategies. This study delivered an array of thirty-nine different – and very diverse – measures. He suggests a number of end state metrics, such as share price, return on assets and return on equity; intermediate metrics might be cost-based and revenue-based metrics, operational efficiency, and competitive advantage.

Examples:

- Blycolin asks its suppliers to ensure compliance to its Code of Conduct (Blycolin, 2012)
- CBRE reports $16 million savings on energy for their clients (CBRE, 2013)
- Johnson Controls aims to work with diverse suppliers, companies owned by minorities or women (Johnson Controls, 2014).

Perspectives

A number of aspects hinder any discussion of the contribution of FM and CREM to CSR. First, there is no single definition of CSR. OECD, Global Compact, ISO 26000, Triple-Ps – they overlap, but there are also differences. Next, the dividing line between CSR and sustainability is somewhat blurred. Some documents (e.g. CFP, 2014) do not clearly differentiate between the two terms. Also, looking at the practice of FM and CREM, current lists of KPIs only partly cover the whole field of CSR. Human rights, abolishing child labour and anti-corruption are all genuine CSR goals, but it may be questioned whether these kinds of goals are in the minds of facility and corporate real estate managers. Besides, FM and CREM may be a department within a company or a company's core business. The CSR reports of FM/CREM suppliers are much more focused on FM/CREM than the CSR report of companies with a FM/CREM department only.

Finally, appropriate knowledge is missing. Though the effect of green buildings on employee health and productivity has been amply documented in academic literature (Heerwagen, 2000), according to Jones Lang LaSalles (2013) the majority of office occupiers do not even know exactly how sustainable their existing accommodation is. CFP (2014) found that almost 50% of the respondents had no overview of investments in sustainability. At an operational level KPIs require thorough knowledge of the firm's performance, but also insight into whether the relationship between FM and CREM on the one hand and CSR on the other hand is causal or correlational. Beyond the operational level this insight is complicated by the strong ethical component of CSR. How do you measure the effect of the absence of unethical behaviour? It is not surprising that there is a lack of quantitative evidence about the relationship between FM/CREM and CSR. To complicate matters even further, an analysis by Redlein et al. (2015) has shown that FM is hardly ever mentioned in CSR reports, even though these reports clearly refer to FM tasks and responsibilities, because companies are not aware of the interrelations between CSR goals and measurements and FM activities. Not surprisingly, CoreNet Global stated the urgency of developing proper metrics for CSR (CoreNet Global, 2012). The field of FM and CREM needs more research into CSR to develop KPIs that will show the specific benefit of both FM and CREM to CSR.

References

Aguinis, H. and Glavas, A. (2012) 'What we know and don't know about Corporate Social Responsibility: A review and research agenda', *Journal of Management*, 38 (4), pp. 932–968.

AirFrance-KLM (2013) *Corporate Social Responsibility Report 2013*, available at www.airfranceklm.com/sites/default/files/.../2013_radd-en.pdf

Bauman, C.W. and Skitka L.J. (2012) 'Corporate social responsibility as a source of employee satisfaction', *Research in Organizational Behaviour*, 32, pp. 63–86.

Baumgartner, R. and Ebner, D. (2010) 'Corporate Sustainability Strategies: Sustainability Profiles and Maturity Levels', *Sustainable Development*, 18, pp. 76–89.

Blome, G. (2012) 'Corporate Social Responsibility in housing management: Is it profitable?' *Property Management*, 30 (4), pp. 351–361.

Blycolin (2012) *Zelfverklaring ISO26000 [Self declaration ISO26000]*, Zaltbommel: Blycolin.

Carroll, A.B. (1999) 'Corporate social responsibility: Evolution of a Definitional Construct', *Business and Society*, 38 (3), pp. 268–295.

CBRE (2013) *Corporate Social Responsibility Report*, available at http://www.csrwire.com/.../1366--CBRE-2013-Corporate-Responsibility-Report?

CFP (2014) *Hoe duurzaam is facility management in Nederland?* [How sustainable is facility management in the Netherlands?], The Hague: FMN.

Cleveland, R. and Duchon, E. (2014) *US Occupier Survey. The corporate view of sustainability*, New York: Cushman and Wakefield, available at http://www.cushmanwakefield.com/.../reports/2014%20Business%20Briefing_SustainabilitySurvey Results_Final.pdf

CoreNet Global (2012) *News Release: Industry Leaders Opinion Poll*, available at http://www.corenetglobal.org/files/about/ISSUES%20ADVOCACY%20-%20Working%20 Environments_FOR%20BOARD%20REVIEW_11%2012_1355152718892_1.pdf

European Commission (2011) *Corporate Social Responsibility: A new definition, a new agenda for action*, Brussels: Memo/11/730.

Evans, S. (2009) *Towards a sustainable industrial system: With recommendations for education, research, industry and policy*, Cambridge: Cambridge University Press.

Facilicom (2014) *MVO verslag* [CSR report], available at http://www.jaarverslag2014. facilicom.nl/mvo-verslag

Feige, A., Wallbaum, H., Janser, M. and Windlinger, L. (2013) 'Impact of sustainable office buildings on occupants' comfort and productivity', *Journal of Corporate Real Estate*, 15 (1), pp. 7–34.

Friedman, M. (1962) *Capitalism and freedom*, Chicago: University of Chicago Press.

Garriga, E. and Melé D. (2004) 'Corporate social responsibility theories: Mapping the Territory', *Journal of Business Ethics*, 53, pp. 51–71.

Heerwagen, J.H. (2000) 'Green Buildings, Organizational Success, and Occupant Productivity', *Building Research and Information*, 28 (5), pp. 353–367.

ING Group (2013) *Sustainability annex 2014*, Amsterdam: ING Group NV.

ISO (2010) *ISO 26000: Guidance on Social Responsibility*, Geneva, Switzerland.

ISS (2014) *Corporate Social Responsibility Report*, Søborg, Denmark: ISS.

Johnson Controls (2014) *The Johnson Controls Way. 2014 Business and sustainability report*, available at http://www.johnsoncontrols.com/.../sustainability/reporting/ business_sustainability.html

Jones Lang LaSalle (2013) *Occupier Special Sustainability*, available at http://www.jll. com/Documents/Sustainability/JLL_Sustainability_Report.pdf

Lavy, S., Garcia, J.A. and Dixit, M.K. (2010) 'Establishment of KPIs for facility performance measurement: Review of literature', *Facilities*, 28 (9/10), pp. 440–464.

Lee, M.-D.P. (2008) 'A review of the theories of corporate social responsibility: Evolutionary path and the road ahead', *International Journal of Management Reviews*, 10 (1), pp. 53–73.

Loew, T., Clausen, J., Hall, M., Loft, L. and Braun, S. (2009) *Case Studies on CSR and Innovation: Company Cases from Germany and the USA Berlin*, Münster, available at http://www.4sustainability.org

Masalskyte, R., Andelin, M., Sarasoja, A.L. and Ventovuori, T. (2014) 'Modelling sustainability maturity in corporate real estate management', *Journal of Corporate Real Estate*, 16 (2), pp. 126–139.

Michel, A. (1993) *Capitalism Against Capitalism*, London: John Wiley and Sons.

Miller, N., Pogue, D., Gough, Q.D. and David, S.M. (2009) 'Green Buildings and Productivity', *Journal of Sustainable Real Estate*, 1 (1), pp. 65–89.

Moon, J. (2007) 'The Contribution of Corporate Social Responsibility to Sustainable Development', *Sustainable Development*, 15 (5), pp. 296–306.

Nappi-Choulet, I. and Décamps, A. (2013) 'Capitalization of energy efficiency on corporate real estate portfolio value', *Journal of Corporate Real Estate*, 15 (1), pp. 35–52.

OECD (2008) *Guidelines for Multinational Enterprises*, available at http://www.oecd. org/corporate/mne/1922428.pdf

Nordic Innovation Centre (2010) *CSR-Driven Innovation – Combining design and business in a profitable and sustainable way*, Nordic Innovation Centre project number: 06315, Oslo, Norway.

Peloza, J. (2009) 'The challenge of measuring financial impacts from investments in corporate social performance', *Journal of Management*, 35 (6), pp. 1518–1541.

Redlein, A., Loeschl, J. and Fuke, F. (2015) 'Corporate Social Responsibility (CSR) and Facility Management (FM) in Europe', *International Journal of Facility Management*, 6 (1), available at http://ijfm.net/index.php/ijfm/article/viewArticle/124

RICS (2013) *Corporate Responsibility Report*, available at http://rics.org/sustainability

Rotmans, J. (2014) *Verandering van tijdperk. Nederland kantelt [A new period: The Netherlands are turning over]*, Aeneas, Boxtel.

Sodexo (2014) *Corporate Social Responsibility Report FY2014*, available at http://www.cr.sodexo.com/.../Sodexo-Corporate-Responsibility-Report-FY2014.pdf

TNT Holding B.V. (2010) *Annual Report 2009*, available at https://www.postnl.nl/.../20100222-tnt-annual-report-2009_tcm10-15612.pdf

Twijnstra Gudde (2012) *Marktonderzoek FM*, Naarden: FMN [Market research FM], available at www.twynstragudde.nl/.../duurzaamheid_de_balans_in_people_planet_profit.pdf

UN (2000) *United Nations Global Compact*, available at https: www.unglobalcompact.org

Visser, W. and Tolhurst, N. (Eds.) (2010) *The World Guide to CSR. A Country-by-Country Analysis of Corporate Sustainability and Responsibility*, Sheffield: Greenleaf Publishing Limited.

Yam, S. (2012) 'The practice of corporate social responsibility by Malaysian developers', *Property Management*, 31 (1), pp. 76–91.

Interview 12: Preben Gramstrup, fm³, Denmark

Preben Gramstrup has over twenty-five years of experience in FM. His educational background is in building engineering and economy. Since 2009 he has run his own one-man consultant company. Previously he was a facilities manager responsible for inhouse organisations in large industrial and finance corporations. He is involved in teaching and facilitating activities for the Danish Facilities Management association.

Added Value in general

Q. Do you commonly use the term "Added Value" in your daily work?

A. *Not so much. I find the term difficult. Compared to whom or what is something "Added"?*

Q   Are there other related terms that you prefer to use rather than "Added Value"?

A. *I prefer to use the term "Value Creation" or the Danish equivalent for "Value Increase" or "Appreciation".*

Q. How do you see the relation between "Added Value" and cost reduction?

A   *Value is not only economic value but can include immaterial values and subjective elements.*

Q. Is "Added Value" mostly treated at a strategic, tactical or operational level?

A. *It starts above the operational level – as a minimum at the tactical level. In general the levels are divided into planning at the strategic level, follow-up at the tactical level and execution at the operational level.*

Q. In which context or dialogue is "Added Value" mostly considered in your experience?

A. *Value Creation or Value Increase has to do with FM's reason for being. It relates to the argumentations for survival. It is an essential part of the dialogue between inhouse FM with their corporate clients, but subsequently it also becomes part of the requirements and expectations in relation to external providers, and of the dialogue internally in the inhouse FM organisation.*

Q. How do you see the relation between innovation and "Added Value" in FM?

A. *Value Increase can be related to innovation, but it can also be based on copying best practice. Innovation as introducing something new does not necessarily create value. It can just be a fad. The introduction of ecological food, which I was responsible for in the financial corporation, is an example of innovation in FM, which increased the value of the canteen offering to the staff at the same cost as the former traditional food – an innovation that was later copied by other companies.*

Benefits and limitations of "Added Value"

Q. What are the benefits of considering and talking about "Added Value"?

A. *It makes you speak the language that top management understands, and makes the dialogue focus on the clients' need. It gives a dialogue from the top down instead of from the bottom up. It creates a real discussion of the value FM delivers and what is necessary, seen from the customers' side. Like the number of stars given to a hotel, it can help to illustrate the different options for quality levels and create alignment of expectations.*

Q. What are the limitations of considering and talking about "Added Value"?

A. *The term can to a certain degree be subjective and it can be difficult to document. It can also be used as a kind of "sugar coating" and spin talk without any real substance.*

Management of Added Value

Q. What are your top five main values to be included in management of accommodations, facilities and services? Could you mention a few examples of concrete FM interventions to attain these added values and KPIs to measure them?

A. *In general it is about how to support the core business in such a way that the customer and end users experience the value. Maslow's pyramid of needs could be a starting point. FM does not create value by supporting the lower levels in the pyramid. These are taken for granted and you will get criticism if they are not fulfilled, but you will not receive any appreciation. That is just doing the work that is necessary. To be appreciated you need to deliver something that is beyond basic expectations. In my job as inhouse facilities manager I always used the criteria of whether changes in FM would create more time or more wellbeing for the staff. This is specified with examples in the table below.*

Prioritised values in FM	Concrete FM interventions	KPIs
1. Create time: Release the core business staff from having to spend time or effort on support related tasks which the FM organisation is better suited to carry out	Meeting facilities that are ready to use	User satisfaction Occupancy of rooms
2. Create wellbeing: Make good working conditions with attractive work environment and internal services	Good food and coffee in the canteen	User satisfaction

Q. Do you measure whether you attain the targeted added values? If so, what Key Performance Indicators do you apply for each of the added values you mentioned before?

A. *User satisfaction is included in the table above. I use a Service Performance Dashboard, which for each service defines some SLAs. For each SLA there are some qualitative and quantitative KPIs. By the use of specific criteria it can be determined whether each KPI is fulfilled, partly fulfilled or not fulfilled, leading to a green, yellow or red "traffic light", which can be aggregated for each SLA and each service. This can be used for operational services like cleaning, for instance, using the standard INSTA 800, but I have also used it for measuring management performance.*

In my former job as inhouse facilities manager in a financial corporation I introduced a Value Account based on the User Value Equation, covering Quality and Process versus Price and Difficulties. This was inspired by Heskett's Customer Value Equation and was only used qualitatively. It was partly introduced as a kind of alternative towards the many accounting focused colleagues in the company, but it was also an attempt to show that FM not only has to do with the economic bottom line.

Q. Do you benchmark your data with data from other organisations?
A. *Yes, particularly user satisfaction and cost pr. units.*

Q. What other methods do you use to document "Added Value"?
A. *For instance with business cases and reporting of finished projects.*

Q. How is "Added Value" included in your communication with your stakeholder?
A. *As a consultant it is a basic condition to show your worth for your client.*

Last comment

Q. Are there other topics that you find important in connection to the concept of "Added Value"?
A. *I feel the term "Added Value" has received too much attention in FM in Denmark. It has been used as a kind of oracle term and everything is possible if it 'only' adds value, regardless of reality. Used more wisely it could help in projects and discussions with non-FM people.*

Part III
Epilogue

17 Tools to manage and measure adding value by FM and CREM

*Jan Gerard Hoendervanger, Feike Bergsma,
Theo van der Voordt and Per Anker Jensen*

Introduction

In Chapter 1 of this book the editors presented a simple Value Adding Management (VAM) process model according to the widely used triplet of input-throughput-output and extended by outcome – impact/added value:

Input → Throughput → Output → Outcome → Impact = Added Value

This model also represents the triplet What → How → Why. In this chapter we present an extended Value Adding Management model and various tools to support the different steps in value adding management processes. This chapter is based on a review of recent literature on aligning corporate real estate and facilities to organisational strategies and primary processes, and a cross-chapter analysis of the twelve value parameters that were discussed in Part II of this book, with possible interventions, tools to measure the added value of interventions and KPIs.

Extended Value Adding Management model

Being a supportive process, value adding FM/CREM processes should be embedded in the strategic management of the organisation and connected to the primary process. Usually processes have a cyclical nature. Evaluation of realised output/outcome/added value may be a starting point for new interventions. In order to integrate VAM of buildings and facilities in business management and to make the VAM model more instrumental, to be able to use it as a decision-support tool, an elaboration of this model and tools are needed.

In the literature many models and tools are presented to support value adding management by optimally aligning buildings, facilities and services to organisational strategies; see Chapter 3. The Center for People and Buildings developed an Accommodation Choice model to support decision makers in the many steps to realise a successful accommodation policy or an improved work environment that fits with the organisational objectives and internal and external constraints and copes with the needs of all stakeholders; see Box 17.1 with Figure 17.1 (Van der Voordt et al., 2012). This comprehensive tool can be perceived as an elaboration of the Plan-Do-Check-Act (PDCA) cycle that Alexander Deming developed in the 1950s to support quality control and total quality management.

Box 17.1 Accommodation Choice Model

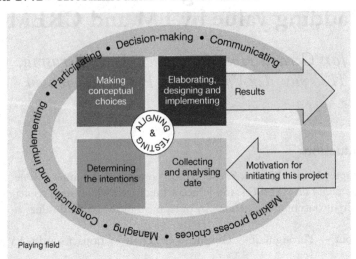

Figure 17.1 Accommodation Choice Model – basic framework

The model starts with the identification of drivers to change and ends with agreed interventions, an implementation plan and a plan for how to manage the building-in-use. Step 1 is the collection and analysis of data about the vision and mission of the organisation, conditions, business processes, expected changes, costs, and experience with the current accommodation. In Step 2 the intentions are specified: what does the organisation want to accomplish with the accommodation? This requires the specification of user profiles and a well-considered choice between one-size-fits-all versus differentiation in the plan in relation to different activity patterns. Step 3 consists of conceptual choices for the accommodation plan, the workplace concept, facilities, services and the aesthetic and technical quality. Also the degree of freedom of choice of teams or individuals has to be defined. Step 4 is the elaboration of conceptual choices in a programme of requirements, design and management plan. The Lynchpin in the centre of the model connects all steps through continuous monitoring, evaluation and coordination. All steps are taken on a playing field with various stakeholders, each with their own needs and priorities. The model is supported with tools such as the work environment diagnosis instrument WODI (see Chapter 5), software to calculate the required number of different types of work places, a Space Utilisation Monitor, and benchmark data on employee satisfaction.

Source: Van der Voordt et al., 2012.

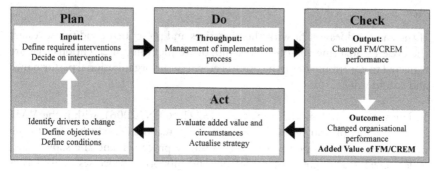

Figure 17.2 Extended Value Adding Management model

Because the PDCA cycle is widely applied, we use this cycle to elaborate the original Value Adding Management model from Chapter 1; see Figure 17.2. The Deming cycle goes back to the Shewhart cycle of Plan-Do-Check-Study (PDCS), developed by Walter Shewhart from the Bell company. The cyclical character of the extended model emphasises that value adding management is – or should be – a continuous process, including four main steps.

Plan

Define the drivers to change, e.g. identify if there is a gap between the desired and actual performance of the organisation and the accommodation, facilities and services; define which interventions may result in improved performance; define the objectives of these interventions in a SMART way (Specific, Measurable, Achievable, Relevant and Time-bound); and define the conditions or prerequisites that should be taken into account. The Plan phase ends with clear decisions about which interventions will be implemented and a plan for how to implement.

Do

Implement the proposed interventions and manage the change process. Decisions to be made regard, for instance, who should be involved in the process and how, time schedules, how to cope with resistance to change, and how to cope with the different needs of different stakeholders.

Check

Measure the costs and benefits of the intervention(s) and the performance of the organisation and its facilities before and after the implementation of the intervention(s). Check if the changed performance fits with the organisational strategy, mission, vison and objectives, and as such adds value to the organisation.

Act

If the change does not fit with the objectives and prescribed conditions, restart the cycle by adapting the objectives and/or (re-)defining interventions that are expected to contribute to further improvement of the organisational performance and to add value.

In line with the FM model from CEN 15221 (CEN, 2006), within the VAM cycle a distinction can be made between the demand side, which refers to the desired organisational performance and primary processes, and the supply side, which refers to FM/CREM performance and supportive processes, as well as in different levels of decision-making: strategic, tactical and operational. An example of value adding management at the strategic level is the introduction of a new workplace policy by implementing activity-based workplaces in order to reduce costs and improve productivity. An example at an operational level is the replacement of a broken window in order to stop draught and leakage. In both cases the whole PDCA cycle may be appropriate.

Plan

Referring to the strategic management model of Johnson et al. (2009), the Plan phase includes 'strategic analysis' and 'strategic choices', whereas 'strategy in action' relates to the Do phase; see Figure 17.3. Strategic analysis is a converging process, starting with a wide-ranging exploration of external developments, internal factors and stakeholders, followed by a narrowing down to the selection of specific added value parameters of FM/CREM that most fit with the goals and interests of different stakeholders. Making strategic choices requires a diverging process, elaborating the strategic focus into more specific strategies, critical success factors, key performance indicators, targets/norms and plans for FM/CREM interventions.

Within the planning process it is recommended to make a clear distinction between the organisational strategy and the FM/CREM strategy. Both require a strategic analysis and both may reveal drivers for change. Linking organisational goals to FM/CREM interventions is a process of alignment between two different playing fields and connecting organisational outcomes to FM/CREM performance. These links are rarely self-evident and require explicit argumentation and consideration in order to avoid sub-optimisation. For example, if an organisation wants to enhance innovation, it seems obvious to invest in a new interior design that may stimulate creativity and support exchange of knowledge. However, reducing real estate costs in order to increase the R&D budget might be more effective. This example illustrates that there may be different ways to use FM/CREM as a means to contribute to one or more organisational goals. For this reason the selection of FM/CREM interventions requires a careful process. This alignment process was illustrated earlier in Figure 3.2 in Chapter 3.

Strategic planning process

Figure 17.3 Visualisation of the three processes within strategic management, based on Johnson et al. (2009)

Tools to identify the need for change, objectives and prerequisites

Analysing the context of value adding management may start with exploring the different roles, interests and power of stakeholders involved, using stakeholder analysis. It is relevant to make a distinction between external and internal stakeholders and the end users of FM/CREM; see also Figures 3.5 and 3.6 in Chapter 3 and Ambrosini et al. (1998). Depending on their strategic role and influence, stakeholders can participate in the planning process.

In practice a SWOT analysis is often applied to analyse the need and direction for change; see Box 17.2. For example, when competitors flood the market with a new product (threat) and the organisation has a strong R&D department (strength), the FM/CREM strategy could be to add value by stimulating innovation (opportunity). When the organisation struggles with low customer satisfaction (weakness) in a growing market (opportunity), the FM/CREM strategy may be to enhance the corporate image by a hospitality policy or to improve its operational excellence by a better price/quality ratio. It is recommended to conduct a SWOT analysis of both the organisation and the FM/CREM processes and products to identify drivers for change *within* the domain of FM/CREM. The need for change can be both internally driven, e.g. a gap between the desired and perceived corporate identity, too much energy consumption, high accommodation costs in comparison to an internal standard or external benchmarks, or externally driven, such as the end of a lease contract or a changing market with more buildings available at a lower rent.

Box 17.2   SWOT analysis

A SWOT analysis is a suitable tool to assess the strengths and weaknesses of the organisation and threats and opportunities from its environment (Hill and Jones, 1989). Steps to take may include:

• Exploration of external opportunities and threats by market research, an assessment of demographic, economic, societal, technological, ecological, market and industry and political trends and developments (DESTEMP, see Keuning et al., 2010, also known as DESTEP or PESTEL), scenario analysis and application of the five forces theory of Porter (1985)

• Internal analysis of the organisation according to the 7-S model (Peters and Waterman, 1982) to explore strengths and weaknesses and to define new possibilities to (re)position FM/CREM to cope with the needs of clients, customers, end users and the environment. Benchmarking and customer satisfaction surveys can help to establish the current position of the organisation

• Assessment of relevant external and internal topics in a SWOT matrix will support the selection of main challenges and opportunities, resulting in a clear view on the need for change.

The value proposition model of Treacy and Wiersema (1995) may provide a useful starting point to relate a corporate strategy to particular FM/CREM value parameters. According to this model each organisation should make a fundamental strategic choice to focus on one out of three different value propositions: product leadership, customer intimacy, or operational excellence. This choice influences the selection of FM/CREM value drivers: product leadership stresses the FM/CREM contribution to innovation, whereas customer intimacy demands a focus on customer satisfaction and operational excellence requires a productivity-oriented approach.

Nourse and Roulac (1993) and Roulac (2001) link nine possible 'driving forces' behind a corporate strategy (e.g. market needs, technology, return on investment) to seven components of competitive advantage (e.g. attracting and retaining customers, efficient business processes), eight strategic accommodation choices (e.g. cost reduction, support of human resources, value creation of real estate) and fourteen operational decisions (e.g. regarding the location, number of m², ICT, ownership and risk management). What they call 'accommodation choices' is rather similar to what we call 'added value parameters'.

A strategic analysis should result in a strategic focus: what objectives should be achieved? It may be helpful to discuss this focus and the underlying analysis with top management (client) and management of business units (customers) to gain a better understanding of business demands and to be able to select suitable objectives for FM/CREM. Cooperation with professionals/colleagues that are responsible for other support functions, such as HRM, ICT, Finance and Marketing could enable an integrated and more client-oriented approach (see for instance Ware and Carder, 2012). These professionals are in a similar position and have to connect the corporate strategy to their own discipline. Moreover, these conversations may reveal opportunities for interdisciplinary collaboration or joint interventions and will help to define conditions for change in a broader perspective. A good example is the introduction of new ways of working, which requires an integrated approach that should be supported by HRM, ICT and FM/CREM.

Tools to define required interventions and to select the most appropriate ones

Defining SMART goals is an important step as it enables the development and selection of interventions that are expected to result in added value. In the second part of the Plan phase, the main question is how to translate the strategic focus and smart goals into appropriate and valuable FM/CREM interventions. Figure 4.1 in Chapter 4 presented a number of optional interventions in order to attain various added values. This figure is elaborated in Table 17.2, which will be discussed in the Check section. To identify the most appropriate interventions it is recommended to create a FM/CREM strategy map. This tool, developed by Kaplan and Norton (2004), may help to identify critical success factors within chains of means/ends, which are crucial for adding value as defined in the strategic focus. Figure 17.4 shows an example of a FM-oriented strategy map from the Danish construction toy company LEGO, which was presented in the first book (Jensen and Katchamart, 2012). An example of how a CRE strategy can be mapped can be found in Lindholm et al. (2006).

The Balanced Scorecard (Kaplan and Norton, 1992) is a widely used tool to link strategic analysis to critical success factors and KPIs. Example of its application in operationalisation of FM strategies can be found in Rasila et al. (2010) and White (2013). For a discussion of BSC related KPIs in different fields, see Schuur (2015). To support organisations in selecting prioritised performance measures and KPIs, a step-by-step plan may be useful including six steps (Riratanaphong and Van der Voordt, 2015):

1 Inventory of KPIs that the organisation currently applies
2 Allocating all KPIs to organisational performance or corporate real estate and facilities performance
3 Classification of all measures in clear categories, for instance according to different value parameters or the four perspectives of the Balanced Scorecard
4 Comparison of possible and currently applied measures and KPIs
5 Reflection on the relevance of KPIs in connection to the vision and mission of the organisation and its main objectives
6 Prioritisation of KPIs that are most relevant and appropriate regarding contextual variables, such as economy and competitive advantage.

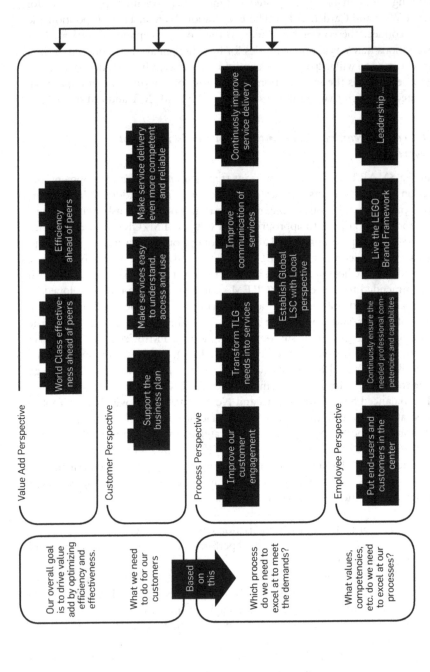

Figure 17.4 LEGO's Facilities Strategy Map 2010 (Møllebjerg, 2010, in Jensen and Katchamart, 2012)

The KPIs can be integrated in a management dashboard to enable continuous monitoring and clear reporting to top management and other stakeholders (Atkins and Brooks, 2015).

The selection of interventions also depends on the future demand and the fit or misfit between the current supply and the future demand. An interesting model in this context is the so-called DAS-frame (De Jonge et al., 2009) that has been developed to support decision makers in Designing an Accommodation Strategy (DAS) that fits with the needs and interests of various stakeholders, now and in the future; see Box 17.3 containing Figure 17.5.

Box 17.3   DAS-frame

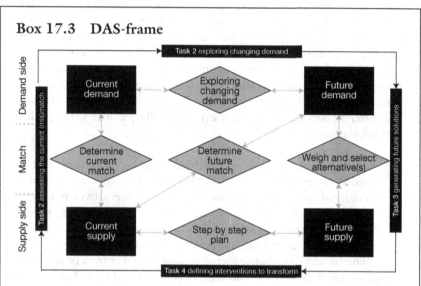

Figure 17.5 Framework for Designing an Accommodation Strategy (adapted from De Jonge et al., 2009, and Den Heijer, 2011)

The DAS framework includes four steps:

1   Assessing the match or mismatch between the current supply and the current demand by a SWOT analysis of the organisation, defining the current need for space and the available space etc., and a SWOT analysis of the current building(s) and facilities, taking into account the needs and interests of different stakeholders.

2   Exploring the changing demand (for instance, using DESTEMP and scenario analysis) and assessing if the current supply fits with the future demand.

3   Designing alternative options for future supply that optimally fit with the current and future demand, and selection of the solution that adds most value, i.e. which results in maximum benefits with feasible costs and risks.

4   Implementation of change to transform the current supply into the desired future supply.

Finally, strategic criteria are a prerequisite to select the most effective FM/ CREM interventions, i.e. the option(s) with highest benefits and lowest costs. Decision support tools such as business cases are often used to select the most appropriate interventions and to support decision making processes; see Box 17.4.

Box 17.4 Business case

A business case enables senior management to take an informed go/ no-go decision about a proposed FM/CREM intervention. Gambles (2009, p.1) defines a business case as "a recommendation to decision makers to take a particular course of action for the organisation, supported by an analysis of its benefits, costs and risks compared to the realistic alternatives, with an explanation of how it can best be implemented". It has become common business practice to ask for a business case as a formal prerequisite to initiating any project that involves considerable impact, investments and risks. While different organisations use different formats, a business case generally includes (Maul, 2011; Hoendervanger et al., 2012):

- *Alternatives:* To enable an informed decision, it is important to consider alternatives and to define why the proposed option is considered to be the best one. Generating alternatives requires a creative process, for which brainstorming, benchmarking, project visits and literature reviews can provide useful input (Tague, 2005). Multiple criteria analysis and preference modelling can be used to define how the alternatives score against several criteria or KPIs and to communicate relevant differences (Barzilai et al., 2010; Arkesteijn et al., 2015).
- *Impact analysis:* In the Plan phase an impact analysis is an ex ante assessment of expected outcomes. The Check phase includes an ex post analysis based on measuring the actual impact. Appropriate KPIs are needed to assess whether an intervention will effectively contribute to adding value and if negative impacts may come to the fore as well. It is also important to include both tangible and non-tangible impacts. A stakeholder analysis is a useful tool to explore impacts in relation to the interests of different stakeholders.
- *Investment analysis:* Decision makers need to assess whether the investment needed to implement the intervention is affordable, feasible and cost-effective. All monetary costs and benefits – during the project and the whole life cycle – should be taken into account. Several tools can be used to make sound financial calculations, e.g. Return On Investment (ROI), payback period, Discounted Cash Flow (DCF), Internal Rate of Return (IRR), Economic Value Added (EVA), and Life Cycle Costing (LCC).

- *Risk analysis:* All possible risks that are related to impacts and return on investment should be analysed. Some risks can be monetised using the formula *risk* = *probability* × *costs*. Others need to be analysed in a qualitative way. A sensitivity analysis or stress test based on different scenarios is recommended. For all major risks, adequate measures should be included (see also Chapter 12).

It is very important but usually also rather difficult to cover both 'hard' factors (e.g. investments, cost savings) and 'soft' factors (e.g. impact on image, possible resistance to change), and to discuss possible trade-offs.

Do

According to the strategic management model of Johnson et al. (2009), the purpose of the Do phase is to put 'strategy in action'. This requires a change process to implement the intended FM/CREM intervention(s). A major challenge is to keep focus on the initial goals regarding adding specific values (as defined and decided in the Plan phase). Implementation processes tend to develop their own dynamics, which can easily shift the focus from long-term strategic organisational goals to the short-term tactical and operational goals of the participants.

Change management has evolved as a specialist discipline and has produced many different tools. As a crucial first step, a tailor-made approach should be designed, which fits the characteristics of the intervention (e.g. complexity, budget, risks and timeframe), the goals (i.e. intended added value) and the social/organisational context. As in the Plan phase, in the Do phase as well it is recommended to conduct a stakeholder analysis to define who is or should be involved in the process, in what way, and what their interests are. These stakeholders may or may not be the same as in the Plan phase. The stakeholder analysis should take into account how different stakeholders perceive change, for instance by using the five colours framework of De Caluwé and Vermaak (2003). This framework links five different change paradigms to five different management process approaches; see Table 17.1.

Since the change approach has to fit with the expectations and needs of different (groups of) participants and characteristics and the goals of the intervention, it is often wise to combine two or more approaches. A blue-print approach to ensure that a refurbishment project will be finished in time and within budget might for instance be combined with a red-print approach for involving users effectively in the design process. The intended added value may play an important role as well. For instance, if the goal of a new work environment is to stimulate social cohesion and interaction, a red-print approach may be most effective, even when this might take more time and effort than a yellow-print or a blue-print approach.

Table 17.1 Five approaches to change management, adapted from De Caluwé and Vermaak (2004)

	Yellow-print	Blue-print	Red-print	Green-print	White-print
Change paradigm	Bring common *interests* together to create a feasible solution, a win-win situation	Think first and then act according to a *plan* to create the best solution	Stimulate *people* in the right way to create a motivating solution, the best 'fit'	Create settings for collective *learning* to create a solution that people develop themselves	Create space for spontaneous *evolution* to create a solution that releases energy
Process approach	Power game	Rational process	Exchange exercise	Learning process	Dynamic process
Typical elements of the approach	Forming coalitions, changing top structures, policy making	Project management, strategic analysis, auditing	Assessment and reward, social gatherings, situational leadership	Training and coaching, open systems planning, gaming	Open space meetings, self-steering teams, appreciative enquiry
Result	Partly unknown and shifting	Described and guaranteed	Outlined but not guaranteed	Envisioned but not guaranteed	Unpredictable

Related to the five colour-print approaches, numerous methods, tools and techniques for change management are available (e.g. De Bruyne, 2007; Cameron and Green, 2015). We refer to two methods that are widely used in practice: PRINCE2, see Box 17.5, a suitable project management method for a blue-print approach; and Lean Management, a comprehensive system for processes optimisation that fits with a green-print approach, see Box 17.6.

Box 17.5 PRINCE2

PRINCE2 (Projects IN Controlled Environments) is a structured project management method that defines a project as "*a temporary organization that is created for the purpose of delivering one or more business products according to an agreed Business Case*" (Buehring, 2011). The method provides a framework based on seven principles:

1 *Continued business justification:* Is there a justifiable reason (business case) for starting/continuing the project?
2 *Learn from experience:* Project teams should continually seek and draw on lessons learned from previous work.

3   *Defined roles and responsibilities:* The project team should have a clear organizational structure and involve the right people in the right tasks.
4   *Manage by stages:* Projects should be planned, monitored and controlled on a stage-by-stage basis.
5   *Manage by exception:* Defined tolerances for each project objective to establish limits of delegated authority.
6   *Focus on products:* Focus on the product definition, delivery and quality requirements.
7   *Tailor to suit the project environment:* PRINCE2 is tailored to suit the project's environment, size, complexity, importance, capability and risk.

These principles are elaborated into seven themes that require continuous attention: 1) business case, 2) organisation, 3) quality, 4) plans, 5) risk, 6) change, and 7) progress; and seven processes which encompass the project lifecycle: 1) starting up a project, 2) directing a project, 3) initiating a project, 4) controlling a stage, 5) managing product delivery, 6) managing stage boundaries, and 7) closing a project.

Source: Buehring, 2011; see also Bentley, 2010 and PMI, 2013.

Box 17.6   Lean Management

Lean Management is a commonly used tool to analyse and optimise processes. To accomplish the shift towards lean thinking, Womack and Jones (1996) presented five famous lean principles:

1   *Specify value.* Lean thinking starts with the aim of defining customer value with a specific product (or service) that offers specific capabilities with a specific price through a dialogue with the specific customer.
2   *Identify the value stream.* The whole value creation process, in which the operations create the value of a specific product or service, is identified in order to pinpoint and eliminate waste.
3   *Flow.* Continuing to remove waste from the value stream and make the value flow between different operations and departments along the entire value stream.
4   *Pull.* Instead of pushing products to the customers, products in the upstream should be produced only when a customer in the downstream asks for them.
5   *Perfection.* The effort for perfection, which is never reached, creates the urge to make improvements: there is no end for waste elimination.

Perfection refers to continuous improvements (in Japanese: *kaizen*) to minimise waste and generate customer value that is one step closer to the real needs and requirements of the customer. Optimisation for maintenance services is an example of improving FM/CREM processes (Jylhä and Junnila, 2013). Lean Management is applicable within a wide range of FM/CREM processes for example workplace and energy management as illustrated by Jylhä and Junnila (2014).
Sources: Womack and Jones, 1996; Jylhä and Junnila, 2013, 2014.

Avoiding or removing resistance to change is usually a major component of any change management approach. According to Kreitner and Kinicki (2007) there is no universal strategy for dealing with resistance, but communication is always essential and should include at least four elements: 1) inform employees about the change ('what'); 2) inform employees about the rationale underlying the change ('why'); 3) organise meetings for answering questions that employees may have; and 4) let employees discuss how the change may affect them.

Check

The Check phase starts during or after the implementation of the selected interventions and measures, if and to what level the objectives have been attained, if the performance of the organisation and FM/CREM has actually been improved, and if the improved output and outcome adds value to the organisations. Table 17.2 gives an overview of many possible interventions and tools to measure the output and outcomes that came to the fore in Part II of this book. Usually various measuring tools are combined in a so-called Post-Occupancy Evaluation (POE), also called evaluation of buildings-in-use (Preiser and Visscher, 2004; Van der Voordt et al., 2012).

A common way to evaluate KPIs is to conduct performance benchmarking internally or with external partners. The benchmarking process can be carried out according to EN15221-7 (CEN, 2012). It is an iterative process consisting of three phases divided into ten steps as shown in Table 17.3. Benchmarking is an important tool to control cost and to find areas of improvement in FM/CREM. However, in practice benchmarking is more often applied to limit the sacrifices by cost reduction than to improve the benefits by creating new and qualitative better solutions, and often applied as a reactive tool instead of a proactive tool. Benchmarking can also be used in the Plan phase and is applicable to both hard (cost) factors and softer factors such as employee satisfaction (Van der Voordt et al., 2012). It is also important to include the why question, i.e. why ones' own organisation or FM/CREM performs better or worse in comparison to other organisations.

Table 17.2 Interventions, assessment methods and KPIs for the twelve value parameters

Chapter Number and Value	Interventions Provide/create:	Tools to measure the impact	KPIs (Top 5)
5. Satisfaction	More suitable spatial layout More collaborative spaces Extra storage space Better indoor climate Modern and attractive interior, furniture and finishing	Employee surveys Observations Interviews Walk-throughs Narratives	Employee satisfaction with: – Workplaces – Collaborative space – Indoor environment – ICT and other equipment – Amenities
6. Image	Move to a new location High quality surroundings Reorganisation of spatial layout Better/more facilities and services "Green" programme	Stakeholder surveys Group discussions Analyses of historical sources Analysis of social media and other ways of communication	Perceptions of – Corporate identity – Corporate value – Corporate brand – Media exposure – Shares and likes on social media

7. Culture	More open settings to support collaboration
	Shared desks/places
	New types of food
	New behavioural rules (e.g. regarding dress code)
		Employee surveys
		Observations
		Interviews
		Workshops
			Perceptions of
			– Corporate culture
			– Match between corporate culture and
			  ○ organisation
			  ○ facilities
			  ○ work environment

8. Health and safety	Higher level of personal control Ergonomic designed furniture Better indoor air quality, acoustics and light More healthy food Healthcare services	Capture and react on complaints Workplace H&S assessment	Sick leave Number of accidents Absence due to accidents Number of complaints about H&S Percentage of (dis)satisfied employees in surveys

Table 17.2 continued

Chapter Number and Value	Interventions Provide/create:	Tools to measure the impact	KPIs (Top 5)
9. Productivity	Higher level of transparency to support collaboration Better opportunities for concentrated work Ergonomic furniture Higher level of personal control of the work environment Better supporting IT facilities and systems Healthy and safe environment	Observations Monitoring of computer activity Counting of output Measuring time spent or saved Employee surveys	Output per employee Quality of output Absence Perceived support of – Individual productivity – Team productivity
10. Adaptability	Surplus of spaces, load-bearing capacity, installation capacity, and facilities Customisability of facilities Disconnection of facilities components Removable and relocatable units and building components	Building performance assessment, i.e. using Flex 2.0 or Flex 2.0 Light Observation of adaptations of the building-in-use	Weighted assessment values, i.e. scores on scales of Flex 2.0 or Flex 2.0 Light
11. Innovation and creativity	Better visibility and overhearing among employees Visual clues Different types of meeting spaces and informal areas Optimised indoor climate ICT that supports virtual knowledge sharing	Spatial network analysis Social network analysis Employee surveys Logbooks on knowledge sharing activities	Level of enclosure/openness Average walking distance Level of personal control with indoor climate Diversity of available workspaces and meeting places Perceived quality of visual clues

12. Risk	Removing hazards Emergency and recovery plans Install security systems Back-up supply systems Insurances	Measuring time of business interruptions Measuring risk expenses • Insurance • Damage prevention • Actual damage	Uptime of critical activities Total risk expenses Total insurance expenses Total damage prevention expenses Total actual damage expenses
13. Cost	Cost saving by • Establishing own FM department • Process optimisation • Outsourcing • Technical upgrade • Utilisation of synergies	Accounting with an appropriate cost structure Measuring space, number of workstations and fte	Cost/m^2, workstation or fte: • Total FM • Space and infrastructure • People and organisation • Space • Workplace
14. Value of assets	Disposal of CRE Sale and lease back Renegotiate rental contract Buy leased property Improve owned CRE by adaptive reuse	Estimate annual potential gross income and annual operational expenses Market valuation Estimate cost of new development	Capitalisation Rent level Market value Cost of new development

Table 17.2 continued

Chapter Number and Value	Interventions *Provide/create:*	Tools to measure *the impact*	KPIs *(Top 5)*
15. Sustainability	Sustainability framework Staff training Reduction of energy consumption Reduction of materials use, increased recycling and improved waste handling Reduction of travel and transport activities	Identify critical success factors from corporate strategy, develop a survey and multi-criteria scoring methodology, and implement a continuous review process	Consumption of primary energy and water CO_2 emissions. Material use, recycling and waste Life cycle cost Access to transport
16. Corporate social responsibility	Higher quality of working environment Employment of challenged workers Promoting public transport Circular purchasing model Collaboration with companies owned by minorities	Depends on corporate CSR policy and target	People: • Diversity of staff • Community satisfaction Planet: • Utilisation of space • Use of resources Profit: • Total FM/CREM cost

Table 17.3 Benchmarking process (based on CEN, 2012)

Phase	Step
Preparing	1. Set objectives (purpose and scope)
	2. Define methodology (indicators and benchmarks)
	3. Select partners (peers and code of conduct)
Comparing	4. Collect data (collect and validate)
	5. Analyse data (determine and normalise)
	6. Determine gaps (compare and explain)
	7. Report findings (communicate and discuss)
Improving	8. Develop action plan (tasks and milestones)
	9. Implement plan (change and monitor)
	10. Process review (review and calibrate)

Regarding KPIs, a distinction should be made between output indicators to measure FM/CREM performance and outcome indicators to measure organisational performance. This distinction is related to the basic distinction in Michael Porter's value chain (Porter, 1985) between support activities and primary activities, which is also reflected in the management model in the European FM standard (CEN, 2006). FM/CREM as support activities deliver value to the primary activities, and the primary activities create value for the organisation by delivering value to external customers and other stakeholders.

Examples of output indicators of benefits, i.e. improved FM/CREM performance, are buildings, facilities and services that better fit with the needs and objectives of the organisation and all stakeholders with respect to:

- Quality of the work environment
- Access to public transport
- Use of space (high occupancy level, low vacancy) and other resources
- Healthy and safe indoor environment
- Adaptability
- Balance between openness and enclosure
- Walking distances
- Personal control of the indoor climate
- Diversity of available workspaces and meeting places
- Quality of visual clues
- Facility costs (investment costs, running costs, life cycle costs).

Examples of outcome indicators of benefits, i.e. improved organisational performance, are a positive contribution of FM/CREM to:

- Job satisfaction, staff commitment and low staff turnover due to high satisfaction with the work environment
- Market share
- Corporate identity, corporate values and corporate brand
- Corporate culture

- Absence due to sick leave
- Number of accidents
- Individual and team productivity, quantitative and qualitative
- Uptime of critical activities
- Consumption of primary energy and water, CO_2 emissions, material use and waste, and high level of recycling
- Attraction and retention of talented staff
- Community satisfaction.

A similar distinction between output and outcome indicators can be made for sacrifices, which can be perceived as the reverse of benefits. Examples of output indicators of sacrifices such as cost and risk of FM/CREM interventions are:

- Downtime of critical activities
- Total expenses of risk and damage prevention and avoidance
- Total insurance expenses
- Total actual damage expenses
- Investment cost and life cycle cost per m², per workstation or per fte of total FM, space and infrastructure, people and organisation, space and work places.

Examples of outcome indicators of sacrifices, i.e. negative effects of FM/CREM interventions on organisational performance are:

- Reduced market share
- Reduced profitability
- Less involvement and commitment of shareholders and stakeholders.

KPIs may regard quantitative numbers that can be compared with objective standards, e.g. the actual m² per person in comparison to a corporate standard, CO_2 emissions in comparison to legislation, or BREEAM, LEED or DGNB scores. However, as we mentioned before in the Plan section, many intangible and "soft" factors can only be measured in a qualitative and sometimes also more subjective way, for instance by measuring the *perceived* support of productivity or the *perceived* support of corporate culture by surveys. To what level the output and outcome has been improved can be measured by calculating the difference between FM/CREM performance and organisational performance before and after the intervention(s).

Whether the increased performance also adds value to the organisation depends on the mission, vision and objectives of the organisation and the trade-off between benefits and sacrifices. For example, if the objective of the organisation is to be as green as possible and to perform in a socially responsible way, a reduction in energy consumption adds value, whereas if the organisation just aims to fit with legislation and the performance assessment in the Plan phase shows that it already fits with the legal requirements, being "more green" does not add value to the organisation (though it is very welcome from a societal point of view!).

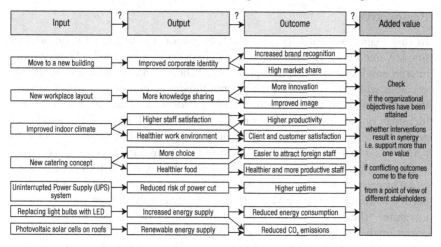

Figure 17.6 Examples of input → output → outcome → added value chains

Furthermore, it is important to check which FM/CREM interventions result in synergy, i.e. improve the outcome regarding more than one value parameter, and which ones may result in conflicting outcomes, e.g. a higher profit but a lower level of employee satisfaction due to a reduction in m² per employee. Figure 17.6 shows examples of input → output → outcome → added value chains to illustrate the complexity of cause-effect relationships between interventions, FM/CREM performance, organisational performance and added value. Assessing the added value of FM/CREM interventions should not only include 'objective' performance measurement and benchmarking, but also a 'subjective' evaluation of whether the improved performance really adds value to the organisation, the clients, customers and end users, and society.

Act

The Act phase is quite similar to the Plan phase but it starts from a different situation. Whereas the Plan phase may start with an analysis of changing internal or external circumstances or a strategic analysis of the strengths and weaknesses of the organisation and FM/CREM products and processes, these factors are already known in the Act phase. When all objectives have been attained and maximum value has been added, the Act phase may include consolidation of the new situation, until new drivers of change come to the fore. If the objectives are not sufficiently or optimally attained, or if too many negative side effects come to the fore, new interventions or broadening or strengthening of earlier interventions should be considered. Another option is to reconsider the objectives. Maybe the aimed performance was not realistic and feasible within the current conditions. If new or revised interventions have to be implemented, the Plan and Do phases start again.

Perspectives

In this chapter we tried to connect existing models and tools to the original simple Value Adding Management model in order to make the VAM cycle more instrumental and practically applicable. Whereas many different tools are available, so far these tools are usually not integrated in a step-by-step approach. Besides, most tools focus on FM/CREM performance (output), with far fewer focusing on assessing the contribution of FM/CREM to organisational performance (outcome). In much research a valuation of the trade-off between benefits and sacrifices in connection to organisational objectives (added value) and interrelationships is often lacking as well (Jensen and Van der Voordt, 2015).

An interesting next step could be to explore the similarities and dissimilarities between various FM/CREM models and generic management models and to integrate "the best of" in a new model or a further improvement and operationalisation of the extended VAM model. This requires intensive collaboration with other support functions and knowledge fields such as HR, ICT, Finance, Marketing and PR. Another next step could be to connect all the tools used to measure FM/CREM and organisational performance and the related KPIs that are presented in Table 17.2 with other lists of KPIs such as the ones mentioned by Lindholm and Nenonen (2006) and Lavy et al. (2010, 2014). A third topic for future research is to further elaborate input → output → outcome → added value relationships and to integrate current qualitative and quantitative data-collection methods to get clear and evidence-based pictures. These three steps may result in a Value Adding Management guidebook, as a third book on how to add value by FM and CREM.

References

Ambrosini, V.R., Johnson, G. and Scholes, K. (1998) *Exploring techniques of analysis and evaluation in strategic management*, Harlow: Prentice Hall.

Arkesteijn, M., Valks, B., Barendse, P. and de Jonge, H. (2015) 'Designing a preference-based accommodation strategy: A pilot study at Delft University of Technology', *Journal of Corporate Real Estate*, (17) 2, pp. 98–121.

Atkins, B. and Brooks, A. (2015) *Total Facility Management*, Chichester: John Wiley and Sons Ltd.

Barzilai, J. (2010) 'Preference Function Modelling: The Mathematical Foundations of Decision Theory', in Ehrgott, M., Figueira, J.R. and Greco, S. (Eds.) *Trends in Multiple Criteria Decision Analysis*, Berline: Springer, pp. 57–86.

Bentley, C. (2010) *PRINCE2: A Practical Handbook*, Oxford: Butterworth-Heinemann.

Buehring, S. (2011) 'What is a project?' https://www.whatisprince2.net/what-is-a-project.php, accessed 3 August 2016.

Cameron, E. and Greene M. (2015) *Making sense of change management: A complete guide to the models and techniques of organizational change*, London: Kogan Page Ltd.

CEN (2006) *Facility Management – Part 1: Terms and definitions.* EN15221-1.

CEN (2012) *Facility Management – Part 7: Guidelines for Performance Benchmarking.* European Standard EN 15221-7. European Committee for Standardization.

De Bruyne, E. (2007) *Effectieve implementatie van kantoorinnovatie. Een literatuurstudie,* Delft: Center for People and Buildings.

De Caluwé, L. and Vermaak, H. (2003) *Learning to change: A guide for organization change agents,* Thousand Oaks: Sage Publications Inc.

De Caluwé, L. and Vermaak, H. (2004) 'Change Paradigms: An Overview', *Organization Development Journal,* 22 (4), pp. 9–18.

De Jonge, H., Arkesteijn, M.H., Den Heijer, A.C., Vande Putte, H.J.M., De Vries, J.C. and Van der Zwart, J. (2009) *Designing an Accommodation Strategy,* Delft: Delft University of Technology, Faculty of Architecture.

Den Heijer, A.C. (2011) *Managing the university campus.* PhD thesis, Delft: Eburon.

Gambles, I. (2009) *Making the business case: Proposals that succeed for projects that work,* Farnham: Gower Publishing Ltd.

Hoendervanger, J.G., Van der Voordt, T. and Wijnja, J. (2012) *Huisvestingsmanagement. Van strategie tot exploitatie,* Groningen: Noordhoff Uitgevers.

Hill, C.W.L. and Jones, G.R. (1989) *Strategic management: An integrated approach.* Boston: Houghton Mifflin.

Jensen, P.A. and Katchamart, A. (2012) 'Value Adding Management: A Concept and a Case', in Jensen et al. (Eds.) *The Added Value of Facilities Management – Concepts, Findings and Perspectives,* Centre for Facilities Management – Realdania Research, DTU Management Engineering, and Polyteknisk Forlag, pp. 104–176.

Jensen, P.A. and Van der Voordt, T. (2015) *How can FM create value to organisations. A critical review of papers from EuroFM Research Symposia 2013-2015 papers,* Baarn: EuroFM publication.

Johnson, G., Scholes, K. and Whittington, R. (2009) *Fundamentals of Strategy,* Oxford: Pearson Education Ltd.

Jylhä, T. and Junnila, W. (2013) 'Learning from lean management – going beyond input-output thinking', *Facilities,* 31 (11/12), pp. 454–467.

Jylhä, T. and Junnila, W. (2014) 'The state of value creation in the real-estate sector – lessons from lean thinking', *Property Management,* (32) 1, pp. 28–47.

Kaplan, R.S. and Norton, D.P. (1992) 'The balanced scorecard – measures that drive performance', *Harvard Business Review,* (70) 1, pp. 70–80.

Kaplan, R.S. and Norton, D.P. (2004) *Strategy maps: Converting assets into tangible outcomes,* Massachusetts: Harvard Business School Publishing.

Keuning, D., Bossink, B. and Tjemkes, B. (2010) *Management: An evidence-based approach,* Groningen: Noordhoff Uitgevers.

Kreitner, R. and Kinicki, A. (2007) *Organisational Behaviour,* New York: McGraw Hill.

Lavy, S., Garcia, J.A. and Dixit, M. (2010) 'Establishment of KPIs for facility performance measurement: Review of literature', *Facilities,* (28) 9, pp. 440–464.

Lavy, S., Garcia, J.A. and Dixit, M.K. (2014) 'KPIs for facility's performance assessment. Part I: Identification and categorization of core indicators. Part II: Identification of variables and deriving expressions for core indicators', *Facilities,* (32) 5/6, pp. 256–274 and 275–294.

Lindholm, A.L., Gibler, K.M., and Leväinen, K.I. (2006) 'Modeling the Value-Adding Attributes of Real Estate to the Wealth Maximization of the Firm', *Journal of Real Estate Research,* (28) 4, pp. 445–473.

Lindholm, A.I. and Nenonen, S. (2006) 'A conceptual framework of CREM performance measurement tools', *Journal of Corporate Real Estate,* (8) 3, pp. 108–119.

Maul, J.P. (2011) *Developing a Business Case: Expert Solutions to Everyday Challenges,* Boston: Harvard Business School Publishing.

Møllebjerg, L. (2010) 'Facility Management Value Add', presentation at *Conference on Key Performance Indicators for FM*, September, Stockholm.

Nourse, H.O. and Roulac, S.E. (1993) 'Linking real estate decisions to corporate strategy', *Journal of Real Estate Research*, (8) 4, pp. 475–494.

Peters, T. and Waterman, J. (1982) *In search of excellence: Lessons from America's Best-Run Companies*, New York: Harper and Row.

PMI (2013) *A Guide to the Project Management Body Of Knowledge*, Pennsylvania: Project Management Institute.

Porter, M.E. (1985) *Competitive advantages: Creating and sustaining competitive performance* London: Free Press.

Preiser, W. and Visscher, J. (Eds.) (2004) *Assessing Building Performance*, Oxford: Elsevier.

Rasila, H., Alho. J. and Nenonen, S. (2010) 'Using balanced scorecard in operationalising FM strategies', *Journal of Corporate Real Estate*, (12) 4, pp. 279–288.

Riratanaphong, C. and Van der Voordt, D.J.M. (2015) 'Measuring the Added Value of Workplace Change. Comparison between Theory and Practice', *Facilities*, (33)11/12, pp. 773–792.

Roulac, S.E. (2001) 'Corporate property strategy is integral to corporate business strategy', *Journal of Real Estate Research*, (22) 1/2, pp. 129–152.

Schuur, R.G. (2015) *Literature Review: Performance Management and Key Performance Indicators (KPIs)*, Atlanta, USA: CoreNet Global.

Tague, N. (2005) *The Quality Toolbox*, Wisconsin: ASQ Quality Press.

Treacy, M. and Wiersema, F. (1995) *The discipline of market leaders*, Massachusetts, USA: Addison Wesley.

Van der Voordt, D.J.M., Ikiz-Koppejan, Y.M.D. and Gosselink, A. (2012) 'Evidence-Based Decision-Making on Office Accommodation: Accommodation Choice Model', in Mallory-Hill, S., Preiser, W.F.E. and Watson, C. (Eds.), *Enhancing Building Performance*, Chichester, UK: Wiley-Blackwell, pp. 213–222.

Van der Voordt, T.J.M., De Been, I. and Maarleveld, M. (2012) 'Post-Occupancy Evaluation of Facilities Change', in E. Finch (Ed.), *Facilities Change Management*, Chichester: Wiley-Blackwell, pp. 137–154.

Ware, J. and Carder, P. (2012) *Raising the Bar: Enhancing the Strategic Role of Facilities Management*, RICS Research Report, London: Royal Institution of Chartered Surveyors.

White, A.D. (2013) *Strategic Facilities Management*, RICS Professional Guidance. London: Royal Institution of Chartered Surveyors.

Womack, J. and Jones, D. (1996) *Lean Thinking: Banish waste and create wealth in your corporation*, New York: Simon & Schuster.

18 Reflections, conclusions and recommendations

Theo van der Voordt and Per Anker Jensen

Reflections

Added value of/adding value by FM and CREM

In this book we used two related terms: 1) the added value of FM/CREM (a noun), defined as a positive trade-off between the benefits and sacrifices of FM/CREM regarding its contribution to organisational objectives, also referred to as the targeted outcome of FM/CREM interventions; and 2) adding value by FM/CREM (a verb), referring to all FM and CREM activities and processes that aim to add value to the organisation. In order to emphasise the management function of FM and CREM, the latter is also called Value Adding Management.

During the development of the book it became clear that we also have to distinguish between adding value to the organisation *by* or *supported by* FM and CREM (outcome) and adding value to FM and CREM processes and products (output), what we called adding value *of* FM/CREM. This was one of the learnings from the collaborative work of the authors of Chapter 11 on Innovation and Creativity (Rianne Appel-Meulenbroek and Giulia Nardelli). They had both done research on innovation in FM/CREM but from a different perspective. They write in their introduction: "On the one hand, dedicated FM and CREM practices regarding workplace management aimed at increasing knowledge sharing among employees and facilitating their creativity were shown to contribute to the added value of FM and CREM by sustaining innovation across all layers of the served organisations ... On the other hand, scholars outlined that innovation *of* the FM and CREM processes and/or services appears to contribute to its added value for the client/end users, as it increases effectiveness and efficiency of FM and CREM practices ... The latter case describes the role of innovation as input and as a way to manage added value (throughput) and, to avoid complexity, will in this chapter be termed innovation *of* FM and CREM practices. In the first case, the emphasis is on innovation and creativity of the organisation as a whole being *supported by* FM and CREM practices, i.e. on innovation as outcome by adding value through FM and CREM activities." Chapter 11 focuses on added value *supported by* FM/CREM in the sections on state of the art and on benefits and costs, and in the sub-section on KPIs, while the sub-section on how to manage focus on management of innovation *of* CREM/FM processes and services.

A similar distinction came to the fore in Chapter 6 on Image. This chapter focuses on how to improve the image and branding of an organisation *by* FM and CREM and also touches briefly on the topic of how to improve the image *of* FM and CREM. Chapter 7 on Culture discussed the reciprocal relation between culture and FM/CREM, i.e. FM/CREM as a means to support the organisational culture versus the organisational culture as one of the factors that influences the design of buildings, facilities and services.

In discussions and communication between different stakeholders it should be very clear whether the focus is on adding value to the organisation *by* FM/CREM, improving the performance *of* FM/CREM, or both.

Value creation

In recent research within service marketing there has been a burgeoning development of the concepts of value and value creation involving customers and users. Among the most influential research here is the work of Vargo and Lusch (2008) about the concept of Service-Dominant Logic. This is seen as an alternative to the traditional goods-dominant paradigm, where value is seen as delivered by a producing company to customers. The Service-Dominant Logic instead regards value as co-created between a service provider and their customers. An even more radical development is represented by the so-called Customer-Dominant Logic, where the service providers only can facilitate value creation, because it is only during the use that *value formation* is taking place and this is dependent on the reality and life experience of the customer (Heinonen et al., 2013). Compared to these new views we consider the understanding of value in this book to mostly be embedded in the traditional view, where value is delivered to the users, but with a strong focus on stakeholder management, which can include user involvement and co-creation of value. However, the new perspectives on value can form an important inspiration for the further development of Value Adding Management.

Relevance of different value parameters

In Part II we have presented comprehensive analyses and suggestions for how to manage and measure twelve different value parameters. The results of our interviews with practitioners from Denmark and the Netherlands presented in Chapter 4 showed that not all of these parameters are seen as important by the interviewees. The focus and priority depends on the context. Which aspects are most important may vary from business to business, between private and public organisations and between commercial versus not for profit activities. According to Carel Fritzsche, priorities depend on geographical location (practice and culture) and position in the S-curve of maturity. Anglo-Saxon countries normally have a strong hierarchical direction and focus on financial performance and shareholder value, while mainland Europe tends to be less hierarchical and more compromise-driven. In industry both Operational Excellence and Safety and Compliance thinking is primary. According to Victor Collado, for manufacturing clients the main priorities are usually business continuity and Health, Safety, Security, Environment and Quality (HSSEQ), whereas in labs quality is the primary driver and in offices corporate image and employee satisfaction are normally more important; cost efficiency is always present. According to Bart Voortman, the desired performance level depends of the context as well, and fluctuates with the economic context and the market.

Based on all interviews, the top three value parameters were Cost, Productivity and Satisfaction. This is very much in line with the purpose of FM in the definition in the draft of the first ISO standard. This defines FM as "organizational function which integrates people, place and process within the built environment with the purpose of improving the quality of life of people and the productivity of the core business" (ISO, 2015). It seems that the measurement of Satisfaction is becoming more and more important in practice as a basis for outcome-based integrated FM contracts, where the payment to the provider is dependent on the results of satisfactions surveys – see, for instance, Interview 1 with Jakob Moltsen concerning Sony Mobile Communications. They not only measure user satisfaction but also buyer and seller satisfaction, which appears to be a very important element in creating dialogue and alignment between customer and provider representatives.

However, we still found it important to treat all twelve parameters, because they are all relevant in the literature, and the importance of the parameters varies between different types of organisations and will change over time. One example is Value of Assets (Chapter 14), which has particular importance for companies that own their real estate, and less importance for companies who rent their space. Most large companies implement a combination of these two solutions, which makes the situation more complex. Production companies usually own their production facilities and headquarters, but they often lease additional office and storage space. Another example is Risk in terms of business continuity (Chapter 12), which is of particular importance for production companies as well as service companies like banks and broadcasting corporation, who are dependent on uninterrupted operation of their digital infrastructure, but less critical for normal office-based companies.

Research-based evidence

The amount of relevant research varies a lot between the twelve parameters. Some parameters are covered by much research-based literature both in general as well as in specific FM and CREM related publications. For instance, this is the case for Satisfaction (Chapter 5) and Productivity (Chapter 9). For these parameters there have been a large number of quantitative studies. For other parameters there is a lot of general literature but only limited research related to

FM/CREM. This is the case for Image (Chapter 6), Culture (Chapter 7), Innovation (Chapter 11) and Corporate Social Responsibility (CSR, Chapter 16). In recent years there has been written a lot about Sustainability (Chapter 15), but this is a very broad topic and it does not have a clear demarcation with CSR. To avoid too much overlap we have focused mostly on environmental aspects in Chapter 15. Adaptability (Chapter 10) has mostly been dealt with in a building design context and not much from a FM/CREM perspective.

Output, outcome and added value

It has become clear during the development of the book that in the current literature and debates on the added value of FM/CREM, quite often no clear distinction is made between added value and related KPIs for the performance of FM/CREM and for the performance of the organisation they serve. The list of KPIs in Tables 17.1a and 17.1b demonstrates that. Therefore, in Chapter 17 we also presented examples of KPIs that we see as related to FM/CREM performance (defined as output indicators), and performance of the organisation (defined as outcome indicators). This distinction was also clearly made in Figure 3.2 in Chapter 3. The valuation and appraisal of increased performance and related sacrifices in connection to organisational objectives defines the level of added value. As such, added value is not an objective phenomenon, but depends on subjective perceptions, management interpretations and negotiations. In Figure 17.5 we have tried to clarify the distinction between output → outcome → added value by presenting various examples of chains of interventions, outputs, outcomes and added value, which may be used as a tool to analyse and identify cause-effect relationships.

It is important for both researchers and practitioners to identify and distinguish outputs, outcomes and added value, and to develop sensible hypotheses about cause-effect relationships. They can then be tested scientifically or less rigidly during ongoing management practice and adjusted as appropriate for the purpose. An example from Figure 17.5 is the chain, where an intervention with improved indoor climate is supposed to lead to the combined outputs of increased staff satisfaction and a healthier work environment, which in turn is supposed to lead to the combined outcomes of higher productivity and increased client and customer satisfaction. This shows the complexity of the concept of satisfaction. In corporate satisfaction surveys staff members are usually asked to evaluate their satisfaction not only with the work environment but also with the management at different levels and with corporate policies. Therefore it seems natural to see staff satisfaction as an output and client and customer satisfaction as an outcome, but other relations could be hypothesised as well.

The difference between output and outcome is also mentioned in recent research on service productivity. Djellal and Gallouj (2013) write: "The time factor of services means that their output should not be considered only statically but also dynamically. It exerts its effects over time and it is therefore necessary to distinguish the short-term service (the output) from its medium- and long-term effects (the outcome). In the case of hospital services, the output means the patient care, while the outcome is the change in the patient's health status, his or her subsequent state of health (a longer life due to the care received)." Another example from healthcare by Lillrank (2015) defines output as what is done to a patient, outcome as what happens to a patient, and value as how a patient benefits. An essential point is that the healthcare system mostly has control over what is done to a patient, less control over what happens to a patient, which for instance depends on behaviour, and very little control over how a patient benefits, which for instance depends on their situation, capabilities and life experience. These discussions illustrate both the relevance and the complexity of the distinction between output, outcome and added value.

Performance measurement and KPIs

In Part II we identified a large number of KPIs to measure the different value parameters. There are several KPIs that can be measured more or less objectively, for instance financial indicators for Cost, uptime for business continuity (Risk), and energy consumption in relation to Sustainability. For Adaptability it is possible to define physical measurements as input and output indicators, but it is very difficult to define a limited number of essential KPIs and to quantify the outcome of physical measures, i.e. their ease of adaptability.

Many KPIs are rather subjective and based on people's perceptions and opinions measured by surveys, often by applying 3-, 5- or 7-point scales, for instance ranging from 1 = disagree or very dissatisfied to 5 = completely agree or very satisfied. This is particularly the case in relation to Satisfaction, but it also applies to Image, Culture, Health and Safety, and Productivity. Besides measuring Satisfaction by questionnaire surveys at regular intervals, there are also examples of more immediate measurements, for instance with "smileys" when leaving a canteen and by "mystery shoppers" in receptions and canteens – see Interview 7 with Liselotte Panduro, ISS.

Conclusions

Value adding management in practice

Our interviews with practitioners show that most practitioners apply the concept of Added Value or Adding Value in daily practice. It is a well-known term among colleagues and clients and may even be their company's reason for being. Added value is often used in contract negotiations with clients and customers to show how FM and CREM can make the core business more effective and make things easier, as part of the in-sale phase in the dialogue with customers, as input to stakeholder dashboards, in business cases, in internal and external benchmarking processes, and as a marketing tool. It is important to make visible that FM and CREM can make a difference and to show what FM and CREM provide and why, for instance regarding the impact of FM interventions on strategic goals, image, return, profit, costs etc. The Added Value concept is related to the term 'value' and has both an economical meaning and meanings referring to feelings and other subjective and qualitative aspects.

According to John Suyker, the use of the Added Value concept has changed over time. In the nineties the focus evolved from bottom line Earnings Per Share (EPS) to Shareholder Value and eventually Economic Value Add (EVA). Nowadays AV relates to a wider scope and variety of financial, economic and softer factors affecting the image and sustainability of the company's future. Value adding management regards optimal performance for lowest costs with clear procedures, appointments, arrangements and measurement of outcomes, and should focus on clear decisions instead of lengthy discussions.

Some practitioners prefer to talk about value instead of cost as value is often perceived as something positive, or use the term 'right costs' to emphasise that an appropriate intervention delivers value for money. Various interviewed practitioners mentioned that the Added Value concept helps practitioners to speak the language that top management understands. It makes them focus on the strategic aspects of FM and gives a more constructive dialogue than focus on cost. They want to be seen as a business partner and a profit centre and not just as a cost centre. By discussing the added value of FM and CREM the dialogue is moved away from SLAs; it makes the customer feel that a practitioner is really interested in his business. Professional FM as a value driver may provide work environments and services that are competitive to tenants' high expectations, resulting in higher rents and longer tenancies.

However, many practitioners also admit to finding added value a complex and ill-defined concept. Added value is perceived differently by different people, at different levels and from different perspectives. According to some interviewed practitioners, added value can only be discussed with people with an academic background and not at an operational level. The most important problem is the difficulty of making added value concrete and operational. It is difficult to document and to measure in economic terms. According to the experience of Liselotte Panduro using the Balanced Scorecard, "the economic and people perspectives are quite easy to document, while the customer and process perspectives are much more difficult to measure". It also depends a lot on what triggers the specific customer and user. Management of expectations is an important aspect of adding value. By exceeding management expectations the FM/CREM position is safeguarded and value add proven. Storytelling is an important way to create understanding and motivation. Sometimes examples of added value convince clients better than explaining theory or presenting a load of data with tables and graphs. Too much pressure placed on economic added value can limit the development of an experimental and creative environment and hinder innovation. Adding value is not often based on an integrated approach and usually focuses on one or a few value parameters.

The interviewed practitioners express a need for a clear operationalisation of added value and appropriate tools to be able to manage and measure the added value of FM/CREM effectively and efficiently – or, in academic terms, in a valid and reliable way. Though the concept is mainly used at the strategic and tactical levels, some practitioners also mentioned a need to develop the right tools to be used at the operational level. However, other practitioners think that on this level FM/CREM is a dissatisfier. FM has to deliver good services in an efficient way, but one should not expect a high added value from basic services on customer satisfaction. According to Bart Voortman, talking about added value at an operational level will result in endless discussions without concrete results. John Suyker also mentioned that talking about AV at the hands-on operational level can be confusing to some individuals, or else complex and counterproductive.

Opening the black box

This book is intended to be research based and practice oriented. From an academic point of view, we may conclude that the state of the art sections regarding the twelve value parameters and the elaboration of the Value Adding Management model and its components in Chapters 1 to 4 and Chapter 17 helped to open the black box of the Added Value concept. We have clarified the links between FM/CREM interventions (input), value adding management processes (throughput), FM/CREM performance (output), organisational performance (outcome) and recognised added value. As such, a huge step has been made towards a clear operationalisation of the Added Value concept and Value Adding Management.

However, though this book shed more light on the what-how-why question – which FM/CREM interventions may improve organisational performance, what are the costs, how to manage change, and how to measure the output, outcome and added value – still much work remains to be done. Adding value is not a simple mechanistic process with a particular intervention resulting in a particular outcome. Rather, it is influenced by many interacting and interrelated factors. Further research is needed to disentangle the complex cause-effect relationships between input, throughput, output, outcome and added value, both to understand these relationships on a conceptual level and, by collecting empirical data, to support evidence-based decision-making.

Regarding practice, we hope and expect that the state of the art sections regarding the twelve value parameters in Part II, the Value Adding Management model, and Table 17.2 with its overview of possible interventions and tools to measure the output and outcome will all be useful to support FM/CREM professionals to convince their clients, customers and end users about the added value of FM/CREM. The VAM-model and the many related tools to manage and measure added value are also intended to be useful means to support top managers and unit managers in identifying the need for change and finding the best possible solutions to increase the organisational performance by increased FM/CREM performance – and thereby adding value to the organisation.

330 *Theo van der Voordt and Per Anker Jensen*

Recommendations

Recommendations for practice

For practice it is recommended to create a sound balance between cost reducing measures and measures that provide benefits for the organisation, clients, customers, end users, external stakeholders and/or society. One of the interviewed practitioners suggested focusing on "cost optimisation" rather than on "cost reduction". Adding Value should be related to the different stakeholders within the company and externally. For instance, Shareholder Value is mainly connected to Return on Investment, low risks, costs and reliabilities. The Board of Management is probably mainly focused on strategic direction, global client-market development, Ebit (Earnings before interest and tax) and overall cash-position. Heads of regional units may equally focus on financial performance (maximum turnover and minimum costs) and on the attraction and retention of talented staff. Site managers may focus on operational issues and employee satisfaction.

It is also important to be clear about and to create transparency in decision making. Søren Andersen and Søren Prahl referred in this context to a "FM landscape" to visualise the scope and responsibilities of FM/CREM. They make use of a "traffic lights" metaphor, with green for areas where FM has the budget and decides, yellow for areas where FM has the budget and others decide, and red for areas where others have the budget and decide. The FM landscape is used internally in the FM organisation and in the dialogue with their customers.

The extended Value Adding Management model may be used as a step-by-step plan to add value to the organisation by appropriate FM/CREM interventions. The many related models and tools to manage adding value by FM/CREM, such as the Accommodation Choice model, the DAS-frame, stakeholder analysis, SWOT analysis of the organisation and FM/CREM products and processes, strategy maps and the Balanced Scorecard, business cases, project management tools such as PRINCE2 and lean management and the KPIs for the twelve value parameters (Table 17.2), can be used to support the PDCA cycle of Plan, Do, Check and Act and to monitor FM/CREM performance and its contribution to organisational performance. One might consider preparing periodical FM/CREM Performance Reviews for the clients and customers with evaluation of recent interventions, proposals for new improvements and strategic advice on developing the property portfolio, facilities and services. This advice can include gain-sharing, dependent on the contract. Another option is to create an Innovation Board with representatives of the provider, clients, customers and end users to decide and monitor innovations and improvements.

Periodical satisfaction surveys among clients, customers and end users, as well as providing a digital portal and a graphical user interface with access to drawings, can be used as input for the discussions. However, responses to satisfaction surveys tend to be too general and dependent on the formulation of questions and the situation in which responses are given. In a recent RICS report it was concluded that relying solely on numerical survey-type information – or conversely on informal conversation-type information – does not enable a true portrayal of the performance of the provision of space and facilities (Tucker, 2014). For this reason it is recommended to use both sets of customer feedback and to compare survey responses with other KPIs.

One of the topics to consider is ownership versus renting premises in connection with market-conforming real estate and more built-to-suit buildings, flexibility, and costs, relating added value to the context (Anglo-Saxon countries versus mainland Europe, USA, Asia) and the specific real estate sector (offices, healthcare, education, industry). Another topic is to make a clear distinction between the market value of CRE and the company value of CRE.

Recommendations for education in FM and CREM

Apart from being research based and practice oriented, the findings presented in this book can be used in educational settings as well: first, to clarify the concepts of Added Value of and Adding Value by FM and CREM and Value Adding Management; second, to disentangle these complex constructs into clear and more or less tangible value parameters; and third, to use the findings in exercises, assignments and BSc or MSc graduation research projects, for instance by organising workshops and asking the students to present examples of adding value by FM and/or CREM per value parameter, or discuss possible value chains like the ones that were presented in Figure 17.5. An interesting subject for graduation research could be to compare the presented insights with current practice: which parameters are applied in practice, which ones are prioritised and why, which interventions come to the fore, and how are the output and outcome being measured in practice? What are the gaps between theory and practice, and how can practice be improved?

Recommendations for further research

Similar questions are relevant for further academic and professional research. Relevant topics for future research include further exploration and quantitative and qualitative data-collection ('evidence') on:

- The twelve value parameters that have been presented in this book
- Additional value parameters, both generic – for instance, how to improve possibilities for financing FM and CREM – and sector-specific – for instance, how to create a healing environment in healthcare facilities, or how to improve educational performance in educational setting by facilities change
- Cause-effect relationships between FM/CREM interventions, ways of implementation, output, outcome and added value, the influence of the internal and external context (staff composition, work processes, branch, maturity of the organisation etc.) and interactions between different values, resulting in synergy or conflicting values

- Implications of time: what are the short-term and long-term impacts of various interventions?
- Adding value by FM/CREM from a historical perspective: how has the concept developed in the last decades?
- The impact of FM at a macro level, for instance the impact of the privatised delivery of post, or the impact of lower space demand on growing vacancy
- Lessons to be learned from other disciplines, such as business management and economics
- The whole cycle of Plan-Do-Check-Act. A critical review of twenty-one research papers showed that so far most research does not measure both benefits and sacrifices of interventions, and neither does it include an ex ante assessment and ex post assessment or assess the impact of the change process (Jensen and Van der Voordt, 2015a and 2015b)
- New tools to measure the output, outcome and added value of FM/CREM choices and interventions, by combining quantitative measurement methods with narratives and the application of less labour-intensive ways to measure performance, such as the use of sensors, apps and other IT devices
- Optimal ways to integrate various tools to measure adding value by FM and CREM in an ex ante assessment – to identify the most appropriate FM/CREM interventions or FM/CREM choices – and Post-Occupancy Evaluation of facility change
- Opportunities and risks of Big Data
- Exploring optimal ways to integrate various tools to manage adding value by FM and CREM in an efficient and effective way.

The current results and findings from further research may be used to develop a Value Adding Management guide book, including many practical examples of value adding management processes, best practices for optimal added value, worst cases of value destruction, and tools to support managing and measuring added value.

References

Djellal, F. and Gallouj, F. (2013) 'The productivity challenge in services: Measurement and strategic perspectives', *The Service Industries Journal*, 33 (3/4), pp. 282–299.

Heinonen, K., Strandvik, T. and Voima, P., (2013) 'Customer dominant value formation in service', *European Business Review*, 25 (2), pp. 104–123

ISO (2015) *Facility Management – Part 1: Terms and definitions*. Draft Standard ISO/DIS 18480-1. International Standard Organization, 16 June.

Jensen, P.A. and Van der Voordt, T. (2015a) 'Added Value of FM – A critical review', conference paper, *European Facility Management Conference EFMC 2015*, Glasgow, 1–3 June.

Jensen, P.A. and Van der Voordt, T. (2015b) *The added value of FM: How can FM create value to organisations. A critical review of papers from EuroFM Research Symposia 2013-2015 papers*. Baarn: EuroFM publication.

Lillrank, P. (2015) 'Why Implementation of Lean Healthcare is Difficult?' *ScAIEM Conference 2015*, Technical University of Denmark.

Tucker, M. (2014) *Balancing the performance scorecard: How to maximise customer feedback in Facilities Management*, London: Royal Institution of Chartered Surveyors (RICS).

Vargo, S. and Lusch, R. (2008) 'Service-dominant logic: Continuing the evolution', *Journal of Academic Marketing Science*, 36 (1), pp. 1–10.

Index

Page numbers in **bold** refer to keywords in tables, and page numbers in *italics* refer to keywords in figures.

Printed in the United States
by Baker & Taylor Publisher Services